GOVERNORS STATE UNIVERSITY LIBRARY

W9-BFU-689

Toxicants in the
Aqueous Ecosystem

Toxicants in the Aqueous Ecosystem

T. R. CROMPTON
Consultant, Anglesey, Gwynedd, Wales

GOVERNORS STATE UNIVERSITY
UNIVERSITY PARK
IL 60466

JOHN WILEY & SONS
Chichester · New York · Weinheim · Brisbane · Singapore · Toronto

Copyright © 1997 by John Wiley & Sons Ltd,
Baffins Lane, Chichester,
West Sussex PO19 1UD, England

National   01243 779777
International   (+44) 1243 779777
e-mail (for orders and customer service enquiries): cs-books@wiley.co.uk
Visit our Home Page on http://www.wiley.co.uk
or http://www.wiley.com

All Rights Reserved. No part of this book may be reproduced, stored in a retrieval
system, or transmitted, in any form or by any means, electronic, mechanical, photocopying,
recording or otherwise, except under the terms of the Copyright, Designs and Patents Act
1988 or under the terms of a licence issued by the Copyright Licensing Agency, 90
Tottenham Court Road, London, UK W1P 9HE, without the permission in writing of the
publisher.

Other Wiley Editorial Offices

John Wiley & Sons, Inc., 605 Third Avenue,
New York, NY 10158-0012, USA

VCH Verlagsgesellschaft mbH,
Pappelallee 3, D-69469 Weinheim, Germany

Jacaranda Wiley Ltd, 33 Park Road, Milton,
Queensland 4064, Australia

John Wiley & Sons (Asia) Pte Ltd, 2 Clementi Loop #02-01,
Jin Xing Distripark, Singapore 129809

John Wiley & Sons (Canada) Ltd, 22 Worcester Road,
Rexdale, Ontario M9W 1L1, Canada

Library of Congress Cataloging-in-Publication Data

Crompton, T. R. (Thomas Roy)
    Toxicants in the aqueous ecosystem / T. R. Crompton.
        p.   cm.
    Includes bibliographical references and index.
    ISBN 0-471-97272-X
    1. Water–Pollution–Toxicology. 2. Aquatic organisms–Effect of
water pollution on. 3. Water–Pollution–Measurement.
    QH90.8.T68C76   1997
    574.5′263–dc21   96-37346
                                                        CIP

British Library Cataloguing in Publication Data

A catalogue record for this book is available from the British Library

ISBN 0 471 97272 X

Typeset in 10/12pt Times by Keytec Typesetting Ltd, Bridport, Dorset
Printed and bound in Great Britain by Biddles Ltd, Guildford, Surrey
This book is printed on acid-free paper responsibly manufactured from sustainable forestation,
for which at least two trees are planted for each one used for paper production.

QH 90.8 .T68 C76 1997

Crompton, T. R.

Toxicants in the aqueous
ecosystem

Contents

Preface

Pollution of the ecosystem has always occurred to some extent or other. For example, over the whole of prehistory and still, to some extent, today the eruption of volcanoes has led to the large-scale contamination of the ecosystem. In more recent times the lead-smelting activities of the Romans caused pollution as is evidenced by the presence of lead-rich layers in peat sediments in Yorkshire, and this is perhaps the earliest example of man-made pollution. Since the start of the Industrial Revolution, pollution of the ecosystem has obviously increased considerably and, despite efforts to control it, it is still doing so. Such inputs of pollution obviously include discharges of industrial waste and sewage directly to rivers and via coastal discharges. The emission of toxic substances to the atmosphere by factory smokestack emissions, incineration plants and fires is another major source of pollution, such emissions inevitably being washed out of the atmosphere by rain to cause pollution of the oceans and land. Another input is the dumping by ships of industrial and sewage wastes into the seas.

Pollution is defined as a change in water quality which causes deleterious effects in the organism community or makes the aesthetic quality of the water unacceptable. Contamination refers to the presence of potentially harmful substances at concentrations which do not cause harm to the environment.

It is becoming increasingly realised that the oceans and rivers, in particular, are not an unlimited reservoir into which waste can be dumped and that control of these emissions is necessary if complete destruction of the environment is to be avoided. Heavy metals are particular offenders in this respect as are organometallic compounds whether the latter are discharged as such directly into the environment or whether, as has been shown in recent years, they are produced by biological conversion of inorganic metallic contaminants such as lead, mercury, tin and arsenic.

There are also many classes of organic pollutants which are encroaching upon the aquatic ecosystem. Organic pollutants, news of which via the media have reached the ears of the public in recent years include crude petroleums, poly-aromatic hydrocarbons, organochlorine and organophosphorus insecticides, polychlorinated biphenyls, chlorinated dioxins, chlorinated aliphatic and aromatic compounds, and nitrosamines. However, the possible organic pollutants number many thousands, only some of which have been studied in detail.

Once a toxic substance enters a river it can cause damage to animal and plant life in the river with possible implications in the survival of fish and crustaceans and also for the health of humans who eat these creatures. Many rivers serve as the inputs to potable water treatment plants and consequently there is a further health implication for humans and animals who drink the water. River waters often carry the pollutants to the oceans where they are added to by pollutants in coastal discharges, atmospheric fallout and shipbound dumping. Again the survival of fish populations and the health of humans become major considerations. Pollutants which discharge directly to land, including sewage and domestic and industrial waste, might enter crops which again become a health hazard to humans and animals. Such pollutants inevitably are washed by rain to a watercourse and eventually reach the sea.

Regulations for controlling the input of pollutants into the environment are slowly being introduced internationally but much still remains to be done. It is the purpose of this book to review the current situation as regards the control of pollution from all sources and discuss the levels of pollutants at present occurring in the aqueous environment, including rivers, lakes, groundwaters, estuary and seawaters and potable water as well as considering the toxic effects of such pollutants and methods of chemical analysis.

The exposure of creatures to known concentrations of toxicants for stipulated periods of time enables the toxicity of the pollutant, as measured by the concentration–time relationship to kill 50% of the creatures (i.e. LC_{50}), or to have an adverse effect on 50% of the creatures (i.e. EC_{50}) to be established. Such 'water analysis'-based methods for assessing the effects of pollutants is discussed in Chapter 2. A further method of assessing the toxicity of pollutants is based on relating the composition of the water in which the creatures live to the concentration of the toxicant found in the animal tissue or, better still, in a particular organ in the animal in which the toxicant preferentially concentrates. Such data can be related to the water composition and the condition of the animal in terms of ill health or mortality and this testing is reviewed in Chapter 2.

In Chapter 3 are discussed the regulations suggested by various European bodies for the control of pollutants in the aqueous ecosystem while in Chapter 4 available toxicity data concerning the effects of various types of pollutants on fish and creatures other than fish are reviewed. These include studies in non-saline and saline waters and cover all the toxic metal pollutants, organic pollutants and organic compounds of arsenic, lead, mercury and tin. Only in the case of polychlorinated biphenyls and carbon tetrachloride has the European Union issued a Directive concerning maximum permitted levels.

In Chapter 5 are discussed specific examples of the effect of dissolved metallic toxicants on freshwater organisms (Section 5.1) and seawater organisms (Section 5.2). Using published LC_{50} and maximum safe concentrations (S_x) data it is possible to draw up for each type of creature 'at-risk' tables from which for any particular water at any particular composition we can compile a list of creatures that will either suffer ill health or die. Examples of clean and dirty rivers are discussed. This chapter concludes with a tabulation of metal contents found in environmental waters throughout the world. Consensus values for toxic metals in seawater are discussed as is a mass balance of metals entering the North Sea from all sources. The metal content of rain is also considered.

Chapter 6, dealing with organometallic and organics in non-saline and saline waters, is similar to Chapter 5. Tabulations are given of the actual levels of such compounds found in a wide range of environmental samples taken throughout the world.

Sediments on the beds of rivers and the oceans have the property of absorbing many toxicants from water in such an amount that the concentration of toxicants in a sediment is many times higher than that in the surrounding water (Chapter 7). Analysis of sediments is therefore a useful means of assessing the levels of pollutants in water over a period of time and is related to ill health or mortality of creatures living in the water. A review is given of levels of inorganic, organic and organometallic toxicants found in such sediments in samples taken from all over the world and an attempt is made to correlate contaminant levels with the health of creatures.

In Chapter 8 are reviewed the levels of metals, organometallics and organics found in the tissues of various types of fish and crustacae as well as phytoplankton and weeds taken at numerous sites throughout the world. In addition, the results are reported for the levels of metals found in the organs of these creatures as in many instances enhanced metal levels occur in particular organs and this facilitates identifying the cause of death. In particular, polyaromatic hydrocarbons, chlorinated biphenyls and 2,3,7,8-tetrachloro dibenzo dioxin are discussed.

A further feature of the book is the discussion in Chapters 5 to 8 of appropriate analytical methods. Detection limits achievable by various techniques are considered and compared with the concentrations likely to be encountered in various types of environmental samples.

In Chapter 9 are summarised concentrations of metals, organometallic compounds and organics found in potable water samples taken throughout the world which illustrate the wide range of concentrations encountered. In particular, haloforms, which are suspected carcinogens, produced during the chlorination of potable water are discussed in some detail. The occurrence of radionuclides is discussed in Chapter 10.

The book concludes with five appendices which give additional information to that in the text on the following:

(1) Ranges of metal concentrations found in environmental waters, freshwaters, marine waters, coastal, bay and estuary waters
(2) Radionuclides in freshwaters and seawaters
(3) Metals, organometallic and organic compounds in sediments
(4) Toxicants in sea organisms and phytoplankton
(5) Composition of potable water.

CHAPTER 1
Toxicity Evaluation—Water Analysis Based

1.1 MEASUREMENT OF LC_{50}

The toxicity of a metal or an organic substance to fish, invertebrates, algae or bacteria is evaluated in toxicity tests in which these are exposed under standard conditions to a range of concentrations of the test substance for a constant period of time and the number of creatures which either die or undergo a particular response (for example, growth reduction) during that period is counted at the end of the test period. A plot of concentration of test substance versus percentage of mortality or percentage of creatures undergoing a particular response enables one to read off the concentration of test substance that causes a 50% effect, as demonstrated in Figure 1.1. Short-term tests are run for 4 days, although, commonly, to obtain more complete information, they may be run for 1, 10, 100 and 1000 days. From the graphs obtained it is then possible by interpolation to read off the median lethal concentration (LC_{50}) or the median effect concentration (EC_{50}), i.e. the concentration which is calculated to cause, respectively, the mortality or a particular response of 50% of the test population. It is also usual to identify the median concentration by the test duration, e.g. 4-day LC_{50} or 100-day EC_{50}. Acute, chronic and lethal and sublethal tests can be performed in a similar manner.*

An example of this type of testing is shown in Figure 1.2 which illustrates LC_{50} results obtained by exposing salmonid and non-salmonid freshwater fish to cadmium in 1-, 10-, 100- and 1000-day tests. Both Figures 1.1 and 1.2 illustrate that, as would be expected, LC_{50} decreases as test duration is increased. Also, for any given test conditions, EC_{50} would always be less than LC_{50}. As might be imagined, the results obtained in such toxicity tests are affected be several factors the more important of which are discussed below.

---

*_Acute toxicity_—the lethal response caused by a short exposure to a substance, at most a few days, commonly 4 days.
Chronic toxicity—deleterious effects (not exclusively fatalities) resulting from prolonged exposure, i.e. more than a few days.
Sublethal toxicity—having a deleterious effect but not causing mortality.

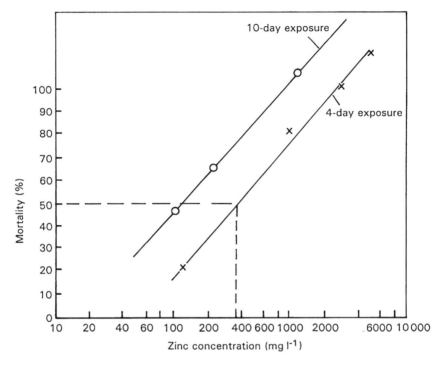

Figure 1.1. Method of obtaining LC_{50} by interpolation. Toxicant: zinc

1.2   FACTORS AFFECTING LC_{50}

1.2.1   TEST-RELATED FACTORS

Space

Adequate space has to be provided for the test creature as overcrowding causes stress and consequently affects the sensitivity of the creatures to toxicants. A minimum of 2 litres of sample is recommended per gram of biomass. Moreover, the sample should be changed every 24 hours and fresh toxicant added.

Water flow through test chamber

The results obtained in static tests are of doubtful value in the case of creatures that normally inhabit flowing rivers. Flow-through tests are inherently more complicated as they involve the provision of a dosing mechanism to maintain a constant concentration of toxicant in the flowing sample. They do, however, have the advantage of providing more constant chemical conditions through the duration of the test. In flow-through tests 2–3 litres of sample containing the

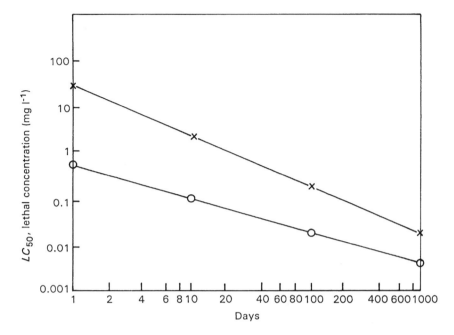

Figure 1.2. Effect of exposure time of fish to cadmium on LC_{50} ✕ non-salmonid fish, ◯ salmonid fish

controlled addition of toxicant should be passed through the test chamber per gram biomass per day. This is equivalent to a 90% replacement of test liquid per day.

In general, as illustrated in Table 1.1, freshwater creatures are more tolerant to toxic metals in static tests than in flow-through tests, all other test conditions being constant.

Table 1.1. Static tests versus flow-through tests in the measurement of LC_{50}

| Element | Species | 4-day $LC_{50}{}^{c}$ (mg l^{-1}) | |
		Static test	Flow-through test
Cadmium	[b]	31.0	4.3
Chromium	[b]	36.2	36.8
Lead	[a]	471	8.0
Silver (as AgNO$_3$)	[a]	0.011	0.009
	[b]	0.010	0.006
Zinc	[b]	12.5	9.2

[a] *Salmo gairdneri.*
[b] *Pimphales promelas.*
[c] pH, hardness and temperature are the same for static and flow-through tests for each element listed.

Temperature

In Figure 1.3 is shown the appreciable effect of sample temperature on 4-day LC_{50} for the elements silver, cadmium and copper. It is clear, therefore, that when evaluating the toxicity of these three elements the test temperature must be carefully controlled and reported with the test result.

Dissolved oxygen content of sample

This should be controlled at a level exceeding 6 mg l^{-1} throughout the test.

Light

A controlled light regime and duration of lighting should be used throughout the test. In general, the absence of light reduces stress, i.e. decreases stress-linked mortality.

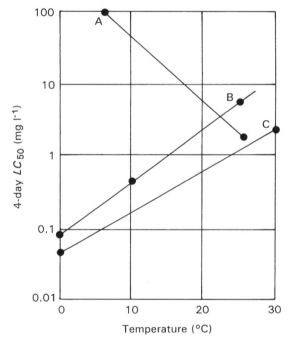

Figure 1.3. Effect of test temperature on 4-day LC_{50}. (a) Silver (as silver nitrate) with salmonid fish, i.e. an increase in temperature increases toxicity. (b) Copper with salmonid fish. (c) Cadmium with non-salmonid fish. In (b) and (c) an increase in temperature reduces toxicity. Chromium, zinc, silver with non-salmonid fish: toxicity independent of temperature. Mercury, cadmium, copper and zinc with freshwater invertebrates: an increase in temperature increases toxicity

Chemical form of toxicant added to sample

Metals vary in their solubility in water depending on their chemical form, the sample pH and the presence in the sample of other chemicals such as phosphates and carbonates. Metal precipitation is expected at higher metal concentrations and higher pH or hardness. The greatest metal solubility is expected in soft acidic waters and it is in these that the best correlation will be obtained between concentration of metal added to sample and LC_{50}.

1.2.2 SAMPLE-RELATED FACTORS

In addition to temperature, the following sample-related factors should be as similar as possible in the test to that in the environmental water from which the samples were taken.

pH

As shown in Figure 1.4, increasing the pH from 6.5 to 8.5 increases the 4-day LC_{50}, i.e. decreases the toxicity of chromium towards freshwater fish and invertebrates by a factor of approximately four. The effect of pH is most

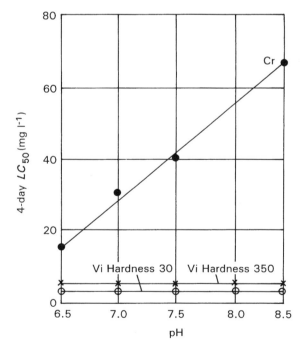

Figure 1.4. Effect of pH on LC_{50}, freshwater fish and invertebrates. Toxicant: chromium

pronounced in hardwaters and is related to an increase in metal solubility at higher pH[1-3]. Sample pH must therefore always be reported when discussing toxicity data.

Hardness

Increase in sample hardness from 10 to 1000 increases 4-day LC_{50}, i.e. decreases toxicity by a factor of between 15 (chromium and zinc) and 120 (nickel and copper) (Figure 1.5) for freshwater fish, while sample hardness has no effect on LC_{50} in the case of arsenic, vanadium and silver (as silver nitrate). Sample hardness must therefore always be recorded when reporting toxicity data.

Salinity

As discussed above, in freshwater samples hardness may be important in reducing the toxicity of a metal as derived from LC_{50} measurements. In the case of estuary or sea water where hardness is not as variable it is salinity that, to a greater or lesser extent, determines the toxicity of the metal. An increase in

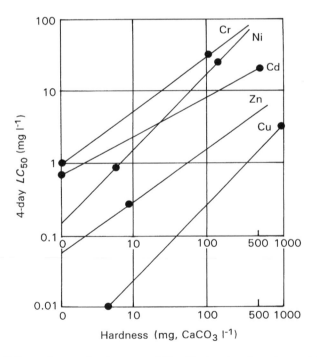

Figure 1.5. Effect of water hardness on LC_{50}. Toxicants: chromium. nickel, cadmium, zinc and copper

sample salinity decreased the toxicity to marine creatures of chromium, copper, cadmium, lead, mercury, nickel and zinc. Typically, an increase in salinity from 10 to 35 g kg^{-1} can decrease toxicity by a factor of 10.

1.2.3 CREATURE-RELATED FACTORS

Age and condition

Young life stages of fish and invertebrates are generally more sensitive to metals than are adults, particularly in the case of marine creatures. Test creatures are to be disease-free.

Feeding

To minimise the toxic effects of animal waste no feed should be given during short-term 24–48-hour tests. Feeding is necessary during long-term tests.

Acclimatisation

The exposure of creatures to toxicants prior to commencing the LC_{50} test for periods of 1 to 1.5 weeks in some instances increases the tolerance of the creature to metals during subsequent LC_{50} tests. For example, pre-exposure of *Jordanella fluoridae* embryos to zinc produced a considerably greater post-hatch tolerance to zinc compared to fry-hatched without pre-exposure. After 30 days[2] exposure to 0.14 μg l^{-1} zinc in the LC_{50} test pre-exposed creatures had nil mortality, while creatures which had not been pre-exposed had 100% mortality during the same period. The benefits of pre-exposure are generally transient.

Reported LC_{50} values for metals obtained from a variety of sources for freshwater and marine fish and invertebrates are quite variable. Typically, a 10-day LC_{50} value for chromium reported by various workers is in the range of 6 to 60 mg l^{-1}, i.e. 33 ± 82% for non-salmonid fish, and in the range 1.5 to 35 mg l^{-1} i.e. 18 ± 93% for salmonid fish. In view of the large number of variables discussed above, this is not surprising. Thus, considering hardness alone it is seen in Figure 1.5 that reported 4-day LC_{50} values for chromium range from 8 at 8 mg l^{-1} hardness to 100 at 500 mg l^{-1} hardness, i.e. a mean of 54 ± 85%. Provided that parameters such as pH, hardness, salinity, test temperature and experimental parameters are reported with the LC_{50} value then data are meaningful and amenable to comparison with results obtained by other workers.

Reported LC_{50} values obtained with fish for a range of elements are summarised in Table 1.2. LC_{50} values obtained in tests of durations 1 to 100 days are included. For salmonid fish mercury and cadmium are among the most toxic elements, while chromium and vanadium are the least toxic.

Table 1.2. Reported LC_{50} values (mg l^{-1}) for various elements: salmonid and non-salmonid fish

Element		As	Zn	V	Ag	Se
Non-salmonids						
Duration of	1	10–140	3.2–100	< 1–1000	15–100	19–152
toxicity test,	10	1.4–16	0.31–14	2.7–37	2.1–11	3–23
days	100	0.16–2.3	0.04–1.4	0.1–1.4	0.19–1	0.35–2.3
	1000	0.05–0.85	0.007–0.16	–	–	
Salmonids						
	1	As	As	As	As	As
	10	above	above	above	above	above
	100					
	1000					

1.3   CUMULATIVE LC_{50} VALUES

Only rarely does a water which is toxic towards fish life contain a single toxicant. If toxic impurities are present in any appreciable amount then it is likely that several of them will adversely affect fishlife. Assuming, as is generally the case, that no synergistic effects exist then the effect of toxicants is additive. The following progressive dilution technique enables the cumulative effect of toxicants on fish to be assessed.

Polluted rivers have been assessed for their toxicity by toxicity tests in flowing water on the river bank using graded dilutions of the river water. Caged fish are exposed to the river water for a number of days and the fish mortality rate and pollutant concentrations measured at daily intervals during this period. Simultaneously, caged fish are exposed to a range of dilutions of the river water and the same measurements repeated. From the results obtained the dilution causing 50% mortality in 2 days is estimated from various fish species at each location.

Figures 1.6(a)–(d) show test duration versus percentage mortality curves obtained from fish in (A) polluted waters and (B) less polluted waters at zero dilution and (×1), ×2, ×5 and ×10 dilutions of river water. From these curves the percentage of mortality occurring after 2 day's exposure for those polluted (A) and less polluted (B) river waters can be obtained. Plots of percentage mortality versus dilution enables the dilution corresponding to 50% mortality to be read off (Figures 1.7(a) and (b)). From these curves it is seen (Figure 1.7(a)) that for the more polluted water sample A 50% mortality results when the original river water sample has been diluted ×4 times and for the relatively unpolluted water sample B (Figure 1.7(b)) only ×2.8 times dilution is required to achieve the same effect. The results of these studies are presented not as a

Table 1.2. (*continued*)

Ni	Cd	Cr	Cu	Hg	Pb
6.6–124	1–53	167–1670	40	0.35–1.2	1.9–35
1.2–23	0.06–2.4	6–60	35	0.03–0.35	0.23–10
0.23–4.3	0.005–0.35	0.15–1.7	5	0.002–0.02	0.03–0.66
0.05–0.81	0.0004–0.028	–	< 1	0.0002–0.002	0.005–0.1
As	0.08–1	23–350	As	As	As
above	0.01–0.19	1.5–35	above	above	above
	0.002–0.035	0.1–2.2			
	0.001–0.005	0.007–0.015			

concentration of pollutants in the rivers but as cumulative fractions of the relevant laboratory-derived 2-day LC_{50} for each species and substance the sum of which is compared with the observed toxicity at each each location.

Thus, considering a simple example, if a relatively toxic river water A before dilution contained 50 mg l^{-1} zinc and 10 mg l^{-1} copper then the ×4 dilution of this causing 50% mortality (Figure 1.7(a)) would contain 12.5 mg l^{-1} zinc and 2.5 mg l^{-1} copper, i.e. river-derived cumulative 2-day LC_{50} = 12.5 + 2.5 = 15 mg l^{-1}. Similarly, if a relatively less toxic water B before ×2.8 dilution (Figure 1.7(b)) contained 8 and 1.5 mg l^{-1} of zinc and copper then the dilution would contain 2.8 and 0.5 mg l^{-1} zinc and copper, i.e. river-derived cumulative 2-day LC_{50} = 2.8 + 0.5 = 3.2 mg l^{-1}. If the laboratory-derived 2-day LC_{50} values for zinc and copper are, respectively, 12 and 6 mg l^{-1}, i.e. cumulative 2-day LC_{50} = 18 mg l^{-1}, then the river-derived cumulative 2-day LC_{50} as a fraction of the laboratory-derived 2-day LC_{50} (i.e. cumulative proportion of laboratory-derived 2-day LC_{50}) is given by:

$$\text{Polluted water A} = \frac{\text{2-day } LC_{50} \text{ river-derived}}{\text{2-day } LC_{50} \text{ laboratory-derived}} = \frac{15}{18} = 0.83$$

$$\text{Less polluted water B} = \frac{\text{2-day } LC_{50} \text{ river-derived}}{\text{2-day } LC_{50} \text{ laboratory-derived}} = \frac{3.2}{18} = 0.18$$

The observed difference between river-derived and laboratory-derived cumulative LC_{50}'s can be ascribed to the effects of factors such as hardness, pH, temperature and dissolved oxygen prior to summation.

This approach has been applied to an assessment of the fishery status of rivers where it has been found that if the sum of the proportions of 2-day LC_{50} exceeds about 0.3 then fish will not survive well enough to support fishing activities[5–8]. Table 1.3 shows this effect for a range of river waters of different total hardness.

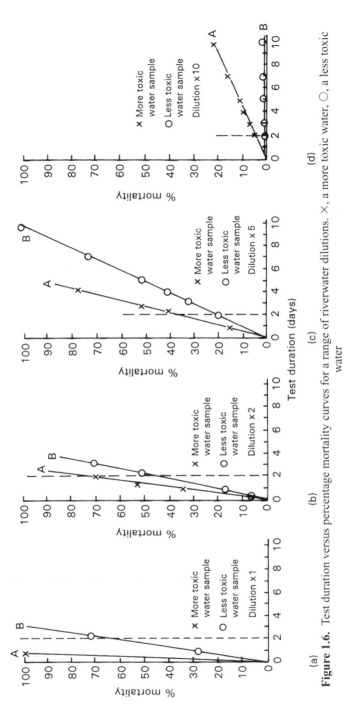

Figure 1.6. Test duration versus percentage mortality curves for a range of riverwater dilutions. ×, a more toxic water, ○, a less toxic water

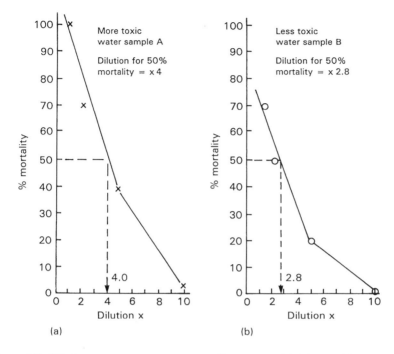

Figure 1.7. Dilution versus percentage mortality curves. (a) More polluted river water. (b) Less polluted river water

Table 1.3. Differences of cumulative proportions of 2-day LC_{50} (laboratory derived) versus fishing status and water hardness

		Median cumulative 2-day LC_{50} ($\mu g\,l^{-1}$)		
Survey	Total hardness	Fishless	Marginal	Fish present
1	11–20	0.45	0.42	0.05–0.25
2	100–170	0.1–0.2	0.16	0.13–0.16
3	100–300	> 0.28	–	< 0.28
4	134–292	> 0.1	–	< 0.1
5	500	> 0.32	0.25–0.32	< 0.25
6	70–745	0.32–2.95	0.3–0.37	0.005–0.02

1.4 CONTINUOUS EXPOSURE (S_x) AND 95TH PERCENTILE EXPOSURE (S_{95}) CONCEPTS

Although the discussion below is concerned with toxic metals similar considerations would apply in the case of organics.

Having available a yardstick by which the toxicity of a given metal to a given creature, i.e. the LC_{50} value, can be evaluated it now remains to assess strategies for measuring the metal concentrations in freshwater and seawater. Clearly, a simple spot measurement of the concentration of a metal in a river water, for example, will not reflect the changes in concentration that occur over a period of time and it is these that will dictate the long-term wellbeing or otherwise of creatures in that water.

Several environmental standard types of approach have been devised for assessing long-term water quality in terms of metal concentration of the environmental water:

(1) A critical metal concentration in the water, which, if exceeded for any period of time, will cause damage to creatures—i.e. environmental change.
(2) A general reference value, reflecting relatively uncontaminated concentrations in creatures for use in identifying areas receiving pollution inputs that may need control.
(3) A maximum safe concentration for continuous exposure for use in calculating discharge limits for toxic metals, which would only be permitted to be exceeded in the immediate vicinity of the discharge.

The use of standard (1) might offer short-term but not long-term protection to the receiving water. The use of (2) in calculating discharge limits would protect receiving water at a prohibitive cost (but may be necessary to ensure no mortalities of creatures at all). Type (3) standards should facilitate adequate discharges of toxicants at a reasonable cost and it is, in fact, this approach that has been adopted by the EU[10–12] and the UK[9] to minimise the deleterious effects of toxicants in discharges on receiving waters. Standards based on (1) would inevitably be lower than standards based on (3), making it necessary to have a much higher level of effluent control at a higher cost.

The toxic effect of a metal on fish is a consequence of not only the concentration but also the duration of the exposure, the adverse effect concentration becoming progressively lower as the period of exposure increases.

Consider, for example, the case of nickel. Various workers have reported LC_{50} values in the following ranges for nickel when non-salmonid fish are subject to toxicity tests of the stated durations (this variability for a constant toxicity test duration may be due to differences in hardness, pH, salinity, temperature, etc. in the various samples tested).

Duration of toxicity test (days)	LC_{50} (mg l^{-1})
1	6.6–124
10	1.24–23
100	0.23–4.3
1000	0.051–0.81

A curve of the type shown in Figure 1.8 can be prepared from these data. In selecting potential values of the standard a boundary line (dotted) drawn to enclose the lower limits of the reported adverse effect concentrations (i.e. conservative estimate) would describe a continuous standard in the form of an equation predicting the maximum acceptable concentration of nickel (with no safety margin) permissible for a specified duration of time. Such real-time management of pollution control is rarely possible and an alternative approach, discussed below, is usually adopted.

Consideration of the relationship in Figure 1.8 would enable most of the statistical values discussed above to be determined, if 100% of the time is assumed to be 365 days, or longer, exposure. Thus for continuous exposure the 365+ days asymptote $S_x = 0.22$ mg l^{-1} (220 μg l^{-1}) nickel would represent the potential standard, after application of a suitable (probably small) safety factor. This long-term standard might be stated as the annual average concentration. However, adoption of this standard would allow higher concentrations to occur for shorter periods and there is a potential risk that these excursions would be sufficiently great as to cause damage to fish. To overcome this the 95% percentile concept has been adopted, i.e. that concentration that could be safely exceeded for 5% of the year (i.e. on 17 days) and this value $S_{95} = 0.9$ mg l^{-1} (900 μg l^{-1}) nickel (excluding safety factors) can be read off from Figure 1.8. Adoption of this approach avoids the need for continuous daily monitoring of nickel concentration stating as it does that the nickel content of the water need be below

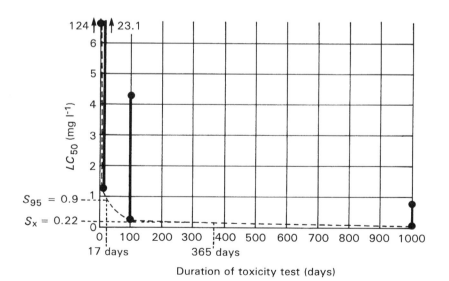

Figure 1.8. Test duration versus LC_{50} plot for nickel

$0.9 \, \text{mg} \, l^{-1}$ ($900 \, \mu g \, l^{-1}$) only on 347 days of the year (365 daily samples) or on 49 weeks of the year (52 weekly samples).

We thus reach the conclusion that for nickel and non-salmonid fish the target standard that would enable fish to survive is that the nickel content of the water should not exceed $0.9 \, \text{mg} \, l^{-1}$ ($900 \, \mu g \, l^{-1}$) (or a slightly lower value if a safety factor is applied) for 95% of evenly spaced out (say, daily or weekly) samples of water taken during a year. It now remains to assess the actual 95th percentile and arithmetic mean nickel concentrations for river or tidal water samples taken at a particular sampling point.

Ideally, the relationship between time and concentrations in river waters would be described by continuous water quality data, but this is rarely available for metals, being replaced by a series of discrete observations of water concentrations. A population of samples of water quality may be summarised to estimate the frequency distribution of observed concentrations (or the probability density of each concentration) or, alternatively, the cumulative probability of not exceeding specified concentrations. It is assumed that the observed water concentrations of nickel adequately represent the natural distributions and, therefore, that the percentage of samples exceeding a specified concentration equate to the proportion of time for which that concentration will be exceeded in the water. On this basis the hypothetical relationship between time and concentration of nickel enables the 95th percentile and the annual average concentration to be calculated.

Suppose that weekly analyses of a river water over 12 months give the results and the distribution of results shown in Table 1.4. A plot of the percentage of time during which nickel contents are within stipulated range versus determined nickel content reveals that for 5% of the time (i.e. 95th percentile) the nickel content is $\geq 1020 \, \mu g \, l^{-1}$ and for 95% of the time it is $\leq 1020 \, \mu g \, l^{-1}$ (Figure 1.9). The arithmetic mean is $448 \, \mu g \, l^{-1}$ (Figure 1.9). The 95th percentile value of $1020 \, \mu g \, l^{-1}$ exceeds the target of $900 \, \mu g \, l^{-1}$ nickel for 95% of the samples taken. Consequently, some mortality of non-salmonid fish would be expected in the circumstances. If all the concentrations quoted in Table 1.4 were halved the 95th percentile value would decrease to $510 \, \mu g \, l^{-1}$ (arithmetic mean $224 \, \mu g \, l^{-1}$) which is less than the standard of $900 \, \mu g \, l^{-1}$ and no adverse effect due to nickel on non-salmonid fish would be expected.

Summarising, the available data for nickel ($\text{mg} \, l^{-1}$) are given below:

Standard		
Upper	*Lower*	*Upper Lower*
S_{95}	S_x	$\dfrac{S_{95}}{S_x}$
$\dfrac{95\%\text{ile}}{900}$	$\overline{220}$	4.09

Table 1.4. Weekly determinations of nickel ($\mu g \, l^{-1}$) in a river water over 12 months

Week no.	Nickel ($\mu g \, l^{-1}$)	Week no.	Nickel ($\mu g \, l^{-1}$)	Week no.	Nickel ($\mu g \, l^{-1}$)	Week no.	Nickel ($\mu g \, l^{-1}$)	Week no.	Nickel ($\mu g \, l^{-1}$)
1	10	11	90	21	420	31	710	41	780
2	120	12	210	22	210	32	60	42	940
3	500	13	900	23	630	33	210	43	160
4	620	14	1210	24	70	34	560	44	200
5	400	15	1410	25	100	35	720	45	220
6	810	16	50	26	150	36	410	46	520
7	420	17	210	27	210	37	510	47	170
8	210	18	430	28	70	38	910	48	400
9	610	19	560	29	420	39	200	49	500
10	700	20	700	30	520	40	200	50	600
								51	710
								52	910

Range of nickel ($\mu g \, l^{-1}$)	Number of samples in this range	Percentage of samples in this range, i.e. percentage of time nickel contents are in stated range
10–90	6	11.5
100–190	6	11.5
200–290	10	19.2
300–390	0	0
400–490	6	11.5
500–590	7	13.5
600–690	4	7.7
700–790	6	11.5
800–890	1	1.9
900–990	4	7.7
1000–1099	0	0
1100–1199	0	0
1200–1299	1	1.9
1300–1399	0	0
1400–1499	1	1.9

River samples S_{95}	Arithmetic	$\dfrac{S_{95}}{m}$
95%ile	mean (m)	
1020	448	2.28

It is the range of ratios of average to 95th percentile obtained for river samples which determines the mode of expression of the standard that is, in fact, most appropriate. Comparison of the observed ratios in rivers for nickel (or any other

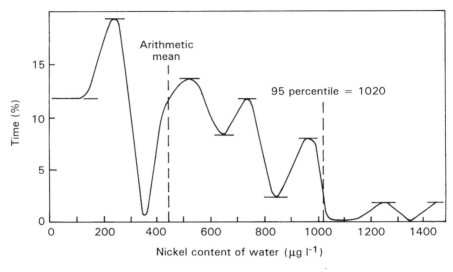

Figure 1.9. 95 percentile. Determination of nickel in water over 12-month period

substance), i.e. S_{95}/arithmetic mean with the ratio of the average standard and the 95th percentile standard, i.e. S_{95}/S_x, will indicate whether the upper or lower standard is appropriate. Where the ratio of standard S_{95}/S_x is less than or equal to the ratio in rivers S_{95}/arithmetic mean then the upper standard (95 percentile) should be selected as the use of the lower or average concentration S_x as the standard would not guarantee against the short-term occurrence of high and damaging river concentrations of metal which exceed the 95 percentile standard. When, as in the case of nickel quoted above, the ratio of standards $S_{95}/S_x = 4.09$ is substantially larger than the ratio in rivers S_{95}/arithmetic mean, $= 2.28$ then the use of the lower or average standard $S_x = 0.22$ will ensure that the 95 percentile, is not transgressed, while providing adequate long-term protection.

The above discussion is concerned with nickel. However, similar considerations can apply to a range of other metals. In Table 1.5 are listed S_x and S_{95} tile values for a range of elements from the least toxic (nickel, $S_x = 220 \mu g\,l^{-1}$, S_{95} tile 900 $\mu g\,l^{-1}$) to the most toxic (cadmium $S_x = 2 \mu g\,l^{-1}$, S_{95} tile $= 6 \mu g\,l^{-1}$ and mercury $S_x = 2 \mu g\,l^{-1}$, S_{95} tile $= 22 \mu g\,l^{-1}$). These data are obtained by plotting the data shown in Table 1.5 in the same manner as is shown in the case of nickel (Figure 1.8). The maximum safe concentrations quoted in Table 1.5 are not amended by safety factors and have not been weighted for the effects of such environmental factors as water hardness, pH, temperature and, in saline waters, salinity. In practice, the available data do not permit this and the effects of experimental factors as demonstrated in short-term acute toxicity tests are extrapolated to long-term exposure.

Table 1.5.

Fish species	Metal	Standard max. safe concentration supporting fish life ($\mu g\,l^{-1}$) i.e. S_x (365 days)	Standard S_{95} ($\mu g\,l^{-1}$) maximum metal concentration permitted for 17 days out of 365 days	$\dfrac{S_{95}}{S_x}$
Non-salmonid	Ni	220	900	4.5
	Se	200	1300	6.5
	V	100	1000–1600	10–16
Salmonid	Cr	100	800	8
Non-salmonid	As	80	600	4.8
Non-salmonid	Ag	70	850	12.1
	Cr	100	1000–3000	10–30
	Zn	23	200	8.7
	Pb	20	100	5
	Cd	4	16	4
Salmonid	Cu	4	17	4.2
Non-salmonid	Hg	2	22	11
Salmonid	Cd	2	6	3

REFERENCES

1. Bowen, H.M.J., in *Environmental Chemistry of the Elements*, Academic Press, London (1979)
2. Waldron, H.A., in *Metals in the Environment*, Academic Press, London (1980)
3. Smith, A.E., *Analyst (London)* **98**, 65 (1973)
4. Brown, V.M., Shurban, D.G. and Shaw, D., *Water Research* **4**, 363 (1970)
5. Alabaster, J.S., Garland, J.H.N., Hart, I.C. and Solbe, J.F.S., An approach to the problem of pollution and fisheries. Symposium of the Zoological Society, London, **29**, 87 (1972)
6. Hart, I.C., The toxicity to fish of some rivers in the Yorkshire Ouse Basin. WPR Report 1299, HMSO, Water Pollution Research Laboratory (1974)
7. Howells, E.S., Howells, M.E. and Alabaster, J.S., *J. Fisheries and Biology* **22**, 447 (1983)
8. Brown, V.M., *Water Research* **2** 723 (1968)
9. Department of the Environment Water and the Environment. The Implementation of Directive 76 (464) EEC on Pollution Caused by Certain Dangerous Substances Discharged in to the Aquatic Environment of the Community: Circular 18/85 September, 1985, HMSO, London (1985)
10. Council of European Communities Directive of 26 September, 1983 on Limit Values and Quality Objectives for Cadmium Discharges: 83/513/EEC; OJL291, 24 October 1983 (1983)
11. Council of the European Communities. Directive on Limit Values and Quality Objectives for Mercury Discharges by the Chlor-alkali Electrolysis Industry: 82/176/EEC; OJL81, 27 March 1982 (1982)
12. Council of European Communities, Directive on Limit Values and Quality Objectives for Measuring Discharges by Sectors other than the Chlor-alkali Industry: 84/156/EEC; OJL74, 17 March 1984 (1984)

CHAPTER 2
Toxicity Evaluation—Animal Tissue Analysis Based

There are several reasons for monitoring the concentration of toxic metals in creatures such as fish and shellfish.

2.1 PROTECTION OF HUMAN HEALTH

This applies to organisms which are harvested for food. Direct analysis of the organisms against accepted standards enables a decision to be made as to whether the organisms are acceptable for human consumption. In the UK, for example[1], regulations exist concerning levels of zinc, chromium, copper, nickel and arsenic in fish and shellfish, and these are based on maximum acceptable intake of these foods for one week. It is stated that the 90th percentile consumption of fish should not exceed 0.79 kg per week, and for shellfish 0.26 kg per week. Table 2.1 shows weekly intakes of metals by consumers observing the recommendations that would result following the consumption of fish containing different levels of total metals. For example, the weekly recommended maximum intake of chromium from fish caught in coastal waters would be 0.237 mg, while that of arsenic would be 11.1–13.2 mg.

2.2 PROTECTION OF ANIMAL SPECIES

Biomagnification and bioaccumulation of metals and organics by fish and creatures other than fish (e.g. crustaceans, molluscs) in non-saline and saline waters will now be considered.

Biomagnification is an increase in concentration of a toxicant moving along a food chain and has been observed for organochlorine pesticides[2] which occur at progressively higher concentrations along the food chain.

Two competing factors operate in *bioaccumulation*, namely the rate of uptake of metals or organics and the rate of loss, and these will govern whether there is a net decrease or an increase in toxicant content of the water creature[3–6].

2.2.1   FACTORS AFFECTING BIOACCUMULATION OF METALS

Bioaccumulation in fish and other creatures is greatest in the following circumstances:

(1) When body weight is lowest, i.e. just after spawning or in younger and smaller creatures
(2) During periods of low rate of growth
(3) In waters of low salinity
(4) In waters of higher temperature
(5) In the absence of competing metals, e.g. bioaccumulation is greater in softwaters than hardwaters
(6) When species is nearer the water surface.

Thus rates of bioaccumulation are greater with creatures of low body weight and rate of growth in surface waters of low salinity and hardness which are at a relatively high temperature.

As a consequence of all these factors having an influence on the extent of bioaccumulation the ratio between reported concentration of metal in water ($\mu g \, l^{-1}$) and in animal or plant life ($\mu g \, kg^{-1}$ dry weight) (i.e. bioaccumulation factor = $\mu g \, kg^{-1}$ in plant or animal/$\mu g \, l^{-1}$ in water) is by no means constant.

Due to bioaccumulation there is an increase in concentration of a toxicant in a particular animal or plant species with time and this has been extensively observed. Metals added to fresh or tidal water tend to be removed by absorption onto particulate matter or by chemical transformation into an insoluble form. Thus sediment concentrations are normally higher than those of the overlying water. At the primary production level macrophytes rooted in these metal-enriched sediments tend to have greater concentrations of metals than the sediment. This is also true for algae whether attached or planktonic, as is illustrated in Table 2.2 for the case of accumulation of metals in sediments in the Severn and Humber estuaries, UK.

Similar bioaccumulation of metal phenomena have been observed in the case of fish and, indeed, bioaccumulation has been studied not only in the whole fish but also in individual fish organs where appreciable differences have been reported between different organs. Van Hoof and Van Son[8] have reported on the extent of bioaccumulation occurring in five different organs taken from rudd (muscle, gill, opercle, liver and kidney). In Table 2.4 are reported concentration factors for four metals (zinc, copper, cadmium and chromium) for organs taken from fish exposed to different levels of the metals for various exposure times between 4 hours and greater than 10 weeks. Of the various organs taken from this particular type of fish it is seen in Table 2.5 that the highest concentration factors always occur in opercle tissue and the lowest in muscle, the other organs being intermediate. It will be noted that higher concentration factors are obtained

Table 2.1.

	Concentration of metal in organisms (mg kg^{-1}), dry weight						Weekly intake of total metals by consumers	
	Cu	Ni	Zn	As	Cr	Total	Maximum intake (recommended kg)	Weight (mg of total metals consumed)
Fish								
Coastal waters	0.5	0.7–1.4	4.6	14.1–16.7	0.3	20.2–23.5	0.79	17.3–18.5
Vicinity of municipal outfall	1–1.9	0.7–1.4	0.8–2.8	10	0.5–1.5	13.0–16.6		10.3–13.1
Remote area	1.2	0.2	2.4	0.5–1.5	0.2	4.5–5.5		3.5–4.3
Shellfish								
Vicinity of municipal outfall	0.4–2.4	2.2–6.5	0.3–0.9	10	1–10	13.9–29.8	0.26	3.6–7.7
Remote area	1	3	0.4	0.5–1.5	0.8	5.7–6.7		1.5–1.7

Table 2.2. Accumulation of metals from Humber and Severn Estuary water into sediment

$$\text{Accumulation factor} = \frac{\mu g\,kg^{-1}\ \text{dry weight in sediment}}{\mu g\,l^{-1}\ \text{in water}}$$

Element	Copper	Lead	Nickel	Zinc	Arsenic	Cadmium
Severn Estuary	15 710–16 300	26 830–67 330	15 280–22 600	13 090–25 640	–	1280–3230
Humber Estuary	57 350–430 000	68 000–136 000	2130–32 000	4060–102 500	3700–37 000	800–4000

Table 2.3. Accumulation of tetramethyl lead in rainbow trout

Exposure (days)	Weight of fish (g)	Fish alive or dead	Water averaged ($\mu g\,l^{-1}$)	Fish ($\mu g\,g^{-1}$) wet weight	Concn factors[a]
1	0.1211	Dead	3.46	0.43	124
2	0.3661	Dead		1.08	312
	0.7982	Dead		2.00	578
3	0.4116	Dead		1.32	382
	0.6300	Dead		2.09	604
7	1.3045	Alive		2.94	850
	1.5466	Alive		3.23	934
	0.8100	Alive		2.25	650
	0.4926	Alive		1.73	500

[a]Concentration factor = concentration of Me_4Pb in fish ($\mu g\,kg^{-1}$)/concentration of Me_4Pb in water ($\mu g\,l^{-1}$)
Source: Chau, Y.K., Wong, P.T.S., Bengert, G.A. and Kramer, V., *Analytical Chemistry* **51**, 186 (1979) © 1979 American Chemical Society.

when the exposure time is extended from 3 to 10 weeks even though the metal concentrations of the water were lower in the 10-week test.

In Figure 2.1 are plotted concentration factors obtained from the opercle versus test duration and concentration of zinc in the water (data taken from Table 2.4). It is seen that the concentration factor increases linearly with increasing exposure time but seems to bear an exponential relationship with metal concentration in the water. A plot of the logarithm of the metal concentration and the exposure time versus concentration factor is linear, as illustrated in Table 2.6 and Figure 2.2.

2.2.2   BIOACCUMULATION OF METALS IN FISH

The bioaccumulation has been measured on copper and zinc in the barnacle *Balanus amphitrite* in estuary water[11]. At concentrations of $1-11$ $\mu g\,l^{-1}$ copper in the water between 39 700 and 625 700 $\mu g\,l^{-1}$ of copper were found in the barnacle tissue giving bioaccumulation factors of 7060 and 384 310. At concentrations of $13-46$ $\mu g\,l^{-1}$ zinc in the water, between 203 600 and 1 937 000 $\mu g\,kg^{-1}$ zinc were found in barnacle tissue giving bioaccumulation factors between 10 660 and 84 600.

Langston and Zhan[12] studied the bioaccumulation of cadmium in the tellinid clam *Macoma balthica* taken at the Whitehaven Cumbria coastline. At 100 $\mu g\,l^{-1}$ cadmium in water the clam picked up 10.150 $\mu g\,kg^{-1}$ cadmium during 29 days' exposure (0.35 $\mu g\,Cd\,g^{-1}\,d^{-1}$) giving a bioaccumulation factor of 101. A bioaccumulation factor of 68 000 has been obtained for iron in kelp (*Ecklonia radiata*) taken in harbour water[13].

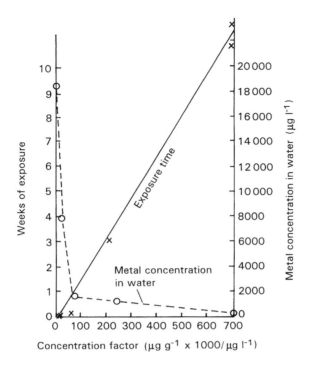

Figure 2.1. Relationship between exposure time and zinc concentration in water and concentration factor obtained for rudd opercle

2.2.3 BIOACCUMULATION OF ORGANIC COMPOUNDS

Bioaccumulation factors of 159, 66 and 17, respectively, have been obtained in exposure of rainbow trout, channel catfish and bluegills to carbendazin[14]. Seawater lampreys were exposed to water containing 50 to 485 μg l^{-1} Kepone for 10 days and gave an average bioaccumulation of 1900[15]. Striped bass (*Morone saxatilis*) exposed to 100 μg l^{-1} Molinate (Ordram) in water for 1 day gave a bioaccumulation factor of 25.3[16]. The crustaceans *Daphnia pulex* and *Palaemon paucideus* upon exposure to 1000 μg l^{-1} fenitrothion for 1–3 days gave, respectively, maximum bioaccumulation factors of 76 and 6[17].

2.2.4 BIOACCUMULATION OF ORGANOMETALLIC COMPOUNDS

The data in Table 2.3 illustrate bioaccumulation of tetramethyl lead present in water at a concentration of 3.46 μg l^{-1} into rainbow trout[7]. The concentration factor ranges from 124 after 1 day's exposure to 800–900 after 7 days' exposure.

Monitoring of bioaccumulation of fresh and tidal waters as trends in spatial monitoring has two purposes:

Table 2.4. Interrelationship between concentration of metal in water, exposure time, and concentration factor in rudd organs

Element	Exposure time	water ($\mu g\,l^{-1}$)	Concentrations in Organs ($\mu g\,g^{-1}$) (or mg kg^{-1})					Concentration factor in organ[a]				
			Muscle	Gill	Opercle	Liver	Kidney	Muscle	Gill	Opercle	Liver	Kidney
Zinc	> 10 weeks[b]	680[b]	16.4	47.9	120.2	29.4	57.0	91	266	667	163	317
	3 weeks[b]	800[b]	24.4	101.9	195.5	104.9	151.7	28	127	244	131	189
	⩽ 24 hours[b]	1600[b]	10.5	38.6	115.3	42.5	154.6	6.5	24	72	26	97
	⩽ 12 hours[c]	7500[bc]	6.6	51.2	90.6	34.1	92.2	0.9	6.8	12.1	4.5	12
	⩽ 4 hours[c]	18 000[bc]	11.2	647.2	174.5	63.5	216.1	0.6	36	9.7	3.5	12
Copper	> 10 weeks[b]	11[b]	0.7	5.5	12.4	6.9	6.0	64	500	11.27	627	545
	3 weeks[b]	50[b]	1.6	8.9	30.9	202	28.5	32	178	618	404	570
	< 12 hours[c]	250[c]	2.3	22.9	52.6	22.3	30.4	9.2	92	210	89	121
	< 12 hours[c]	1200[c]	2.2	29.3	72.1	31.1	39.0	1.8	24	60	26	32
	< 12 hours[c]	1600[c]	4.0	43.2	104	39.8	100	2.5	27	65	25	62
Cadmium	> 10 weeks[b]	3[b]	0.3	2.6	9.5	5.0	4.2	100	868	3166	1666	1400
	3 weeks[b]	250[b]	0.41	2.5	8.7	9.6	15.7	1.6	10	35	38	55
	< 12 hours[c]	1100[c]	0.6	3.9	6.0	4.1	14.4	0.5	3.5	5.4	4.5	13
	< 12 hours[c]	4000[c]	0.5	10.4	20.7	3.8	12.8	0.12	2.6	5.2	0.95	3.2
	< 12 hours[c]	11 000[c]	3.2	87.9	29.2	12.3	28.2	0.29	8.0	2.65	1.1	2.5
Chromium	> 10 weeks[b]	3	< 0.2	< 0.2	< 0.2	< 0.2	< 0.2	< 66	< 66	< 66	< 66	< 66
	3 weeks[b]	16	< 2	< 2	< 2	< 2	< 2	< 125	< 125	< 125	< 125	< 125
	< 12 hours[c]	20	0.5	4.9	8.3	5.6	10.3	125	245	415	280	515
	< 12 hours[c]	80	0.8	48.2	26.0	18.4	23.8	10	602	325	230	297
	< 12 hours[c]	145	0.6	30.6	19.6	15.2	27.8	41	211	135	105	192

[a] Concentration factor $= \dfrac{\text{Concentration in organ (mg kg}^{-1}\text{ or }\mu g\,g^{-1})}{\text{Concentration in water (}\mu g\,l^{-1})} \times 1000$

[b] Subacute toxicity tests, i.e. no fish mortalities.

[c] Acute toxicity tests, i.e. 100% fish mortality.

Source: Van Hoof, F. and Van Son, M., *Chemosphere* **10**, 1127 (1981) with kind permission of Elsevier Science Ltd.

Table 2.5. Summary of concentration factors obtained for different organs taken from rudd at different metal concentrations in water and different exposure times

Exposure time (weeks)	Metal concentration in water (μg l^{-1})		Highest concentration factor[a]	Lowest concentration factor[a]
3	Zn	800	244 (opercle)	28 (muscle)
	Cu	50	618 (opercle)	32 (muscle)
	Cd	250	35/38 (opercle/muscle)	1.6 (muscle)
10	Zn	180	667 (opercle)	91 (muscle)
	Cu	11	1127 (opercle)	64 (muscle)
	Cd	3	3166 (opercle)	100 (muscle)

[a] $\dfrac{\mu\text{g l}^{-1} \text{ in tissue} \times 1000}{\mu\text{g l}^{-1} \text{ in water}}$

Source: Van Hoof, F. and Van Son, M., *Chemosphere* **10**, 1127 (1981) with kind permission of Elsevier Science Ltd.

Table 2.6. Dependence of concentration factor obtained for the opercle on product of log concentration factor and exposure time

Element	Exposure time of rudd (weeks)	Concentration of metal in water (μg l^{-1})	(Log concentration) $\times$ exposure time (a)	Observed concentration factor (b) $\dfrac{\text{mg kg}^{-1} \times 1000}{\mu\text{g l}^{-1}}$	Slope a/b
Zinc	10	180	22.55	667	0.034
	3	800	8.71	244	0.036
	0.143 (1 day)	1600	0.46	72	
	0.0715 (0.5 day)	7500	0.25	12.1	
	0.024 (4 hr)	18 000	0.10	9.7	
Copper	10	11	10.41	1127	0.0092
	3	50	5.10	618	0.0082
	0.0715 (0.5 day)	250	0.17	210	
	0.0715 (0.5 day)	1200	0.22	60	
	0.0715 (0.5 day)	1600	0.23	65	

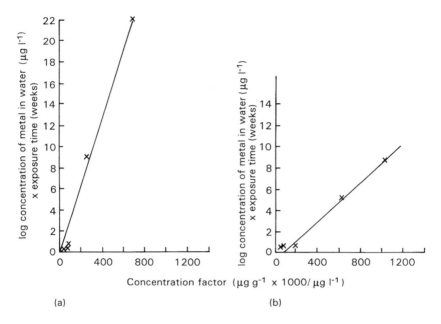

Figure 2.2. Linear relationship between concentration factor

$$\frac{(\mu\mathrm{g\,g}^{-1} \text{ in fish} \times 1000)}{\mu\mathrm{g\,l}^{-1} \text{ in water}} \text{ and product of concentration}$$

in water ($\mu\mathrm{g\,l}^{-1}$) and exposure time (weeks); (a) zinc, (b) copper

(1) Macroscale, i.e. the identification of potentially unknown areas of elevated concentration and assessment of the extent of the zone of contamination
(2) Monitoring of bioaccumulation in fresh and tidal waters as trends in time. These need to be maintained to identify trends in contamination, especially near effluent discharges in order to identify stability, improvement or deterioration in contaminant levels.

Spatial and time monitoring programmes of the types discussed above will also give information needed to assess the risk to top predators in a particular ecosystem.

The design of such programmes is typified by the US Mussel Watch Programme[9] which takes into account the following factors:

- *Species studied*: *Mytilus edulis* mussel was used in this programme as this creature had already been studied for factors affecting accumulation.
- *Time of year*: Late winter was chosen as metal content is stable (i.e. avoiding post-spawning maximum)
- *Size or age*: Dominant size of population sampled to avoid effect of age and size.

- *Position on shore*: Collected on rocky shores to avoid contamination by soft sediments at level on shore exposed for approximately 6 hours each tidal cycle, i.e. 3–4 hours after high tide.
- *Sample size*: Minimum 25 animals to allow statistical assessment.
- *Sampling*: Transported alive in polyethylene bags regularly drained from free water. Placed in clean water for 24 hours prior to analysis to ensure gut contents are eliminated. Analysis on homogenised individual animals and shell dimensions recorded.

This scheme is designed to detect a 10% change in metal concentrations in *Mytilus* mussels with a confidence of 90%.

In one such study mussels from a clean environment were suspended in cages at several locations in the Firth of Forth. A small number were removed periodically, homogenised and analysed for methylmercury. The rate of accumulation of methylmercury was determined and, by dividing this by mussel filtration rate, the total concentration of methylmercury in the seawater was calculated.

The methylmercury concentration in caged mussels increased from low levels (less than 0.01 $\mu g\,g^{-1}$) to 0.06–0.08 $\mu g\,g^{-1}$ in 150 days (Figure 2.3), giving a mean uptake rate of 0.4 $ng\,g^{-1}$ daily, i.e. a 10 g mussel accumulated 4 ng daily. The average percentage of total mercury in the form of methylmercury increased

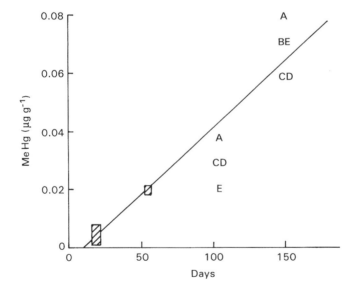

Figure 2.3. The increase with time of methylmercury concentration in caged mussels at positions A–E. Methylmercury was not detectable (0.01 $\mu g\,g^{-1}$) after 20 days and animals from all five positions contained 0.02 $\mu g\,g^{-1}$ after 55 days' exposure, as shown by the shaded rectangles. The case at position B was not sampled at 106 days' exposure. (Reproduced from Davies, I.M., *et al.*, *Marine Chemistry* **7**, 11 (1979) Elsevier Science Publishers, by courtesy of authors and publishers)

from less than 10% after 20 days to 33% after 150 days. This may be compared with analyses of natural intertidal mussels from the area in which the proportion of methylmercury was higher in mussels of lower (less than 10 μg g^{-1}) than of higher total mercury concentrations.

Davies, Graham and Pirie[10] calculated the total methylmercury concentration in the seawater as 0.06 μg l^{-1}, i.e. 0.1–0.3% of the total mercury concentration as opposed to less than 5–32 ng l^{-1} methylmercury found in Minamata Bay, Japan. The bioaccumulation factor μg kg^{-1}/μg l^{-1} of methylmercury in mussels ranged from 17 (one day's exposure) to 1333 (150 days' exposure).

A potentially valuable consequence of this type of bioassay is that estimates of the relative abundance of methylmercury may be obtained at different sites by the exposure of 'standardised' mussels, as used in their experiment, in cages for controlled periods of time, and by the comparison of the resultant accumulations of methylmercury.

REFERENCES

1. Mance, G., Brown, V.M., Gardiner, J. and Yates, J., Proposed Environmental Quality Standards for List II Substances in Water; Chromium. Technical Report TR207, The Water Research Centre, Marlow, Bucks, UK (1984)
2. Phillips, D.J.H., *Quantitative Aquatic Biological Indications*, Applied Science Publishers, London (1980)
3. Bohn, A., *Marine Pollution Bulletin* **6**, 87 (1975)
4. Leatherland M. and Burton J.D., *J. Marine Biological Association* **54**, 457 (1974)
5. Prosi, F., *Schmermetallbleastung in den Sedimenten der Elsenz und ihre Austrilkung auf Unische Organismen*, dissertation, University of Heidelberg (1977)
6. Davis, J.J., Perkins, R.W., Palmer, R.F., Hanson, W.C. and Clive, J.F. Radioactive materials in aquatic and terrestial organisms exposed to reactor effluent water. In 2nd UN International Conference on the Peaceful Uses of Atomic Energy 1968
7. Chau, Y.K., Wong, P.T.S., Bengert, G.A. and Kramer, V., *Analytical Chemistry* **51**, 186 (1979)
8. Van Hoof, F. and Van Son, M., *Chemosphere* **10**, 1127 (1981)
9. Baydon, C.F., *J. Marine Biological Association* **57**, 675 (1977)
10. Davies, I.M., Graham, W.C. and Pirie, S.M., *Marine Chemistry* **7**, 11 (1979)
11. Anil, A.C. and Wagh, A.B., *Marine Pollution Bulletin* **19**, 177 (1988)
12. Langston, W.J. and Zhou, M., *Marine Environmental Research* **21**, 225 (1987)
13. Higgins, H.W. and Mackey, D.J., *Australian Journal of Marine and Freshwater Research* **38**, 307 (1987)
14. Palawski, D.U. and Knowles, C.O., *Environmental Toxicology and Chemistry* **5**, 1039 (1986)
15. Mallett, J. and Barron, N.G., *Archives of Environmental Contamination and Toxicology* **17**, 73 (1988)
16. Tjeerdema, R.S. and Crosby, D.G., *Aquatic Toxicology* **9**, 305 (1987)
17. Takimoto Y., Okshima M. and Miyamoto, J., *Ecotoxicology and Environmental Safety* **13**, 126 (1987)

CHAPTER 3
Control of Pollution Regulations

National (UK) and European (EU) control of the disposal of pollutants to rivers and the oceans is regulated by the following bodies:

	UK	EU
Water pollution control	National river authorities	
Industrial effluents	(eg/NRA)	
Liquid effluents		EU regulations to surface water and groundwater
Solid effluents		
Landfill	Local authorities (WA consulted at planning stage)	EU regulations
Sea dumping	Ministry of Agriculture, Fisheries and Food (MAFF)	EU regulations, also Paris and Oslo Conventions
Special storage	–	–
Atmospheric emissions	Air Inspectorate (HMIP and HMIPI)	EU regulations
Radioactive substances	Nuclear Inspectorate	
Pollution arising from shipping	Marpol	Marpol
Pollution in the Mediterranean	–	Mediterranean Action Plan

Some basic differences exist between the philosophy adopted by the UK and the EU regarding means of controlling pollution.

3.1 EU PHILOSOPHY

The EU defines pollution 'as the discharge by man, directly or indirectly, of substances or energy into the aquatic environment, the results of which are such as to cause hazards to human health, harm to living resources and to aquatic ecosystems, damage to amenities or interference with other legitimate uses of water'. The main EU Directives do not aim to specify any standards but create a framework within which detailed controls for each substance could be developed in subsequent Directives which do quote limit values[1-13]. The limit value is defined as the maximum permissible effluent concentrations of toxic substances expressed as a monthly average. These monthly figures are to be reported to the EU by member states. The EU limit value is equated to the best practical and economic method of effluent treatment. Thus the EU Directives are concerned only with analysis of the effluent and not with the analysis of the receiving water, river or ocean. While concentration maxima for toxicants are not specified for the receiving water, it is required to show that contamination of sediments and biota in the receiving ecosystem is stable and does not increase. Thus the limit values published for mercury and cadmium (Table 3.1) apply to the effluents not the receiving waters and are effectively the maximum permissible effluent concentrations reported as monthly averages. The only requirement of this Directive concerning mercury and cadmium contents of the receiving environment is that the yearly average values should not exceed the stipulated values.

Control by limit values includes no requirement for regular monitoring of the receiving environment or for the environmental concentrations of the controlled substance to be within acceptable limits (i.e. less than the environmental standard in the Directive). This conformity with the limit value may not in practice actually ensure environmental protection. It is possible that the expenditure necessary to comply with the limit value in such circumstances would be a waste in terms of environmental benefit.

The limit value approach takes no account of the effects of such environmental factors as water hardness and temperature and its relative stringency will vary depending upon the water chemistry and geographical location of the receiving water's bodies. Also, for mercury and cadmium the limit values quoted in Table 3.1 for estuaries are less stringent than those for coastal waters when, in fact,

Table 3.1. Limit values (monthly average concentrations) and environmental standards (annual average concentrations) in EU Directives for mercury[2,3], and cadmium[1]

	Limit value for effluents ($\mu g \, l^{-1}$)[a]	Environmental standard ($\mu g \, l^{-1}$)	
Mercury	(all industrial sectors)		
Freshwater	50	1	total
Estuaries	50	0.5	dissolved
Coastal waters	50	0.3	dissolved
		0.03	$mg \, kg^{-1}$ in fish
Cadmium	(most industrial sectors)		
Freshwater	200	5	total
Estuaries	200	5	dissolved
Coastal waters	200	2.5	dissolved

[a] Equating to best practical method of effluent treatment having regards to economics of treatment.

toxicity data indicate that metal toxicity increases as salinity decreases in estuaries.

If the dilution required for the limit value to achieve the environmental standards is considered, then for freshwater such dilution will only be available in the larger rivers, which emphasises the uncertainty of the environmental protection afforded by compliance with the limit value.

3.2 CURRENT EU DIRECTIVES

The EU has, to date, published 300 standards covering determinands (some bacteriological) (see Table 3.2). These standards are published in the form of Directives. The Directives have no direct legal status with individual member states but they are legally binding agreements between the member states who can refer disputes to the European Court.

3.3 UK PHILOSOPHY

The UK has advocated an alternative method of pollution control by quality objectives achieved on the basis of the best practical environmental option (BEPO) as expounded by the Royal Commission on Environmental Protection. This approach is based on the identification of an environmentally acceptable concentration of toxicant in the receiving water identified by the 95th percentile concentration (S_{95}) and a long-term exposure standard (S_x) as defined in Chapter 1. This environmentally safe concentration is identified to ensure the continued use of fresh or saline waters for specific purposes. Discharge limits are then calculated to ensure that the safe concentration is only exceeded in the immediate zone of mixing of the effluent with the receiving water. This approach ensures environmental protection, but could mean that a discharge may require more or less stringent effluent standards than specified by the EU limit values.

In the UK approach, when determining consent conditions for metal-bearing effluents, account is taken of all known downstream uses of the receiving water and the most stringent standards for the composition of the receiving water applicable from Table 3.3 are used for each metal to ensure that the relevant standards are not being exceeded. These standards are, when necessary, weighted for water hardness and salinity. This approach has some distinct advantages and many disadvantages over the EU approach and debate on the relative merits and demerits of both approaches continues unabated.

As mentioned above, the UK best practical environmental option (BEPO) is based on the identification of an environmentally acceptable concentration of toxicant in the receiving water identified by a 95th percentile concentration (S_{95}) and a long-term (4-year) exposure standard (S_x). The figures given in Table 3.3

Table 3.2. EU Directives

Directive No.	Environmental water quality	Substances covered
79/923/EEC (30.10.79)	Shellfish[4]	
76/160/EEC (8.12.75)	Bathing waters[5]	
78/639/EEC (18.7.78)	Fresh water[6]	Cu, Zn
75/440/EEC (16.6.75)	Surface water abstracted for public supply[7]	As, Cd, Cr, Pb, Hg, Zn
Effluent controls		
76/464/EEC (4.5.76)	Dangerous substances directive[8]	Lists 1[a] and 11[b]
80/68/EEC (17.12.79)	(a) Groundwater[13]	Lists 1[a] and 11[b]
82/176/EEC (27.3.82)	(b) Mercury from chloralkali process[1]	Hg
84/156/EEC (17.3.84)	(c) Mercury from other sources[2]	Hg
83/513/EEC (26.9.83)	(d) Cadmium[3]	Cd
84/491/EEC (17.10.84)	(e) Lindane[9]	Lindane
Waste from titanium dioxide industry		
78/176/EEC (20.2.78)	(a) Waste disposal[10]	
82/883/EEC (3.12.82)	(b) Surveillance and monitoring[11]	Ti, Cr, Cd, Hg, V, Ni, Zn, Cu, Pb
Potable water industry		
80/778/EEC (15.7.80)	Drinking water for human consumption[12]	Al, As, Cd, Cr, Cu, Pb, Hg, Ni, Sb, Ag, Zn

[a]List 1 (black list) comprising:

Organohalogen compounds
Organophosphorous compounds
Organotin compounds
Carcinogenic compounds
Mercury and its compounds
Cadmium and its compounds
Persistent mineral oils and petroleum hydrocarbons
Persistent synthetic substances (organics) which float or sink in water

[b]List 11 (grey list) comprising:

Substances on List 1 for which limit values have not been determined.
Individual substances and categories of substances belonging to the following families and groups:
    Zn, Cu, Ni, Cr, Pb, Se, As, Sb, Mo, Ti, Sn, Ba, Be, B, U, V, Co, Tl, Te and Ag.
Biocides and their derivatives
Substances having taste and smell
Toxic or persistent organic compounds of silicon
Inorganic phosphorus compounds in elementary phosphorous
Non-persistent mineral oils and petroleum hydrocarbons
Cyanides, fluorides
Substances affecting dissolved oxygen balance of water, e.g. ammonia and nitrates.

Table 3.3. Example of UK control by quality objective approach annual average ($\mu g\,l^{-1}$) dissolved metals in receiving water = S_x 95th percentile concentration ($\mu g\,l^{-1}$) = S_{95}

Element	Lead		Chromium		Zinc		Copper		Nickel		Arsenic	
	S_x	S_{95}	S_x	S_{95}	S_x	S_{95}	S_x	S_{95}	S_x	S_{95}	S_x	S_{95}
Use: protection of fish												
Salmonids												
Hardness 0–50	–	–	5	–	10[a]	30	1	5	50	–	50	–
50–100	10	–	10	–	50[a]	200	6	100	100	–	50	–
100–150	10	–	20	–	75[a]	300	10	150	150	–	50	–
150–200	20	–	20	–	75[a]	300	10	150	150	–	50	–
200–250	20	–	50	–	75[a]	300	10	200	200	–	50	–
250+	20	–	50	–	125[a]	500	28	200	200	–	50	–
Coarse fish												
Hardness 0–50	50	–	150	–	75[a]	300	1	5	50	–	50	–
50–100	125	–	175	–	175[a]	700	6	22	100	–	50	–
100–150	125	–	200	–	250[a]	1000	10	40	150	–	50	–
150–200	250	–	200	–	250[a]	1000	10	40	150	–	50	–
200–250	250	–	250	–	250[a]	1000	10	40	200	–	50	–
250+	250	–	250	–	500[a]	2000	28	112	200	–	50	–
Use: protection of other forms of freshwater life												
Hardness 0–50	5	–	5	–	100	–	1	–	8	–	150	–
50–100	60	–	10	–	100	–	6	–	30	–	150	–
100–150	60	–	20	–	100	–	10	–	50	–	150	–
150–200	60	–	20	–	100	–	10	–	50	–	150	–
200–250	60	–	50	–	100	–	10	–	100	–	150	–
250+	60	–	50	–	100	–	28	–	100	–	150	–
Use: protection of saltwater fish and shellfish												
Taken	25	–	15	–	140	–	5	–	30	–	25	–
from table	20	100	100	800	23	200	4	17	220	900	80	600

[a]Total zinc.

are weighed for the above water hardness and are based on practical measurements of lethal effect LC_{50} for various time intervals and for waters of different hardness.

As discussed in Chapter 1, measurement for different toxicants' lethal effect (LC_{50}) for different time intervals, e.g. 1-, 4-, 10-, 100-, 500- and 1000-day test durations enables the parameters, namely S_{95} and S_x, to be calculated. Plots of LC_{50} versus test duration enables the 365-day asymptotic concentrations S_x to be obtained for each toxicant, i.e. the annual average concentration. Such plots enable the 95% percentile S_{95} concentration to be obtained, i.e. that concentration which can be safely exceeded for 5% of one year (in 17 days out of 365). In Table 1.5 (Chapter 1) are quoted S_x and S_{95} values for a range of toxicants.

Consider the cases of the toxicity of chromium and copper to salmonid fish, summarised in Table 3.4. For chromium the S_x value derived from measurement of lethal effect LC_{50} for various time intervals is 100 $\mu g\,l^{-1}$ (Table 1.5). The maximum concentration recommended in the UK BEPO approach (Table 3.3) ranges from 5 $\mu g\,l^{-1}$ at a water hardness of 0–50 mg l^{-1} calcium carbonate to 50 mg l^{-1} at a hardness exceeding 250 mg l^{-1} calcium carbonate. Thus the BEPO figures have a built-in safety factor ranging from $20(=100/5)$ at a hardness of less than 50 to a value of $2(=100/50)$ at a hardness exceeding 250.

For toxicity of copper to salmonids the S_x and S_{95} values derived from the measurement of lethal effect LC_{50} for various time intervals are, respectively, 4 and 17 $\mu g\,l^{-1}$ (Table 1.5). The S_x and S_{95} concentrations recommended in the UK BEPO approach for waters with a hardness of less than 50 are 1 and 5 $\mu g\,l^{-1}$, i.e. safety factors, respectively, of $4(=4/1)$ and $3.4(17/5)$. For waters with hardnesses exceeding 50 there is no safety factor. For example, at a hardness exceeding 250 the safety factor for S_x and S_{95} quoted in the BEPO list are $0.14(=4/28)$ and $0.085(=17/200)$. This is presumably because it is known that

Table 3.4. Chromium and copper S_x and S_{95} safety factors for salmonid fish in the UK BEPO[a] approach to quality objectives

Element	Calculated from LC_{50}—exposure time plots		UK control by quality objective approach				
	S_{xA}	S_{95A}	Hardness	S_{xB}	S_{95B}	$\dfrac{S_{xA}}{S_{xB}}$	$\dfrac{S_{95A}}{S_{95B}}$
Chromium	($\mu g\,l^{-1}$)		(mg l^{-1})	($\mu g\,l^{-1}$)			
	100	800	0–50	5	–	20	–
			150–200	20	–	5	–
			250+	50	–	2	–
Copper	4	17	0–50	1	5	4	3.4
			150–200	10	150	<1	–
			250+	28	200	<1	–

[a]BEPO: Best practicable environmental option.

the acute toxicity of copper decreases sharply in waters of high water hardness[14-19].

REFERENCES

1. Council of the European Communities Directive of 20 September 1983 on Limitations and Objectives for Cadmium Discharges. 83/513/EEC; OJL 291 24 October 1983 (1983)
2. Council of the European Communities Directive on Limit Values and Quality Objectives for Mercury Discharges by the Chloralkali Electrolysis Industry. 82/176/EEC; OJL 81 27 March 1982 (1982)
3. Council of the European Communities Directive on Limit Values and Quality Objectives for Mercury Discharges by Sectors other than the Chloralkali Industry. 84/156/EEC; OJL 74 17 March 1984 (1984)
4. Council of the European Communities Directive on the Quality Required of Shellfish Waters. 30 October 1979. 79/923/EEC; OJL 281 10 November 1979 (1979)
5. Council of the European Communities Directive of 8 December 1975 concerning the Quality of Bathing Water. 76/160/EEC; OJL 31 5 February 1976 (1976)
6. Council of the European Communities Directive of 18 July 1978 on the Quality of Freshwater Needing Protection or Improvement in Order to Support Fish Life. 78/659; OJL 222 14 August 1978 (1978)
7. Council of the European Communities Directive of 16 June 1975 concerning the Quality Required of Surface Water Intended for the Abstraction of Drinking Water in the Member States. 75/440/EEC; OJL 194 25 July 1975 (1975)
8. Council of the European Communities Directive of 4 May 1976 on Pollution Caused by Certain Dangerous Substances Discharged into the Aquatic Environment of the Community. 76/464/EEC; OJL 129 18 May 1976 (1976)
9. Council of the European Communities Directive on Limit Values and Quality Objectives for Discharge of Hexachlorocyclohexane, and in Particular Lindane. 84/491/EEC; OJL 274 17 October 1984 (1984)
10. Council of European Communities Directive on Waste from the Titanium Dioxide Industry, 20 February 1978. 78/176/EEC; OJL 54 25 February 1978 (1978)
11. Council of European Communities Directive of 3 December 1982 on Procedures for the Surveillance and Monitoring of Environments Concerned by Waste from the Titanium Dioxide Industry. 82/883/EEC: OJL 378 31 December 1982 (1982)
12. Council of European Communities Directive of 1 July 1980 Relating to the Quality of Water Intended for Human Consumption. 80/778/EEC: OJL229 30 August 1980 (1980)
13. Council of European Communities Directive of 17 December 1979 on the Protection of Groundwater against Pollution caused by certain Dangerous Substances. 80/68/EEC; OJL 20, 20 January 1980 (1980)
14. Alabaster, J.S. and Lloyd, R. (eds), *EIFAC Water Quality Criteria for Freshwater Fish*, UN Food and Agriculture Organisation, Butterworths, London (1980)
15. Environmental Protection Agency, *Ambient Water Quality, Criteria for Copper*, EPA-440/5-80-036, Washington, DC (1980)
16. De, Mayo A. and Taylor, M.C., *Guidelines for Surface Water Quality Criteria, Vol. 1. Inorganic Chemical Substances, Copper*, Environment Canada, Ottawa (1981)
17. Spear, P.A. and Pierce, P.G., *Copper in the Aquatic Environment. Chemistry, Distribution and Toxicity*, National Research Council of Canada, Ottawa No. 16454 (1979)

18. Nriagu, J. (ed.) In *Copper in the Environment. Part 11, Health Effects*, John Wiley, New York (1979)
19. Mance, G., Brown, V.M. and Yates, J., *Proposed Environmental Quality Standards for List 11 Substances in Water. Copper*, Technical Report TR 210, Water Research Centre, Marlow, Bucks (1984)

CHAPTER **4**
Toxicity Data

This chapter reviews the available information on the toxicity towards fish and creatures other than fish of various types of pollutants—metallic, organometallic and organic. Most of this information is concerned with the concentrations of these substances in the water to which the creatures are exposed whether it be freshwater or seawater. Information has also been reported on the the concentrations of toxicants found in tissues of creatures that are known to have been killed by pollutants, i.e. a study of acute exposures resulting in fish killed. Although there is a vast amount of literature available concerning the presence of metals in fish it deals almost exclusively with levels in muscle tissue or in whole fish after chronic exposure. Van Hoof and Van Son[1] pointed out that investigation of the cause of fishkills by water analysis alone has serious drawbacks since in many cases the causative agent may have been diluted, biodegraded or volatilised to a level which does not allow an unambiguous interpretation at the time of sampling. Eventually it may have been displaced from the site where killed fish are localised. The work of Van Hoof and Van Son on copper, cadmium, zinc and chromium is discussed where relevant in this chapter.

Mount and Stephen[2] developed an autopsy technique for zinc-caused fish mortalities and found that the ratio of opercle/gill zinc concentrations gave valuable information for discrimination between acute and chronic exposure. Mount and Stephen found that cadmium intoxications in the bluegill sunfish (*Lepomis macrochirus*) and catfish (*Ictalurus nebulosus*) could be demonstrated through analysis of gill tissue. Martin *et al.*[3] and Kariya *et al.*[4] found similar results for copper in five different fish species. These findings were not confirmed by the work of Brings *et al.*[5], who found no significant differences between copper tissue levels in *Ictalurus nebulosus* exposed to acute lethal and subacute non-lethal concentrations.

4.1 TOXICITY OF METALS

The toxicity of metals towards fish and invertebrates is now discussed in order of increasing toxicity. For comparison, each element is identified by the mean (S_x) and 95% percentile (S_{95}) concentrations discussed in Chapters 1 and 2. Lethal

LC_{50} values for fish and invertebrates are given in Table 4.1 as a function of the type of water (non-saline or saline), type of creature and the exposure time.

4.1.1   FRESH (NON-SALINE) WATERS

Nickel ($S_x = 220\,\mu g\,l^{-1}$, $S_{95} = 900\,\mu g\,l^{-1}$, non-salmonid and salmonids)

Fish

The long-term LC_{50} value ($500\,\mu g\,l^{-1}$) is similar for salmonid and non-salmonid fish. Impaired reproducibility of non-salmonids occurs at $100\,\mu g\,l^{-1}$ nickel for 4 days' exposure or $50\,\mu g\,l^{-1}$ nickel for 15 days' exposure or $110\,\mu g\,l^{-1}$ nickel for 100 days, compared with lethal concentrations of, respectively, 35 000, 8000 and $2200\,\mu g\,l^{-1}$ (Table 4.1).

Invertebrates

Low concentrations of nickel ($15\,\mu g\,l^{-1}$) impair the reproduction of invertebrates.

Selenium ($S_x = 200\,\mu g\,l^{-1}$, $S_{95} = 1300\,\mu g\,l^{-1}$, non-salmonids)

In recent years, the physiological role of selenium as a trace element has created considerable speculation and some controversy. Selenium has been reported as having carcinogenic as well as toxic properties. Other authorities have presented evidence that selenium is highly beneficial as an essential nutrient[6,7].

Fish

Exposure of salmonid fish for 250 days to $40-50\,\mu g\,l^{-1}$ selenium reduced blood volume[8]. Salmonids and non-salmonids are equally sensitive to selenium. Depending on its concentration, selenium can cause mortalities in spawning, cataract development and reduced larval survival in fish (see Section 5.1.1, Table 5.6).

Invertebrates

The toxicity of selenium depends on its chemical form. Thus the 2-day LC_{50} for selenite and selenate are, respectively, 1100 and $5300\,\mu g\,l^{-1}$[9]. The 2-day LC_{50} for selenium in the case of *Daphnia magna* in non-saline waters is $680\,\mu g\,l^{-1}$ (adults) and $750\,\mu g\,l^{-1}$ (juveniles)[10].

   Depending on its concentration selenium can have an adverse effect on growth and reproduction in invertebrates[230] (see Section 5.1.1, Table 5.7).

Table 4.1. Concentration of elements in freshwater causing mortalities (LC_{50}) for salmonid and non-salmonid fish and invertebrates

Element	S_x (μg l^{-1}) (Table 1.5)	S_{95} (μg l^{-1}) (Table 1.5)	Fish Exposure time (days)	Fish LC_{50} (μg l^{-1})	Invertebrates Exposure time (days)	Invertebrates LC_{50} (μg l^{-1})
Ni	220	900 (n/s,s)	Long term 100 15 4	500 (s,n/s) 2200 (s,n/s) 8000 (s,n/s) 35 000 (s,n/s)	–	–
Se	200	1300 (n/s)	4 (8) 10	2900–3060 300	2 2 2	1100 (asSeO3) (9) 5300 (asSeO4) (9) 680 (adult) (10) 750 (juvenile) (*Daphnia magna*)
V	100	1000–1600 (n/s)	7	2400–3000 (s) (saltwaters) 2900–5000 (s) (hardwaters)	–	–
Cr	100	800 (s)	100 10 4 60 4	1150 (s) 18 300 (s) 3300–65 000 (s) 200 (n/s) 25 000–169 000 (n/s)	3	30–80 (as CrVI) (crustaceans)
As	80	600 (n/s,s)	4	14 400 (12)		
Ag	70	850 (n/s)	Short term	10–10 000 (as AgNo$_3$)	3–5	1000
Zn	23	200 (n/s)	500–1000	260 (23, 24) (soft waters)	4	70 (*Daphnia magna*)

continued overleaf

Table 4.1. (*continued*)

Element	S_x (µg l⁻¹) (Table 1.5)	S_{95} (µg l⁻¹) (Table 1.5)	Fish Exposure time (days)	Fish LC_{50} (µg l⁻¹)	Invertebrates Exposure time (days)	Invertebrates LC_{50} (µg l⁻¹)
Pb	20	100 (n/s)	500–1000	1050 (23, 24) (hardwaters)	4	10 000 (annelids, insect larvae)
			4	2600 (juvenile) 2400 (adult) Rainbow trout (*Salmo gairdnera*) (14)	60	200–600 (snail, *Ancyclus fluviatus*) (14)
			4	13 300–33 000 (*Tilapiazilli*) (15)	60	2000 (amphipod, *Allorchestes compressa*) (17)
			4	2600–52 000 (*Clarias lazena*) (15)		
Cu	4	17 (s)	90	5500 (16)		Similar to fish
			40	900 (16)		
			4	1500 (16)	3	24 (17) (crustaceans)
			72	80 (s)	4	400–2000 (17, 18) (molluscs)
			30	200 (s)	1	500 (19) (adult clams)
			6	250–400 (s) (16)	40	140 (19) (juvenile clams)
					8	5000 (19) (juvenile clams)

Cd	4	16 (n/s)	100	180 (n/s)	4	480 (amphipod *Allorchestes compressa*)
	2	6 (s)	10	4000 (n/s)	4	680 (crustaceans)
			700	2 (s)	4	780 (amphipod, *Allorchestes compressa*) (29)
			4	<10 000 (*Notropis lutrensis*) (48)		
				<10 000	25	10 (*Daphnia magna*) (53)
			4	(fathead minnow, *Pimephales promelas*) (48)		
			4	12 600 *Punctius conchonus* (45)		
			4	350 000 (male)	3	0.2 (crayfish)
				371 000 (female) (*Herbistes reticulatis*) (19)	2	110 (slipper limpet, *Cripidula fornicata*) (17)
Hg	2	22 (n/s)	30	2 (asMe Hg (25))		
Al	–	–	4	3800 (rainbow trout, *Salmo gairdneri*) (55)		
Fe	–	–	–	–	4	25 610–43 100 (isopod, *Asselus aquaticus*) (58)
NH$_4$	–	–	4	2170 (fathead minnow, *Pimephales promelas*) (60)	–	–

n/s: non-salmonids
s: salmonids
References in parentheses.

Vanadium ($S_x = 100 \, \mu g \, l^{-1}$, $S_{95} = 1000-1600 \, \mu g \, l^{-1}$, non-salmonids

Fish

The lowest adverse effect concentration (LC_{50}) observed for this element is
1130 $\mu g \, l^{-1}$. Toxicity increases at higher pH values. The chemical form of
vanadium also affects toxicity, vanadate being the most toxic and vanadium
pentoxide the least[49].

Chromium ($S_x = 100 \, \mu g \, l^{-1}$, $S_{95} = 800 \, \mu g \, l^{-1}$ (salmonids), and $S_x = 100 \, \mu g \, l^{-1}$, $S_{95} = 1000-3000 \, \mu g \, l^{-1}$ (non-salmonids)

Fish

The 4-day LC_{50} for salmonids is appreciably lower than that for non-salmonids,
(Table 4.1), i.e. salmonids are more sensitive to chromium. Concentrations of
chromium as lows as 13 $\mu g \, l^{-1}$ for 60 days adversely affect growth in fish,
720 $\mu g \, l^{-1}$ for 60 days reduces growth of non-salmonids and 2300 $\mu g \, l^{-1}$ for 10
days or 100 $\mu g \, l^{-1}$ for 100 days reduces growth in non-salmonids. Chromium is
more toxic to fish at lower pH values.

Invertebrates

Insect larvae are least affected and crustacae most affected by chromium.
Invertebrates are more sensitive than fish. Crustaceans are very sensitive, having
a 3-day LC_{50} of 30–80 $\mu g \, l^{-1}$ as hexavalent chromium. Trivalent chromium is
believed to be less toxic than the hexavalent form.

In Table 4.2 are presented results obtained in subacute and acute toxicity tests
carried out by exposing rudd to water containing various levels of chromium,
present as potassium dichromate, for various times. One hundred per cent
mortalities of rudd occurred when the chromium content of the water was
somewhere within the range 20–80 $\mu g \, l^{-1}$ during a 12-hour exposure. Chromium
levels were found in all organs of rudd, and the most elevated values were detected
in gill tissue in all fish killed. Chromium levels found in the experiment using the
highest chromium concentration (145 mg Cr l^{-1})) were lower than those found
after exposure to 80 mg Cr l^{-1}, probably because of the shorter exposure time
before death. Chromium levels in all organs of killed fish differ significantly from
exposed surviving fish, which have higher chromium concentrations in opercle,
kidney and liver than in gill tissue. Similar results were reported by Van der Putte
et al.[11] after exposing rainbow trout to hexavalent chromium.

Arsenic ($S_x = 80 \, \mu g \, l^{-1}$, $S_{95} = 600 \, \mu g \, l^{-1}$, non-salmonids)

Large amounts of arsenic enter the environment each year because of the use of
arsenic compounds in agriculture and industry as pesticides, feed preservatives,

Table 4.2. Toxicity of chromium (as potassium dichromate)

	Control fish	Subacute exposure	Acute exposure		
Composition of water (μg l^{-1})	3	16	20	80	145
Composition of tissue (μg l^{-1}) (dry weight basis) Organ					
Muscle	< 0.2	< 2	0.5	0.8	0.6
Gill	< 0.2	< 2	4.9	48.2	30.6
Opercle	< 0.2	< 2	8.3	26.0	19.6
Liver	< 0.2	< 2	5.6	18.4	15.2
Kidney	< 0.2	< 2	10.3	23.8	27.8
Mortality (%) during exposure time	0	0	0	100	100
(Weeks)	> 10	3			
(Hours)			< 12	< 12	< 12

Source: Van Hoof, F. and Van Son, M., *Chemosphere* **10**, 1127 (1981) by kind permission of Elsevier Science Ltd.

herbicides, insecticides, feed additives and wood preservatives. The main amount is used as inorganic arsenic (arsenite, arsenate) and about 30% as organoarsenicals such as monomethylarsinate and dimethylarsinate as agricultural chemicals. Arsenic is known to be relatively easily transformed between organic and inorganic forms in different oxidation states by biological and chemical action. As the toxicity and biological activity of the different species vary considerably, information about the chemical form is of great importance in environmental analysis.

Of the two oxidation states, As(III) and As(V), the latter is the more common in an oxidising environment and is more toxic.

Arsenic has some similar toxic properties to lead, mercury and cadmium as regards bonding to sulphur and inhibiting enzyme action such as pyruvate dehydrogenase. The order of toxicity of arsenic compounds is arsines (As(III)) > arsenite (As(III)) > arsenate (As(V)) and arseno-organic-acids (As(V)). Arsenic, which is found mainly in the liver, kidneys, lungs and intestinal walls, is readily absorbed if water soluble.

Fish

Exposure of fish to 4000 μg l^{-1} arsenic for 30 days reduces fish growth[12]. Organoarsenic compounds are less toxic to fish than inorganic arsenic.

Invertebrates

Arsenic is relatively non-toxic to these creatures, needing 1000 μg l^{-1} to cause mortalities in short-duration tests (Table 4.1). Insect larvae are the least sensitive and crustaceans are the most sensitive to arsenic.

Silver ($S_x = 70 \, \mu g \, l^{-1}$, $S_{95} = 850 \, \mu g \, l^{-1}$, non-salmonids)

Fish

The toxicity of silver reduces depending on its chemical form in the order nitrate, chloride, iodide, sulphide and thiosulphate. Thus eggs and larvae of *Pimephales prometas* exposed to 650 $\mu g \, l^{-1}$ silver nitrate or 11 000 $\mu g \, l^{-1}$ silver sulphide for 30 days caused 20% mortalities[13]. In general, non-salmonids are more sensitive to silver than salmonids.

Invertebrates

Adverse effects have been observed at 10 $\mu g \, l^{-1}$ silver.

Zinc ($S_x = 23 \, \mu g \, l^{-1}$, $S_{95} = 200 \, \mu g \, l^{-1}$, non-salmonids)

Fish

Exposure of fish to 210–520 $\mu g \, l^{-1}$ zinc for 30–140 days affected non-salmonid fish growth while 200–300 days' exposure to 180 $\mu g \, l^{-1}$ zinc affected reproduction. Salmonids are more sensitive to zinc than non-salmonids in short-term exposures but have a similar sensitivity in long-term exposures. The 4-day LC_{50} value in non-saline water obtained for rainbow trout (*Salmo gairdneri*) exposed to zinc was 2600 $\mu g \, l^{-1}$ (juveniles) or 2400 $\mu g \, l^{-1}$ (adults)[15] (Table 5.6). Corresponding 4-day LC_{50} values in non-saline waters obtained for *Titapiazilli* and *Clarias lazena* were 13 300–33 000 $\mu g \, l^{-1}$ and 26 000–52 000 $\mu g \, l^{-1}$ [16] (see Section 5.1.1., Table 5.6).

In Table 4.3 are presented results obtained in subacute and acute toxicity tests carried out by exposing rudd to water containing various levels of zinc for various times[16]. One hundred per cent mortalities of rudd occurred when the zinc level of the water was somewhere in the range 1600–7500 $\mu g \, l^{-1}$ for up to 12 hours' exposure.

The zinc values found in all tissue were significantly different from control values after exposure to 18 000 $\mu g \, l^{-1}$ zinc l^{-1}. In this case all fish died within 4 hours and values found in gill tissues are clearly higher than in other tissues. These findings were not entirely confirmed after exposure to 7500 $\mu g \, l^{-1}$ zinc l^{-1}, in which case all fish died within 12 hours, although zinc levels in gills and opercle are significantly higher than control levels, and after exposure to 1600 $\mu g \, l^{-1}$ zinc l^{-1} which caused no kill after 24 hours. In all experiments kidney zinc levels were significantly higher than control values, suggesting that this organ might in this case also give supplementary information about acute

Table 4.3. Toxicity of zinc

	Control fish	Subacute exposure	Acute exposure		
Composition of water ($\mu g\, l^{-1}$)	180	800	1600	7500	18 000
Composition of tissue ($\mu g\, l^{-1}$) (dry weight basis) Organ					
Muscle	16.4	22.4	10.5	6.6	11.2
Gill	47.9	101.9	38.6	51.2	647.2
Opercle	120.2	195.5	115.3	90.6	174.5
Liver	29.4	104.9	42.5	34.1	63.5
Kidney	57.0	151.7	154.6	92.2	216.1
Mortality (%) during exposure time	Nil	Nil	Nil	100	100
(Weeks)	> 10	3	–	–	–
(Hours)	–	–	$\geqslant 24$	$\leqslant 12$	$\leqslant 4$

Source: Van Hoof, F. and Van Son, M., *Chemosphere* **10**, 1127 (1981) by kind permission of Elsevier Science Ltd.

exposure. After exposing rudd to 800 $\mu g\, l^{-1}$ zinc for three weeks zinc in the opercle gave higher values than kidney and gill tissue, suggesting that the opercle gives the most valuable information concerning non-lethal exposure. A study of the toxicity of zinc sulphate to rainbow trout dealt only with acute toxicity as measured by fish mortality[35].

Invertebrates

Four days' exposure to 70 $\mu g\, l^{-1}$ zinc caused a 50% mortality in *Daphnia magna*. The 4-day LC_{50} of annalids and insect larvae was 10 000 $\mu g\, l^{-1}$. Higher water hardness reduces toxicity of zinc to some gastropods.

Sixty-day LC_{50} values obtained in non-saline waters obtained for zinc were in the case of *Ancylus Ruviatis* snails 200–600 $\mu g\, l^{-1}$ depending on snail size[14] and in the amphipod *Allorchestes compressa* 2000 $\mu g\, l^{-1}$ [18] (see Section 5.1.1, Table 5.7). Depending on its concentration, zinc can reduce the reproductive capacity of invertebrates[18] (see Section 5.1.1, Table 5.7).

Lead ($S_x = 20\ \mu g\, l^{-1}$, $S_{95} = 100\ \mu g\, l^{-1}$, non-salmonids)

Fish

LC_{50} values differ little between salmonids and non-salmonid fish. Improved reproducibility of non-salmonids occurs at 70 $\mu g\, l^{-1}$ for 40 days' exposure or 400 $\mu g\, l^{-1}$ for 90 days' exposure[296]. The 4-day LC_{50} values obtained for *Lebistes reticulatis* in non-saline waters are 1 620 000 $\mu g\, l^{-1}$ (males) and 1 630 000 $\mu g\, l^{-1}$ (females)[19]. (See Section 5.1.1, Table 5.6).

Invertebrates

Crustaceans and gastropods have a similar sensitivity to lead in long-term exposure tests.

Copper ($S_x = 4\ \mu\mathrm{g\,l^{-1}}$, $S_{95} = 17\ \mu\mathrm{g\,l^{-1}}$, salmonids)

Fish

The reported lowest adverse effect concentration for copper is 2 $\mu\mathrm{g\,l^{-1}}$ [20,21]. The 6-day LC_{50} value is 250–400 $\mu\mathrm{g\,l^{-1}}$ [22]. Doses for copper with salmonid fish range from 200 $\mu\mathrm{g\,l^{-1}}$ for 30 days to 80 $\mu\mathrm{g\,l^{-1}}$ for 72 days (lethal dose) (Table 4.1) to 100 $\mu\mathrm{g\,l^{-1}}$ for 30 days to 30 $\mu\mathrm{g\,l^{-1}}$ for 72 days (reduced growth). An increase in water temperature reduces the toxicity of copper to fish. Depending on copper concentration copper can cause a decreased emergence success of fish eggs, reduced fish activity, necrosis, stress[23] and interference in nervous functions[24] (see Section 5.1.1, Table 5.6).

Invertebrates

Crustaceans are the most sensitive to copper and molluscs the least sensitive[25–27] (Table 4.1). The life stage of the invertebrate is important in determining the toxicity of copper[28]. Thus, adult clams had a 1-day LC_{50} of 500 $\mu\mathrm{g\,l^{-1}}$ while juveniles had an 8-day LC_{50} of 5 $\mu\mathrm{g\,l^{-1}}$ and from 140 $\mu\mathrm{g\,l^{-1}}$ for 40 days to 30 $\mu\mathrm{g\,l^{-1}}$ for 72 days (impaired reproducibility).

In Table 4.4 are presented results obtained in subacute and acute toxicity tests

Table 4.4. Toxicity of copper

	Control fish	Subacute exposure	Acute exposure		
Composition of water ($\mu\mathrm{g\,l^{-1}}$)	11	50	250	1200	1600
Composition of tissue ($\mu\mathrm{g\,l^{-1}}$) (dry weight basis) Organ					
Muscle	0.7	1.6	2.3	2.2	4.0
Gill	5.5	8.9	22.9	29.3	43.2
Opercle	12.4	30.9	52.6	72.1	104
Liver	6.9	20.2	22.3	31.1	39.8
Kidney	6/0	28.5	30.4	39.0	100
Mortality (%) during exposure time	Nil	Nil	100	100	100
(Weeks)	> 30	3	–	–	–
(Hours)	–	–	< 12	< 12	< 12

Source: Van Hoof, F. and Van Son, M., *Chemosphere* **10**, 1127 (1981) by kind permission of Elsevier Science Ltd.

carried out by exposing rudd to water containing various levels of copper for various times[1]. One hundred per cent mortalities occurred when the copper concentration of the water was somewhere in the range of 50–250 $\mu g\,l^{-1}$ for up to 12 hours' exposure.

In acute exposures where 100% mortalities of fish occurred in a few hours the highest concentrations of copper were found in opercle and kidneys. The high concentration of copper found in the opercle may be partly due to adsorption.

The amphipod *Allorchestes compressa* in non-saline waters had a 4-day LC_{50} of 40 $\mu g\,l^{-1}$. Depending on its concentration, copper can reduce the survival rate of juvenile invertebrates[30] (see Section 5.1.1, Table 5.7).

Cadmium ($S_x = 4\,\mu g\,l^{-1}$, $S_{95} = 16\,\mu g\,l^{-1}$ (non-salmonids) and $S_x = 2\,\mu g\,l^{-1}$, $S_{95} = 6\,\mu g\,l^{-1}$ (salmonids)

Cadmium is increasingly recognised as an important environmental pollutant with toxic effects on human and animal life at relatively low levels[31–33]. Environmental concentrations of cadmium are of serious concern because cadmium accumulates in the human body throughout life, from 1 $\mu g\,l^{-1}$ at birth to about 30 mg, with about one third to be found in the kidneys[34]. Based on animal studies, cadmium is preferentially retained by the kidney and liver[35].

In view of the known accumulation of cadmium in biological tissues, a detailed study has been carried out to determine the rate of uptake in the common bluegill (*Lepomise macrochirus Raf.*) exposed to known amounts of cadmium in a carefully controlled aquatic environment. An important objective was to evaluate the relative rates of uptake in vital organs including heart, skin, muscle, gut, gill, kidney, liver and/or bone.

Another study examined the chronic toxicity of cadmium (also copper and zinc) mixtures at sublethal concentrations to the fathead minnow using mortality, physical characteristics and reproduction as bioassay methods[37]. While these studies increase our understanding of biological effects for relatively concentrated heavy metal pollutants in aquatic systems, they provide no evidence for actual rate of accumulation of toxic metals nor the distribution of these in vital organs. Moreover, only rarely in natural waters do the concentrations of toxic metals attain the levels used in most acute and chronic bioassay studies. Thus, experimental evidence for heavy metal accumulation and distribution in organisms exposed to environmentally unrealistic levels of heavy metal pollutants in natural waters should make possible more reliable and general predictions for the long-term effects of such pollutants.

In humans, cadmium accumulates in the liver and kidneys, the average level in wet kidney tissue being 25–50 $\mu g\,g^{-1}$, a level of 200 $\mu g\,g^{-1}$ producing irreversible kidney damage.

Zinc and calcium

Both protect against cadmium poisoning. A calcium-deficient diet enhances cadmium accumulation. The contributing factors to Itai-Itai disease were high cadmium intakes ($>$ 600 μg for most sufferers), low calcium diet and a lack of vitamin D. Itai-Itai disease is always accompanied by renal dysfunction. The levels of cadmium were high in the bones of Itai-Itai sufferers, 1.0–1.4% (ash weight). A low molecular weight protein in the liver, metallothein (MW-7000) approximately one third of which is cysteine, bonds to heavy metals (especially cadmium and mercury) and protects against toxic metals. A sample of metallothein was found to contain 4.2% cadmium.

Hypertension has been attributed to cadmium although the topic is controversial. Respiratory and pulmonary damage is reported to occur from the breathing of cadmium vapour or particles. Cadmium, unlike mercury and lead, does not affect the central nervous system, it cannot cross the placental membrane and the mammary gland is an effective barrier.

Fish

A concentration of 4–13 μg l^{-1} cadmium for 30 to 60 days causes growth reduction of fish[37]. Non-salmonid species are one tenth as sensitive to cadmium as salmonids[38–43]. Young life stages are more susceptible than adults. Impaired reproducibility of non-salmonids occurs at a concentration of cadmium of 15 μg l^{-1} over 100 days (or 240 μg l^{-1} cadmium over 10 days).

Depending on concentration, cadmium can reduce the ability of the fish to withstand heat stress[44], cause branchial lesions and mucus secretion[46], and reduce alkaline phosphatase activity[46,47] (see Section 5.1.1, Table 5.6).

Four-day LC_{50} values for cadmium in non-saline water for *Notropis lutrenis*, fathead minnow (*Pimephales promelas*) and *Punctius conchonus* were respectively $<$ 10 000 μg l^{-1} [48], and 12 600 μg l^{-1} [49]. The corresponding values obtained for *Lebistes reticulatus* were 350 000 μg l^{-1} (males) and 371 000 μg l^{-1} (females) (see Section 5.1.1, Table 5.6).

Invertebrates

Some species are more sensitive to cadmium than others. Thus *Daphnia magna* was adversely affected by 5–7 μg l^{-1} cadmium over 4–20 days[43,50] while *Gammarin pulex* crustacean had a 4-day LC_{50} of 680 μg l^{-1}.

Crustacae are more sensitive to cadmium and insect larvae the least sensitive. A rise in temperature increased toxicity and increased pH reduced toxicity[51]. In Table 4.5 are presented results obtained in subacute and acute toxicity tests carried out by exposing rudd to water containing various levels of cadmium for various times.

Table 4.5. Toxicity of cadmium

	Control fish	Subacute exposure	Acute exposure		
Concentration of water (μg l^{-1})	3	250	1100	4000	11000
Composition of tissue (μg l^{-1}) (dry weight basis) Organ					
Muscle	0.3	0.41	0.6	0.5	3.2
Gill	2.6	2.5	3.9	10.4	87.9
Opercle	9.5	8.7	6.0	20.7	29.2
Liver	5.0	9.6	4.1	3.8	12.3
Kidney	4.2	13.7	14.4	12.8	28.2
Mortality (%) during exposure time	Nil	Nil	100	100	100
(Weeks)	> 10	3	–	–	–
(Hours)	–	–	< 12	< 12	< 12

Source: Van Hoof, F. and Van Son, M., Chemosphere **10**, 1127 (1981) by kind permission of Elsevier Science Ltd.

One hundred per cent mortalities occurred when the cadmium content of water was somewhere in the range 250–1100 μg l^{-1} for up to 12 hours' exposure.

Exposure to increasing concentrations results in elevations of cadmium levels in gill and kidney tissues which are statistically significant against control levels at all exposure levels. Cadmium accumulation in gill tissue has also been observed by Mount and Stephen[2] in bluegills and brown bullheads after acute lethal exposure and Sangalang and Freeman[52] in brook trout after chronic sublethal exposure. Cadmium levels in gills from killed fish are significantly different from levels in exposed surviving fish. Subacute exposures (250 μg Cd l^{-1}) caused the most significant increases in kidney and liver.

The 4-day LC_{50} value for cadmium obtained for the amphipod *Allochestes compressa* in non-saline water was 780 μg l^{-1} [29] (see Section 5.1.1, Table 5.7). *Daphnia magna* in non-saline water had a 25-day LC_{50} value of 10 μg l^{-1} [53]. Depending on its concentration, cadmium can reduce the survival rate of invertebrates[54] (see Section 5.1.1, Table 5.7).

Mercury ($S_x = 2 \mu$g l^{-1}, $S_{95} = 22 \mu$g l^{-1}, non-salmonids)

In recent years mercury has been recognised as a toxic contaminant in the environment. The toxicity of mercury is related to its chemical form. Liquid mercury appears to have little effect, but mercury vapour is readily adsorbed producing brain damage. Mercury I salts are relatively toxic compared to mercury II salts because of their low solubility.

Mercury present in fish occurs almost entirely as methylmercury. The WHO recommends a maximum daily intake of mercury by humans from all sources of 43 μg day^{-1} of which no more than 29 μg day^{-1} should be methylmercury. In lakes and streams mercury can be collected in the bottom deposits where it may remain for long periods of time.

Fish

Thirty days' exposure of fish to $1 \mu g \, l^{-1}$ inorganic mercury causes weight reduction and poor spawning. Three $\mu g \, l^{-1}$ mercury as methylmercury chloride caused 88% mortality in fish[297]. The toxic effects of organomercury compounds are similar to those of inorganic mercury. Depending on its concentration and time of exposure, mercury will inhibit gonadal growth in fish (see Section 5.1.1, Table 5.6).

Invertebrates

Most invertebrates are very sensitive to mercury, e.g. crayfish have a 3-day LC_{50} of $0.2 \mu g \, l^{-1}$. Slipper limpets (*Cripidula fornicata*) have a 2-day LC_{50} value of $1100 \mu g \, l^{-1}$ [17] in non-saline water (see Section 5.1.1, Table 5.7).

Aluminium

Fish

Rainbow trout (*Salmo gairdneri*) when exposed to aluminium in non-saline waters gave a 4-day LC_{50} value of $3800 \mu g \, l^{-1}$ [55]. Depending on its concentration pH and water hardness, aluminium can cause growth suppression[300], reduced survival rate[56] and delayed hatching[55] as well as stress and embryo larval mortalities[57] (see Section 5.1.1, Table 5.6).

Iron

Invertebrates

The freshwater isopod *Asselus aquaticus* when exposed to iron had a 4-day LC_{50} value of $25\,610–43\,100 \mu g \, l^{-1}$ [58] (see Section 5.1.1, Table 5.7).

Ammonium

Fish

Rainbow trout (*Salmo gairdneri*) when exposed in non-saline water to ammonium ions had a 4-day LC_{50} value of $530 \mu g \, l^{-1}$ [59]. Fathead minnow (*Pimphales promelas*) under similar conditions had a LC_{50} value of $2170 \mu g \, l^{-1}$ [60]. Depending on its concentration and exposure time, ammonium ion can cause liver and thyroid degeneration, hyperactivity and mortalities in fish[61] (see Section 5.1.1, Table 5.6).

Invertebrates

Depending on its concentration, ammonium ion can reduce the survival rate[62] and growth[63] of invertebrates (see Section 5.1.1, Table 5.7).

Cobalt

Invertebrates

Depending on its concentration, cobalt can decrease the muscle glycogen level in invertebrates[64] (see Section 5.1.1, Table 5.7).

4.1.2   SALINE WATERS (ESTUARIES, BAYS, COASTAL AND OPEN SEA)

The effect of metal concentrations on marine life, particularly invertebrates, has been examined by various workers and is summarised below.

Nickel

Fish

This element is relatively non-toxic to fish. It is less toxic in saline water (4-day LC_{50} 35 000 $\mu g\,l^{-1}$) than in non-saline water (4-day $LC_{50} = 10\,000$ $\mu g\,l^{-1}$). Toxicity is greater at higher water temperatures[65].

Invertebrates

Nickel is relatively non-toxic to marine organisms (4-day LC_{50} 10 000 $\mu g\,l^{-1}$). Planktonic crustacae and bivalve mollusc larvae are more sensitive (4-day LC_{50} 50–600 $\mu g\,l^{-1}$).

Selenium

In recent years the physiological role of selenium as a trace element has created considerable speculation and some controversy. Selenium has been reported as having carcinogenic as well as toxic properties; other authorities have presented evidence that selenium is highly beneficial as an essential nutrient[6,7]. Its significance and involvement in the marine biosphere is not known. A review of the marine literature indicates that selenium occurs in seawater as selenite ions ($SeO_3{}^{2-}$) with a reported average of 0.2 $\mu g\,l^{-1}$ [66].

Selenium is particularly toxic to invertebrates (4-day $LC_{50} = 2900$ to $> 10\,000\ \mu g\,l^{-1}$). The lowest observed adverse effect concentration was $200\ \mu g\,l^{-1}$.

Vanadium

Vanadium has a tendency towards concentrating in the environment for reasons not yet understood. Environmental mobilisation of vanadium and its compounds occurs by a number of means in the net transport of vanadium into the oceans. Some of these transport processes include terrestrial run-off, industrial emissions, atmospheric wash-out (vanadium in the air comes only from industry, as there are no significant natural sources), river transport, and oil spills, resulting in a complex ecological cycle. There has been discussed the possibility of vanadium deposition due to oil spillage, but no evidence is yet available to confirm the release of vanadium from oil. Since crude oils are rather rich in vanadium (50–200 ppm), it is not inconceivable that some vanadium should be released upon the contact of oil with seawater. The LC_{50} of vanadium is greater than $10\,000\ \mu g\,l^{-1}$.

Chromium

The relative LC_{50} values of annelids, molluscs, crustacae and fish when exposed to trivalent chromium (exposure period not stated) are, respectively, $2200–8000\ \mu g\,l^{-1}$, $14\,000–105\,000\ \mu g\,l^{-1}$, $2000–98\,000\ \mu g\,l^{-1}$ and $12\,400–91\,000\ \mu g\,l^{-1}$. Reduction in salinity from 35 to 15 $g\,kg^{-1}$ reduces the 4-day LC_{50} from $640\,000$ to $190\,000\ \mu g\,l^{-1}$.

Arsenic

The chemical form of arsenic in marine environmental samples is of interest from several standpoints. Marine organisms show widely varying concentrations of arsenic[67–69] and knowledge of the chemical forms in which the element occurs in tissues is relevant to the interpretation of these variable degrees of bioaccumulation and to an understanding of the biochemical mechanisms involved. Different arsenic species have different levels of toxicity[70] and bioavailability[71], and this is important in food chain processes, while physiochemical behaviour in processes such as adsorption onto sediments also varies with the species involved[72]. It has been shown that inorganic arsenic (III and V), monomethylarsenic (MMA) and dimethylarsenic (DMA) acids are present in natural waters[73], biological materials[74] and sediments[75].

The Arsenic in Food Regulations 1959[76] state that foodstuffs must not contain more than 1 $mg\,kg^{-1}$ of total arsenic. Certain exceptions are listed which include fish and edible seaweed and their products, where arsenic contents of above

1 mg kg^{-1} are present naturally. The UK total diet survey suggests that at least 75% of total arsenic ingested originates from fish and shellfish. It is accepted that the arsenic in fish and shellfish is mainly organically bound, hence if any of the more toxic inorganic arsenic is present this is of greater interest. If the levels of the total inorganic arsenic approach 1 mg kg^{-1}, the proportion of arsenic (III) relative to arsenic (V) also assumes importance as the latter is considered to be more toxic than the former[77].

Crab are susceptible to arsenic at the larval life stage[78]. Crustacea are the most sensitive to arsenic and annelids are the least sensitive. For instance, toxicity decreases in the order AsV $>$ organic arsenic $>$ AsIII. Fish are less susceptible (4-day $LC_{50} = 15\,000 - 28\,000 \ \mu\text{g l}^{-1}$) than invertebrates (4-day $LC_{50} = 4000 \ \mu\text{g l}^{-1}$).

Silver

Fish embryos and eggs are relatively sensitive to silver, and 9-day exposure to $90 \ \mu\text{g l}^{-1}$ silver had no adverse effect[84]. At $180 \ \mu\text{g l}^{-1}$ silver growth deformities and 30% mortality were observed[85]. Increase in salinity reduced the toxicity of silver[86].

Zinc

Marine life is relatively resistant to zinc at all life stages. Crustacea, bivalve molluscs and worms undergo damage or fatalities upon 1 to 2 weeks' exposure to $340 \ \mu\text{g l}^{-1}$ zinc[87,88]. Decrease in salinity increased the toxicity of zinc to invertebrates and fish sevenfold[89–91]. Fiddler crabs *Uca annulipes* and *Uca triangularis* when exposed to zinc gave a 4-day LC_{50} value of $66\,420 - 76\,950 \ \mu\text{g l}^{-1}$.

Lead

Mollusc larvae are particularly sensitive to lead, abnormal development occurring upon 2 days' exposure to $400 \ \mu\text{g l}^{-1}$ lead[78].

Copper

The importance of complexing agents in the mineral nutrition of phytoplankton and other marine organisms has been recognised for more than 20 years. Complexing agents have been held responsible for the solubilisation of iron and therefore its greater biological availability[92]. In contrast, complexing agents are assumed to reduce the biological availability of copper and minimise its toxic effect. Experiments with pure cultures of phytoplankton in chemically defined media have demonstrated that copper toxicity is directly correlated to cupric ion in activity and independent of the total copper concentration. In these experiments, cupric ion (Cu^{2+}) concentrations can be varied in media containing a wide range of total concentrations through the use of artificial complexing agents. When Cu^{2+} concentration was calculated for earlier experiments with phytoplankton in defined media it appeared that Cu^{2+} was toxic to a number of phytoplankton species in concentrations as low as 10^{-6} μmol l^{-1}. Since copper concentrations in the world oceans typically range from 10^{-4} to 10^{-1} μmol l^{-1} complexing agents and other materials affecting the solution chemistry of copper must maintain the Cu^{2+} activity at sublethal levels.

Copper may exist in particulate, colloidal and dissolved forms in seawater. In the absence of organic ligands or particulate and coloidal species, carbonate and hydroxide complexes account for more than 98% of the inorganic copper in seawater[93,94]

The young life stage of crustacea and molluscs are more sensitive to copper (2-day $LC_{50} = 300$ μg l^{-1}) than adults (2-day $LC_{50} = 30\,000$ μg l^{-1}). This also applies to fish[75]. Low salinity increases the toxicity of copper. Bivalve molluscs are the most copper-sensitive species yet examined, undergoing reduced growth in the presence of $3-10$ μg l^{-1} copper for prolonged periods. Beyond 500 days' exposure mortalities occurred[97]. Fiddler crabs *Uca annulipes* and *Uca triangularis* when exposed to copper gave 4-day LC_{50} values of $12\,820-14\,810$ μg l^{-1}.

Cadmium

Reduced salinity and higher water temperature both increase the toxicity of cadmium to invertebrates[98,99]. Planktonic crustaceans have a 4-day LC_{50} of $60-380$ μg l^{-1} [100]. Young life stages of invertebrates are sensitive to cadmium[57]. Adult crustaceans are also susceptible to 60 μg l^{-1} of cadmium causing 30% mortality in 60 days' exposure[101]. Fish are relatively resistant to cadmium with a 4-day LC_{50} of $6400-16\,400$ μg l^{-1}. Fiddler crabs *Uca annulipes* and *Uca triangularis* when exposed to cadmium gave 4-day LC_{50} values of $43\,230-48\,210$ μg l^{-1}.

Mercury

The toxicities of organic and inorganic mercury to marine fauna are similar. Thus fish embryos undergo damage when exposed to 67 μg l^{-1} mercury for 4 days[102] and poor hatching when exposed to 32 μg l^{-1} mercury for 32 days.

Crustacea and molluscs are sensitive to mercury as are fish (e.g. crab $LC_{50} = 8$ μg l^{-1} [103]). Fiddler crabs *Uca annulipes* and *Uca triangularis* when exposed to mercury gave a 4-day LC_{50} value of $2750-2830$ μg l^{-1}.

Summary

In Tables 4.6 and 4.7, respectively, are compared the short-term 4-day LC_{50} values obtained for various metals in non-saline and saline waters with typical concentrations of these elements which have been found in natural waters. Such typical concentrations are summarised in Appendix 1. When the concentration of a metal in an environmental water is lower than the 4-day LC_{50} value then less than 50% mortalities occur in this period.

The higher the 4-day LC_{50} value relative to the observed concentration in the environmental water, the fewer the mortalities. Thus if 4-day LC_{50} is 3000 μg l^{-1} and 5 μg l^{-1} are present in environmental water, then few or zero mortalities will occur. If 1000–2000 μg l^{-1} of the metal is present in environmental water then some mortalities ($<$ 50%) and adverse effects will take place.

4.1.3   NON-SALINE FRESHWATERS

Applying this treatment to the results in Table 4.6 it is seen that the following creatures will undergo extensive mortalities (50–100%) upon short exposure to the quoted concentrations of metals:

Some types of fish exposed to 32 μg l^{-1} silver
Daphnia magna exposed to 630 μg l^{-1} zinc
Crustacea exposed to 200 μg l^{-1} copper
Crayfish exposed to 1.3 μg l^{-1} mercury.

Consideration of the ratio of 4-day LC_{50} to environmental concentrations of metals reveals that when these elements occur at the higher end of their observed concentration range in the environment then some mortalities ($<$ 50%) and certainly adverse effects are likely to occur:

Crustacea in the presence of 23 μg l^{-1} chromium
Fish and amphipods in the presence of 200 μg l^{-1} copper
Fish in the presence of 3600 μg l^{-1} aluminium
Some types of fish in the presence of 630 μg l^{-1} zinc.

4.1.4   SALINE FRESHWATERS

Fewer types of creatures will undergo mortalities in saline waters due to the lower environmental concentrations of metals that occur, compared to concentrations present in non-saline inland waters. Thus the only observed case where the environmental concentration exceeds the 4-day LC_{50} value is that of crabs exposed to 15.1 μg l^{-1} mercury in seawater. Other cases where low percentage mortalities or adverse effects might occur (i.e. the 4-day LC_{50}/environmental concentration ratio is low) include bivalve molluscs and plankton, also

Table 4.6. Effect of short-term (4-day) exposure of creatures to typical concentrations of metals found in freshwaters (non-saline)

Element	Creature	4-day LC_{50}[a]	Typical concentration of metal ($\mu g\,l^{-1}$) in freshwater (see Appendix 1)	
			Max	Min
Nickel	Fish	3060 max	40	1.5
		2900 min	40	1.5
Vanadium	Fish	5000 max	24	0.1
		2900 min	24	0.1
Chromium	Fish	65 000 max	23	0.05
		3300 min	23	0.05
	Crustacea	80 max	23	0.05
		30 min	23	0.05
Arsenic	Fish	14 400	490	0.42
Silver	Fish	6700 max	32	0.3
		7 min	*32*	0.3
Zinc	Fish	52 000 max	630	0.86
		2400 min	630	0.86
	Daphnia magna	70	*630*	0.86
	Annelid	10 000	630	0.86
Lead	Fish	1500	60	0.13
	Invertebrates	1500	60	0.13
Copper	Fish	400 max	200	0.48
		250 min	200	0.48
	Crustacea	24	*200*	0.48
	Mollusc	2000 max	200	0.48
		400 min	200	0.48
	Amphipod	480	200	0.48
Cadmium	Fish	371 000 max	5	0.013
		< 10 000 min	5	0.013
	Amphipod	780	5	0.013
Mercury	Crayfish	0.2	*1.3*	0.009
	Slipper limpet	1100	1.3	0.009
Aluminium	Fish	3800	3600	14
Iron	Isopod	43 100 max	5000	1
		28 610		

[a] Max to min range depending on creature type.

Greater than 50% mortalities on 4 days' exposure in the case of (italic): some types of fish when exposed to 32 $\mu g\,l^{-1}$ silver, *Daphnia magna* exposed to 630 $\mu g\,l^{-1}$ zinc, crustacea when exposed to 200 $\mu g\,l^{-1}$ copper, and crayfish when exposed to 1.3 $\mu g\,l^{-1}$ mercury.

planktonic crustacea in the presence of 5.3 $\mu g\,l^{-1}$ nickel and young crustacea and molluscs in the presence of 20 $\mu g\,l^{-1}$ copper.

It will be noted that in all the above considerations only the 4-day LC_{50} test is discussed. This parameter gives the concentration of the test metal in the test water that will kill 50% of the creatures under test in 4 days. Obviously, if the duration of the LC_{50} test is increased then the particular creature under discussion will only tolerate a lower concentration of the test metal over the extended period for 50% fatalities to occur. Thus, as shown in Figure 1.2, the LC_{50} value of 0.7 $\mu g\,l^{-1}$ obtained for salmonid fish in non-saline waters when exposed in the short term to cadmium for 4 days is approximately 15 times greater than the value of 0.05 $\mu g\,l^{-1}$ obtained in a long-term 365-day exposure.

Exposure of the fish to various concentrations of cadmium in non-saline waters would have the following results:

Environmental concentration of cadmium ($\mu g\,l^{-1}$)	*Exposure time* (days)	
	4	*365*
	% mortality	*% mortality*
0.005	Very low	< 50
0.05	Very low	50
0.5	< 50	> 50
0.7	50	> 50
5	> 50	> 50

Thus long-term exposure is more likely to produce mortalities at lower exposure concentrations. This is illustrated in Table 4.8 where it is seen, for example, that exposure to 780 $\mu g\,l^{-1}$ and 52 $\mu g\,l^{-1}$ cadmium for 4 days and 365 days, respectively, would kill 50% of amphipods. In the case of environmental waters containing 5 $\mu g\,l^{-1}$ cadmium, adverse effects on amphipods and possibly a small number of fatalities are more likely to occur during the long-term exposure. In the case of zinc short-term exposure under stated conditions would lead to more than 50% mortalities in the case of *Daphnia magna* and long-term exposure would, in addition, have a similar effect on certain types of fish.

4.2 ORGANIC COMPOUNDS

Compared to the metals and organometallic compounds there are fewer published toxicity data on the effects of organic compounds on freshwater or oceanic sea creatures. It is seen in Tables 4.9 and 4.10 that 4-day LC_{50} values obtained with fish and creatures other than fish exposed to various organic compounds

Table 4.7. Effect of short-term (4-day) exposure of creatures to typical concentrations of metals found in saline waters

| Element | Creature | 4-day LC_{50} [a] | Typical concentration of metal ($\mu g\,l^{-1}$) in saline water (see Appendix 1) | | | |
| | | | Open seawater | | Coastal, bay and estuary waters | |
			Max	Min	Max	Min
Nickel	Fish	35 000	1.58	0.099	5.3	0.2
	Marine organisms	10 000	1.58	0.099	5.3	0.2
	Planktonic	600 max	1.58	0.099	5.3	0.2
	Crustacea	50 min	1.58	0.099	5.3	0.2
	Bivalve mollusc	600 max	1.58	0.099	5.3	0.2
		50 min	1.58	0.099	5.3	0.2
Selenium	Invertebrates	> 10 000 max	0.029	0.001	0.4	0.4
		2900 min	0.029	0.001	0.4	0.4
Vanadium	Fish	> 10 000	2.0	0.45	5.1	< 0.01
	Invertebrates	> 10 000	2.0	0.45	5.1	< 0.01
Chromium	Annelid	8000 max	1.26	0.005	3.3	0.15
		2200 min	1.26	0.005	3.3	0.15
	Mollusc	105 000 max	1.26	0.005	3.3	0.15
		14 000 min	1.26	0.005	3.3	0.15
	Crustacea	640 000 max	1.26	0.005	3.3	0.15
		2000 min	1.26	0.005	3.3	0.15
	Fish	190 000 max	1.26	0.005	3.3	0.15
		12 400 min	1.26	0.005	3.3	0.15
Arsenic	Fish	28 000 max	–	–	1.04	1.0
		15 000 min	–	–	1.04	1.0
	Invertebrates	4000	–	–	1.04	1.0
Zinc	Fiddler crab	76 950 max	10.9	0.05	250	0.007
		66 420 min	10.9	0.05	250	0.007
Copper	Crustacea (young)	300	8.6	0.006	20	0.065
	Mollusc (young)	300	8.6	0.006	20	0.065
	Crustacea (adult)	30 000	8.6	0.006	20	0.065
	Mollusc (adult)	30 000	8.6	0.006	20	0.065

		4-day LC_{50}	Max	Min	
Cadmium	Fish	16 400 max	0.3	5	0.013
	Fiddler crab	6400 min	0.3	5	0.013
		48 210 max	0.3	5	0.013
		42 230 min	0.3	5	0.013
Mercury	Crab	8	0.078	15.1	0.00002
	Fiddler crab	2830 max	0.078	15.1	0.00002
		2750 min	0.078	15.1	0.00002

[a] Max to min range depending on creature type.
Greater than 50% mortalities on 4 days' exposure in the case of (italic) crab exposed to 15.1 $\mu g\,l^{-1}$ mercury.

Table 4.8. Effect on mortalities of duration of exposure to metals in water

		Short-term exposure (4 days)			Long-term exposure (1 year)		
		4-day LC_{50} ($\mu g\,l^{-1}$)	Metal concentration in freshwater ($\mu g\,l^{-1}$) (see Appendix 1)		365-day LC_{50} ($\mu g\,l^{-1}$)[b]	Metal concentration in freshwater ($\mu g\,l^{-1}$) (see Appendix 1)	
			Max	Min		Max	Min
Cadmium	Fish (max)[a]	37 100	5	0.013	24 733	5	0.013
	(min)[a]	< 10 000	5	0.013	< 666	5	0.013
	Amphipod	780	5	0.013	52	5	0.013
Zinc	Fish (max)[a]	52 000	630	0.86	3466	630	0.86
	(min)[a]	2400	630	0.86	160	630	0.86
	Daphnia magna	70	630	0.86	4.6	630	0.86
	Annelid	10 000	630	0.86	666	630	0.86
Mercury	Crayfish	0.2	1.3	0.009	0.013	1.3	0.009
	Slipper limpet	1100	1.3	0.009	73	1.3	0.009

[a] Depending on type.
[b] Assumed 365-day LC_{50} = 4-day $LC_{50}/15$ (see Figure 1.2 which applies to cadmium), assumed similar relationship for mercury and zinc. Greater than 50% mortalities in italic, i.e. LC_{50} < concentration of metal in freshwater—short-term (4-day) exposure, Daphnia magna exposed to 630 $\mu g\,l^{-1}$ zinc, crayfish exposed to 1.3 $\mu g\,l^{-1}$ mercury—long-term (1 year) exposure fish and Daphnia magna exposed to 630 $\mu g\,l^{-1}$ zinc, crayfish exposed to 1.3 $\mu g\,l^{-1}$ mercury.

Table 4.9. LC_{50} values of organic compounds in non-saline and saline waters

Compound	Organism	Water type	LC_{50} value	LC_{50} test duration (days)	Reference
Diethyl hexyl phosphate	Daphnia and fish	ns	10–1000 mg kg⁻¹	4	273, 278
1-octanol	Rainbow trout (*Salmo gairdneri*)	ns	15.84 mg l⁻¹	4	274
Sodium decyl sulphonate	Lugworm (*Arenicola marina*)	ns	15.2 mg l⁻¹	4	275
Triton X-100			15.2 mg l⁻¹	4	275
Sodium dodecyl benzene sulphonate			12.5 mg l⁻¹	4	275
Ethylene dibromide	*Hydra oligatis*	ns	50 mg l⁻¹	3	276
Methylene dichloride	Juvenile fathead minnow (*Pimphales promelas*)	ns	502 mg l⁻¹	2	277
1,2,4-trichlorobenzene	Fathead minnow (*Pimphales promelas*)	ns	(a) 7.8 mg l⁻¹ (b) 2.76 mg l⁻¹	4	278
1,4-dichlorobenzene		ns	1.10 mg l⁻¹	4	278
1,2,3,4-tetrachlorobenzene		ns	4.2 mg l⁻¹	4	278
Pentachlorophenol	*Selenastrum capricornatum*	ns	(a) 0.11–0.15 mg l⁻¹ softwater (b) 0.76 mg l⁻¹ hardwater	4	279
	Roach (*Rutilus rutilus*)	ns	0.028 mg l⁻¹	4	280
	Rainbow trout (*Salmo gairdneri*)	ns	0.09 mg l⁻¹	4	274
2,4-dichlorophenol	Rainbow trout (*Salmo gairdneri*)	ns	4.64 mg l⁻¹	4	274
2,4,6-trichlorophenol	Roach (*Rutilus rutilus*)	ns	0.05 mg l⁻¹	4	280
2,3,4,6-tetrachlorophenol	Roach (*Rutilus rutilus*)	ns	0.071 mg l⁻¹	4	280

Compound	Species		Concentration		Ref
Polychlorobiphenyl (Arochlor 1254)	Rainbow trout (Salmo gairdneri)	ns	30 mg l⁻¹	4	281
Endosulphan	Crab (Oziotelphusa Senex Senex)	s	6.2 mg l⁻¹ (sublethal) 18.62 mg l⁻¹ (lethal)	4	170, 171
Kepone	Lamprey (Petromyzon marinus)	s	414–444 mg l⁻¹	4	282
Carbaryl	Lugworm (Arenicola marina)	ns	7.2 mg l⁻¹	3	278
	Catfish (Clarius batrachus)	ns	46.9–107.7 mg l⁻¹	4	283
Parathion ethyl	Lugworm (Arenicola marina)	ns	2.7 mg l⁻¹	3	278
Mirex	Rainbow trout (Salmo gairdneri)	ns	5.0 mg l⁻¹	4	284
Malathion	Rainbow trout (Salmo gairdneri)	ns	1.73 mg l⁻¹	4	285
Roundup herbicide	Rainbow trout (Salmo gairdneri) Chinook Coho salmon	ns	7.4–12 mg l⁻¹	4	286
Rodeo herbicide	Rainbow trout (Salmo gairdneri) Chinook Coho salmon	ns	580 mg l⁻¹	4	286
Bromacil	Fathead minnow (Pimphales promelas)	ns	182 mg l⁻¹	4	287
Diuron	Fathead minnow (Pimphales promelas)	ns	14.2 mg l⁻¹	4	287
Lindane	Teleost fish (Anguilla anguilla)	ns	0.32–0.68 mg l⁻¹	4	288
Hexazinone	Juvenile pacific salmonid	s	276 mg l⁻¹	4	289
Methylenebis Thiocyanate	Chlorella pyrenozdosa	ns	0.042 mg l⁻¹	4	290
	Poecilia reticulata	ns	0.39 mg l⁻¹	4	290
Cyanogen chloride	Daphnia magna	ns	0.065 mg l⁻¹ (adult) 0.029 mg l⁻¹ (juvenile)	2	291

continued overleaf

Table 4.9. (*continued*)

Compound	Organism	Water type	LC_{50} value	LC_{50} test duration (days)	Reference
3,4-dichloro-aniline	Fathead minnow (*Pimphales promelas*)	ns	6.99–8.06 mg l⁻¹	4	292
Pronone 109	Fish	s	904 mg l⁻¹	4	289
Varpar L	Fish	s	1686 mg l⁻¹	4	289
3-fluoromethyl 4-nitrophenol	Walleye (*Stiz ostedium vitreum*)	s	LC_{25} 4.1 mg l⁻¹ (gametes) 2.6 mg l⁻¹ (eggs)	0.5	293
3 fluoromethyl 4-nitrophenol	Larval sea lamprey (*Petromyzon marinus*)	s	$LC_{99.9}$ 1 mg l⁻¹	8-hour	293

ns: non-saline
s: saline

Table 4.10. Relative 4-day LC_{50} values for organic and organometallic compounds in water creatures

Most toxic				Least toxic
LC_{50} 0.01–10 µg l⁻¹	0.01–1 mg l⁻¹	1–10 mg l⁻¹	10–100 mg l⁻¹	100–1000 mg l⁻¹
Organotin compounds	Pentachlorophenol 2,4,6-trichloro phenol 2,3,4,6-tetra chlorophenol Lindane, methylene *bis* thiocyanate, cyanogen chloride	1,2,4-trichloro-benzene, 1,4-dichloro benzene 1,2,3,4-tetra chlorobenzene, 2,4-dichlorophenol carbaryl, parathion, mirex, malathion, Roundup, 3,4-dichloroaniline 3-fluoro-4-methyl nitrophenol	1-octanol Sodium decyl sulphonate, Triton X-100, sodium dodecyl benzene sulphonate, ethylene dibromide, polychlorobiphenyls, endosulphan, carbaryl diuron	Methylene dichloride Kepone Rodeo bromacil hexazinone, Pronone 109, Varpar L

range from $10\ \mu g\,l^{-1}$ (i.e. very toxic) to $106\ \mu g\,l^{-1}$ (i.e. less toxic). Further information on the toxicity of organic compounds towards fish and creatures other than fish is discussed in Chapter 6. As seen in Table 4.10, of the organic compounds discussed the chlorophenols, chloroaromatics and chlorinated insecticides are the most toxic ($LC_{50} < 10\ \mu g\,l^{-1}$) while compounds such as alcohols and surface-active agents are the least toxic.

In Table 4.11 are compared the 4-day LC_{50} values for a range of organics and creatures in river water. It is seen that, of the substances tested, pentachlorophenol is by far the most toxic and when its concentration in environmental waters is high it can cause fatalities in fish.

Various classes of compounds such as polyaromatic hydrocarbons, polychlorinated biphenyls and organochlorine insecticides, organophosphorus insecticides and chlorinated aliphatics have been the cause of great concern in recent years and have received some attention. The available information on the toxicity of organic compounds is discussed below, but clearly much further work remains to be done in this area.

Table 4.11. Comparison of 4-day LC_{50} values for fish and other creatures and environmental concentrations of organic compounds in river waters (in order of increasing toxicity)

Organic compound	Creature	4-day LC_{50} ($\mu g\,l^{-1}$)	Environmental concentration ($\mu g\,l^{-1}$) (see Table 6.2)	
			Max	Min
Most toxic				
Pentachlorophenol	Roach	28	*250*	0.1
	Rainbow trout (*Salmo gairdneri*)	90	*250*	0.1
	Selenastrum capricornatum	110–150	*250*	0.1
Lindane	Teleost fish (*Anguilla anguilla*)	320–680	0.01	0.001
Malathion	Rainbow trout (*Salmo gairdneri*)	1730	0.032	0.027
Alkylbenzene sulphonate	Lugworm (*Arenicala marina*)	12 500	600	10
Non-ionic detergents	Lugworm (*Arenicala marina*)	15 200	70	8
(Triton X-100) *Least toxic*				

Results in italic: environmental concentration exceeds 4-day LC_{50}, i.e. greater than 50% mortalities of test creature.

4.2.1   ALIPHATIC HYDROCARBONS

In many areas oil has become the most frequently encountered water pollutant, and oil pollution incidents are becoming more numerous. This reflects the expanding and widespread consumption of petroleum products, a consumption that will continue to increase in the foreseeable future. Oil pollution has harmful effects on aquatic life and lowers the aesthetic appearance of an inland water. Occasionally, it necessitates the closure of waterworks intakes. Although marine oil pollution has received much attention in recent years, this has not been the case with oil pollution of inland waters, and for some time now there has been the need for an assessment of the analytical and related problems in this field. These problems are discussed below. Particular attention is given to the identification of the polluting oil.

Wherever oil is produced, stored, transported by vehicle or pipeline or consumed there exists a potential source of oil pollution, either directly by surface drains and surface run-off or indirectly by seepage into the ground. Unlike some pollutants, oil pollution is generally unpredictable as to location and time, and usually exists as a surface phenomenon. Heavy pollution is obviously unwelcome, but even thin ephemeral films representing only small amounts of oil may cause complaints and require investigation if continually present on surface water.

Practically the total range of petroleum products is encountered as pollutants on inland water. Crude oils are very seldom found inland, but petrol, paraffin, gas oils, heating fuels, lubricating oils, transformer oils and cutting fluids have given widespread problems. In most areas, heating fuel, due to its widespread industrial and domestic utilisation, and diesel fuel are the most commonly occurring oil pollutants. Petrol, although used in greater quantity than most other petroleum products, does not often pollute inland waters. Probably this is due to its high volatility on water surfaces, the strict regulations concerning its storage and the general public awareness of its dangers. Its relatively high water solubility may be a lesser factor. Lubricating oils give pollution problems especially in highly industrialised areas. Since lubricating oils are seldom stored or used in large quantities, pollutions tend to be of a smaller nature, but are often responsible for intermittent surface films on inland waters, which are regarded as insignificant. However, the increase in concern over the aesthetic appearance of our inland waters may soon render this significant. Heavy fuel oils are occasionally met as pollutants, and owing to their high viscosity can cause extensive soiling of banks and riparian structures as well as being extremely troublesome to remove.

Hydrocarbons can occur in seawater as a result of oil spills at sea. These can have adverse effects on fish and invertebrates which include allometric growth inhibition[104], mortality and reduced yolk reserves in fish[105] and reduced pumping activity and increased mucus production in marine bivalves[106]. This is discussed further in Chapter 6. Oil dispersants have been used to attempt to disperse oil spillages. These dispersants themselves can have adverse effects on fish and invertebrates such as allometric growth inhibition and reduced growth rate[107] (see Section 6.1.1, Tables 6.1 and 6.2).

4.2.2 POLYAROMATIC HYDROCARBONS

Many polyaromatic hydrocarbons in trace quantities have been shown to be directly carcinogenic to mammals (Table 4.12). These are attributed to particular materials which may be present in water samples and are also water-soluble to some extent, so that their occurrence in the environment has caused widespread concern. At least a hundred compounds of this type have been detected and characterised in environmental samples. The basic molecular structure consists of benzene rings either fused together or bridged by methylene side chains. Alkyl substituents also occur.

These compounds can be produced by biochemical degradation of other organic compounds under suitable conditions. They may occur in the environment from the combustion of materials such as wood or leaves. Other sources of aromatic materials from which polyaromatic hydrocarbons may be derived include crude oil which can contain 20% by weight of dicyclic and higher polyaromatic hydrocarbons and high-grade petrol, the aromatic content of which is over 50%. Unsaturated fatty acids, terpenoids, and steroids may also be potential polyaromatic hydrocarbon precursors.

The behaviour and effects of anthropogenic polycyclic aromatic hydrocarbons in aquatic biota in chronically and acutely polluted waterways have been intensely studied for many years[108–14]. Although molluscs have been shown to accumulate polyaromatic hydrocarbons the question of whether the concentrations of potentially toxic and carcinogenic polyaromatic hydrocarbons are magnified through the food chain is not yet resolved.

The analytical chemistry of polyaromatic hydrocarbons in tissues can provide an important part of the answer to the biomagnification question, but it must be improved by new technology and the modification of existing analytical procedures to the point where unambiguous, detailed, and reproducible data can be obtained on a routine basis. Furthermore, few papers dealing with the analytical methodology for determining polycyclic aromatic sulfur heterocycles and polycyclic aromatic nitrogen heterocycles in fish tissues are to be found in the literature. Considering that the heterocyclic fractions are at least as biologically active as the polyaromatic hydrocarbons[115–21] it is clearly desirable that techniques be developed which will provide accurate and quantitative and qualitative data on the sulphur and nitrogen heterocycles in aquatic biota. Polyaromatic hydrocarbons have been shown to reduce the reproduction rate of plaice in seawater[122] (see Section 6.3.1).

Table 4.12. PAHs commonly found in water

Structure	IUPAC name	Mol. wt	Relative carcinogenicity	Abbreviation
	Benzo(ghi)perylene	276	−	B(ghi)P
	Chrysene	228	−	Ch
	Fluoranthene	202	−	Fl
	Indeno(1,2,3-cd)pyrene	276	+	IP
	Phenanthrene	178	?	Ph
	Perylene	252	−	Per
	Pyrene	202	−	Pyr
	Anthracene	178	?	An
	Benzo(a)anthracene	228	+	B(a)A
	Benzo(b)fluoranthene	252	++	B(b)F
	Benzo(j)fluoranthene	252	++	B(j)F
	Benzo(k)fluoranthene	252	−	B(k)F
	Benzo(a)pyrene	252	+++	B(a)P
	Benzo(e)pyrene	252	+	B(e)P

+++, active; ++, moderate; +, weak; ?, unknown; −, inactive.

4.2.3 CHLORINATED INSECTICIDES

Persistent chlorinated hydrocarbons of agricultural and non-agricultural interest, such as 1,1,1-trichloro-2,2-bis-(p-chlorophenyl)ethane (DDT), polychlorinated biphenyls (PCBs) and hexachlorobenzene now have a global distribution and can be detected in wildlife samples in variable amounts. PCBs together with 1,1-dichloro-2,2-bis-(p-chlorophenyl)-ethylene (DDE) are the main types of chlorinated hydrocarbons found in Norwegian avian fauna and in fish along the Norwegian coast[123-5].

In Friefjorden, a fiord in south-east Norway, heavy local contamination with chlorinated hydrocarbons of industrial origin has been detected. The contaminants most often found in fish in this area are hexachlorobenzene, octachlorostyrene and decachlorobiphenyl. In addition, complex mixtures of PCBs and chlorinated naphthalenes have been detected[126,127]. Decachlorobiphenyl has previously been found in arctic fox (*Alopex lagopus*) from Svalbard[128] and octachlorostyrene was first detected in birds in the Netherlands[129-31].

In a monitoring programme over the last six years, the above chlorinated hydrocarbons have been determined in samples from cod (*Gadus morhua*). Lindane and Endosulphan and Trichlorophon insecticides have been shown to cause erratic swimming behaviour, hyperventilation and mortalities in invertebrates (see Section 6.1.1).

4.2.4 POLYCHLORINATED BIPHENYLS

Since their introduction polychlorinated biphenyls have caused much ecological damage and are harmful to humans. One aspect of these toxicants is that they have been shown to have a very severe adverse effect on wildlife by causing thin eggshells and consequently a poor reproductive rate in the laying season. Deleterious effects on seals have also been observed.

Polychorinated biphenyls have until recently been used extensively as cooling media in electrical transformers and also in railway engine repair shops. PCBs are marketed as Arochlors by Monsanto. All Arochlors are characterised by a four-digit number; the first two digits represent the type of molecule (e.g. 12 represents biphenyl, 54 terphenyl and 25 and 44 are mixtures of biphenyl and terphenyl) and the last two digits give the percentage by mass of chlorine, e.g. Arochlor 1260 is a carbon system with 60% m/m of chlorine. The compositions of two further Arochlors are given below:

Composition	Arochlor 1016 (16% mm chlorine)
Biphenyl	% w/w
2	0.03
4	1.1
2,4[1]	0.4

$2,5,2^1-$	12.7
$4-4^1$	3.4

Arochlor 1254 (50% m/m chlorine)
% w/w

$2,3,2^1,5^1$	5.3
$2,5,2^15^1$	10.3
$2,5,3^1,4^1$	3.3
$2,4,5,2^1,5^1$	11.7
$2,3,4,2^1,4^1,5^1$	4.9
$2,4,5,2^1,4^1,5^1$	5.3

Polychlorinated biphenyls are sold under a variety of trade names, of which Arochlor is one. The following is a list of principal trade names used for PCB-based dielectric fluids which are usually classified as Apkarels: Aroclor (UK + USA), Pyoclor (UK), Inertren (USA), Pyanol (France), Clophen (Germany), Apirolio (Italy), Kaneclor (Japan), Solvol (USSR).

Arochlor causes severe weight reduction and liver degeneration in rainbow trout (*Salmo gairdneri*)[133] and inhibition of reproduction at the $50-100 \ \mu g \, l^{-1}$ level in non-saline waters in *Daphnia puliccaria*[134] (see Section 6.1.1, Tables 6.1 and 6.2).

4.2.5 HERBICIDES

The use of certain herbicides in or near to water will give rise to rapid decomposition of the affected vegetation, which in turn can cause deoxygenation of the water.

The most obvious method of entry of herbicides into river water is by their direct application to the water in order to control aquatic vegetation. Where emergent vegetation is being sprayed some of the material may be sprayed directly onto the water surface and some may run off plants into the river. Any herbicide reaching the soil or the banks close to the water may or may not be available for leaching into the watercourse. If the herbicide remains in the plants after their death then it may enter the water when they decompose.

The above observations are also pertinant to field-applied herbicides which may enter the water by spray drift, leaching from or erosion of the soil or via rotting vegetation or silage. The quantities reaching the water by leaching will depend upon the herbicide, rainfall and soil type. The terrain may also be important in that it will affect the pattern of leaching or run-off. The period that the herbicide persists in the soil is also important in that it will affect how long pollution is likely to continue. All these factors apply at the same time, making each herbicide application an individual event and generalisations must be treated with caution.

The question of accidental spillage of a concentrate or a diluted spray into water must also be considered as well as malpractices such as dumping of excess of chemicals, washing out of empty containers in ponds and rivers, and improper disposal of containers. The herbicides may also be present in industrial or agricultural effluents.

Factors which will reduce the concentration of herbicides downstream and must be taken into account are:

(1) The stability of the compound towards chemical and biological degradation, and its removal from the water by volatilisation
(2) The absorption into the bottom muds, suspended material and living organisms.

2,4-dinitrophenoxy acetic acid herbicide has been shown to cause death to macro invertebrates during 12 months' exposure in non-saline water[125] (see Section 6.1.1, Table 6.2).

Diflubenzuron causes mortalities at 50 μg l^{-1} and moulting delay at 5 μg l^{-1} in larval horseshoe crabs (*Limulus polyphemus*) in seawater[136] (see Section 6.3.1).

4.2.6   *p*-DIOXINS AND DIBENZOFURANS

Polychlorinated dibenzo-*p*-dioxins (PCDDs) polychlorinated dibenzofurans (PCFDs), and *ortho*-unsubstituted polychlorinated biphenyls (non-*ortho* polychlorinated biphenyls) are three structurally and toxicologically related families of anthropogenic chemical compounds that have in recent years been shown to have the potential to cause serious environmental contamination[137–41]. The substances are trace-level components or byproducts of several large-volume and widely used synthetic chemicals, principally PCBs and chlorinated phenols[44,142] and can also be produced during combustion processes and by photolysis[146,147]. In general, polychorinated dibenzo-*p*-dioxins, polychlorinated dibenzofurans and non-*ortho* polychlorinated biphenyls are classified as highly toxic[148], although the toxicities are very dependent on the number and positions of the chlorine substituents[149]. About ten individual members of a total of 216 polychlorinated dibenzo-*p*-dioxins, polychlorinated dibenzofurans and non-*ortho* polychlorinated biphenyls are among the most toxic man-made or natural substances to a variety of animal species[131]. The toxic hazards posed by those chemicals are exacerbated by their propensity to persist in the environment[139] and to readily bioaccumulate[62,63,150] and although the rate of metabolism and elimination is strongly species-dependent[151–153] certain highly toxic isomers have been observed to persist in the human body for more than 10 years[154].

The majority of scientific concern for the hazards of these compounds has been directed towards the disposition in the environment of the single most toxic isomer, 2,3,7,8-tetrachlorodibenzo-*p*-dioxin (2,3,7,8-TCDD)[133–42]. More recently, however, investigations into the formation and occurrence of polychlorinated dibenzofurans suggest that this family of toxic compounds may also commonly occur at comparable or greater levels and could possibly pose a greater hazard than polychlorinated dibenzo-p-dioxins. Polychlorinated dibenzofurans are often found as co-contaminants in and are more readily produced from pyrolysis of polychlorinated biphenyls[47,155-8]. Most important, polychlorinated dibenzofurans produced from pyrolysis of polychlorinated biphenyls are predominantly the most toxic isomers, those having a 2,3,7,8-chlorine substitution pattern[140]. A number of recent fires involving electrical transformers and capacitators have demonstrated the potential for formation of hazardous levels of polychlorinated dibenzofurans from pyrolysis of polychlorinated biphenyls[157-61].

In the light of these findings and because of a dearth of data pertaining to the occurrence of these compounds in the environment polychlorinated dibenzofurans and non-*ortho* polychlorinated dibenzo-p-dioxins were included as target compounds in a survey of important US rivers and lakes for polychlorinated dibenzo-p-dioxins. The decision to include as many polychlorinated dibenzo-p-dioxins isomers as possible was based on several facts: (1) several other polychlorinated dibenzo-p-dioxins isomers are also extremely toxic[149]; (2) pentachlorophenol, a large-volume fungicide and wood preservative, contains relatively high levels of hexa-, hepta- and octachlorodibenzodioxins and essentially no tetrachlorodibenzo-p-dioxins[111,112,133]; and (3) incineration of materials containing chlorophenols readily produces mixtures of polychlorinated dibenzo-p-dioxins but 2,3,7,8-tetrachloro dibenzo-p-dioxins is a minor component. On the other hand, the highly toxic 1,2,3,7,8-pentachloroisomer is a major component of polychlorinated dibenzo-p-dioxin incineration products of pentachlorophenol[44,163].

Component-specific analyses can be a crucial link to the sources of contamination because different sources of polychlorinated dibenzo-p-dioxins and polychlorinated dibenzofurans usually produce mixtures of distinctly different relative component abundances[44]. On the other hand, the preferential accumulation of certain isomers in animals may prevent source identification from analyses of biological samples.

4.2.7 NITROSAMINES

Many N-nitrosamines are toxic and carcinogenic, and furthermore the carcinogenic action exhibits a high degree of organ specificity. Nitrosamines are formed by interaction between a nitrite and an amine with varying ease, depending on the nature of the amine and the prevailing conditions. The reaction is not restricted to secondary amines but also occurs with primary and tertiary amines and even quaternary ammonium salts. Thus, the precursors are widespread as naturally occurring compounds and nitrosamines are generated in many commercial and industrial processes. It is therefore conceivable that trace amounts may be present in air and water in the vicinity of industrial sites.

Mills and Alexander[161] have discussed the factors affecting the formation of dimethylnitrosamine in samples of water and soil. Dimethylnitrosamine was formed as readily in sterilised samples as in non-sterile ones, indicating that although microorganisms can carry out an enzymatic nitrosation in some soils and waters, dimethylnitrosamine can be formed by an non-enzymatic reaction, even at near-neutral conditions. The presence of organic matter appears to be important in promoting nitrosation in the presence of the requisite precursors.

4.2.8   OTHER INSECTICIDES

Other types of insecticides cause adverse effects in fish and invertebrates in non-saline waters (see Section 6.1.1, Tables 6.1 and 6.2).

Permethrin

This insecticide causes nerve poisoning and blocking of anaerobic and aerobic metabolism in the snail *Hymnaea acuminata*[164].

Phosphamidon

Phosphimidon causes glycogen depletion in muscles, i.e. reduced mobility, in freshwater prawn *Macrobrachum lamarrei*[165].

Fenitrothion

Fenitrothion causes growth abnormalities of follicle and epithelium in freshwater murrel *Channa punctatis*[166].

Carbaryl

Carbaryl above certain concentration levels reduces the survival time of catfish *Clarias leatrachus*[167].

Malathion

Above certain concentration levels malathion reduces the survival time of freshwater teleosts *Channa punctatis*. It also produces mortalities upon 5 days' exposure at 44 000 $\mu g\,l^{-1}$ in toad embryos (*Bufo arenarum*)[168].

Bromacil

Bromacil reduces the growth and survival time and deforms fry in fathead minnow (*Pimephales promelas*)[169].

Endosulphan

This insecticide in seawater increases the body weight, haemolymph volume and hydration at sublethal concentrations (6200 μg l^{-1}) and decreases these parameters in concentrations above the lethal level (18 600 μg l^{-1}) when the crab *Oziotelphusa senex* is exposed to endosulphan[170,171].

4.2.9   ORGANIC ESTERS

Di-2-ethyl hexylphthalate

This ester causes increased surfacing behaviour in *Daphnia magna*[172] as well as deterioration of reproductive capacity and the immune system and carcinogenic activity in Daphnia and fish (see Section 6.2.1). Di-2-ethyl hexyl phthalate causes mortalities of young baltic herring (*Clupea lapengus*) and Atlantic cod (*Gadus marina*)[173].

Acrylates and methacrylates

These esters cause respiratory and metabolic inhibition and neurotoxicity in juvenile fathead minnow (*Pimphales promelas*)[174].

4.2.10   SURFACE-ACTIVE AGENTS

Sodium decyl sulphate, Triton X-100 non-ionic detergents and sodium dodecyl benzene sulphonate surface-active agents have adverse effects on gills and epidermic receptors in the non-saline water lugworm *Arenicola marina L*[175].

4.2.11   PHENOL

Phenol above certain concentrations in non-saline waters causes immobilisation, paralysis and mortality in *Ascellus aquaticus*[176].

4.2.12   PENTACHLOROPHENOLS

Pentachlorophenols above certain concentrations in non-saline waters cause low survival rates (e.g. 31.6 h at 90 μg l^{-1}) in rainbow trout (*Salmo gairdneri*)[177].

4.2.13   ANILINE

Aniline above certain concentrations (1000 μg l^{-1}) causes inhibition of embryo development in the South African clawed toad (*Xenobus laevis*)[178].

4.2.14 *p*-CHLOROANILINE

p-Chloroaniline above certain concentrations (100 000 $\mu g\,l^{-1}$) kills embryos in the South African clawed toad (*Xenobus laevis*)[178].

4.2.15 METHYL BROMIDE

Methyl bromide at concentrations above 100 000 $\mu g\,l^{-1}$ for a period of 1 to 3 months causes paralysis in the guppy *Poecilia reticulata* and at concentrations above 1800 $\mu g\,l^{-1}$ for 4 days causes degenerative changes in the gills as well as oral mucosa[179].

4.2.16 TETRACHLORO-1,2-BENZOQUINONE

This compound when present in seawater causes skeletal abnormalities in the fourhorn skulphin (*Hyoxecephalus quadricornis*)[180].

4.2.17 3-FLUORO-METHYL-4-NITROPHENOL LAMPRICIDE

This compound in seawater damages the eggs and fry of walleye (*Stizostedium vitreum*)[181].

4.3 ORGANOMETALLIC COMPOUNDS

The four types of organometallic compounds that occur in the environment and which have been most extensively studied are those of arsenic, lead, mercury and tin. These can originate in the ecosystem either as man-made pollutants or by microorganism-induced biomethylation of metals in sediments, fish or marine invertebrates. The toxic effects of organometallic compounds on fish and creatures other than fish are discussed in Sections 6.21 and 6.4.1.

4.3.1 ORGANOARSENIC COMPOUNDS

Organoarsenic species are known to vary considerably in their toxicity to humans and animals[182]. Large fluxes of inorganic arsenic into the aquatic environment can be traced to geothermal systems[183], base metal smelter emissions and localised arsenite treatments for aquatic weed control. The methylated arsenicals have entered the environment either directly as pesticides or by the biological transformation of the inorganic species[184,185].

It has been shown that arsenic is incorporated into both marine and freshwater organisms in the form of both water- and lipid-soluble arsenic compounds[186]. Studies to identify the chemical forms of these arsenic compounds have shown the presence of arsenite (As(III)), arsenate (As(V)), methylarsonic acid, dimethylarsinic acid and arsenobetaine[73]. Methyl arsenicals also appear in the urine and plasma of mammals, including humans, by biotransformation of inorganic arsenic compounds[73].

The biological methylation of inorganic arsenic by microorganisms such as moulds and bacteria present in sediment sludges and muds has been established although there is no unequivocal evidence of the proposed pathways[187−94].

Organoarsenical pesticides such as sodium methanearsonate and dimethylarsinic acid are used in agriculture as herbicides and fungicides. It is possible that these arsenicals enter soil, plants and consequently humans. On the other hand, arsenic is a ubiquitous element on earth, and the presence of inorganic arsenic and several methylated forms of arsenic as monomethyl-, dimethyl- and trimethylarsenic compounds in the environment has been well documented[195]. The occurrence of biomethylation of arsenic in microorganisms[196], soil[197], animals and humans[295] has also been demonstrated. Therefore, further investigation of the fate of arsenicals in the physical environment and living organisms requires a knowledge of their complete speciation.

4.3.2   ORGANOLEAD COMPOUNDS

The use of tetraalkyl leads as antiknock additives/octane enhancers for automotive gasolines has been reduced due to environmental considerations in several countries. However, the complete elimination of tetraalkyl lead additives is unlikely.

Organolead compounds are generally more toxic than inorganic lead compounds[198], and the toxicity of the alkylated lead compounds varies with the degree of alkylation, with tetraalkyl lead being the most toxic[199].

The highly polar dialkyl and trimethyl lead compounds in particular have a high toxicity to mammals[200] and are formed as a result of the degradation of tetraalkyl lead in aqueous medium[201].

Tetramethyl and tetraalkyl lead compounds are considerably more toxic than inorganic lead (1000 times)[202] or di- or tri-methyl or diortriethyl lead compounds[199].

The high toxicity of tetraalkyl leads is attributed to their ability to undergo the following decomposition in the environment[201]:

$$R_4Pb \rightarrow R_3Pb^+ \rightarrow R_2Pb^{2+} \rightarrow Pb^{2+}$$

The formation of alkyl lead salts, probably associated with proteins, arising in tissues from rapid metabolic dealkylation of tetraalkyl lead compounds is of toxicological importance in evaluating exposure to tetraalkyl leads. The toxic effect of tetraalkyl leads to mammals has been attributed to the formation of trialkyl lead compounds in body fluids and tissues.

Wong et al.[203], Reisinger et al.[204] and others[163−5] have demonstrated that microorganisms in lake sediment can transform inorganic and organic lead compounds into volatile tetraalkyl lead. The possibility of biomethylation of lead or organolead ionic species by microorganisms, reversing the decomposition mechanism given above, may add to the problem of lead toxicity already faced by humans, although the area is presently much disputed[204].

Organically bound lead is a minor but important contribution to total lead intake by humans and animals. Alkyl lead salts such as trialkyl lead carbonates, nitrates and/or sulphates can be formed in tissues by the rapid metabolic dealkylation of tetraalkyl lead compounds.

Recently a renewed interest in the speciation of lead in environmental samples has resulted from several diverse lines of investigation. Organolead compounds have been detected in cod, lobster, mackerel, and flounder meal (10–90% of the total lead burden[208], and in freshwater fish[209,210].

Fairly high concentrations of tetraethyl lead (30 ppm) have been detected in mussels collected at a buoy near the S.S. *Cavtat* incident where a shipload of tetraethyl lead was sunk[208] in the Adriatic Sea. High organolead concentrations, mainly of tetraethyl lead, were also found in mussels in other parts of the Italian seas. The presence of tetraethyl lead in aquatic organisms may indicate that the alkyl lead compounds are not immediately metabolised by living organisms and may remain in their authentic forms in the living tissues for a long time[211]. The occurrence of tetralkyl lead compounds in aquatic biota is highly significant because of the possibility of their incorporation into the food chain.

A steady input of organoleads into the environment results from the continued use of tetralkyl leads as antiknock additives. In addition, evidence for the chemical[204] and biological alkylation of organolead salts or of lead (II) salts has been obtained[203,209,212].

Although organoleads may make only a small contribution to the total lead intake of an organism. it has been demonstrated that trialkyl lead salts arising in tissues from the degradation of tetralkyl leads are important in lead toxicity. The conversion of R_4Pb to R_3Pb^+ occurs rapidly in liver homogenates from rats to rabbits. Acute toxicities of tetralkyl leads and of trialkyl lead salts are similar and are at least an order of magnitude greater than dialkyl lead salts or inorganic lead salts. Relatively little is known of either the effect of chronic exposure to small amounts of such compounds or the levels of organic lead compounds, such as the tetralkyl leads, in biological and food material. Dialkyl lead salts cause symptoms of toxicity similar to those produced by inorganic lead salts and they exhibit an affinity for thiol compounds. Triorganolead salts inhibit oxidative phosphorylation.

Speciation of alkyl lead compounds, including molecular and ionic, volatile and solvated forms, has become immensely important and in urgent demand in studies related to toxicity and environmental consequences. The highly polar dialkyl and trialkyl forms in particular are more important species because of their high toxicity to mammals and the consequence of the formation as a result of degradation of tetralkyl lead in aqueous medium[201].

4.3.3   ORGANOMERCURY COMPOUNDS

Organomercury compounds are more toxic than metallic mercury[213] and inorganic mercury forms, when present in the environment, may cause serious illness in extremely polluted areas. Methylmercury has been stated to be neurotoxic. Due to its chronic toxicity and its tendency to bioaccumulate, mercury is of prime interest. Being extremely volatile in the organic and elemental forms, mercury is well dispersed in the atmosphere.

The interest in mercury contamination, and particularly in organomercury compounds, is a direct reflection of the toxicity of these compounds to humans. Some idea of the proliferation of work can be derived from the reviews of Krenkel[214], Robinson and Scott[215] (460 references) and Uthe and Armstrong[216] (283 references).

All forms of mercury are potentially harmful to biota, but monomethyl and dimethyl mercury are particularly neurotoxic. The liphophilic nature of the latter compounds allows them to be concentrated in higher trophic levels and the effects of this biomagnification can be catastrophic[217]. Certain species of microorganisms in contact with inorganic mercury produce methylmercury compounds[189]. Environmental factors influence the net amount of methylmercury in an ecosystem by shifting the equilibrium of the opposing methylation and demethylation processes. Methylation is the result of mercuric ion (Hg^{2+}) interference with biochemical C-1 transfer reactions[191]. Demethylation is brought about by non-specific hydrolytic and reductive enzyme processes[218-20]. The biotic and abiotic influences that govern the rates at which these processes occur are not completely understood.

Although much of the early work on cycling of mercury pollutants has been performed in freshwater environments, estuaries are also subject to anthropogenic mercury pollution[221]. A strong negative correlation exists between salinity of anaerobic sediments and their ability to form methylmercury from Hg^{2+}. As an explanation for this negative correlation the theory was advanced that sulphide, derived by microbial reduction of sea and salt sulphate, interferes with Hg^{2+} methylation by forming mercuric sulphide which is not readily methylated[222-5]. There are several reports in the literature concerning the methylation of Hg^{2+} by methylcolbalamin[194,226,227].

The synthesis of methylmercury compounds from inorganic mercury by microorganisms, mould and enzymes in freshwater sediments has been investigated by some workers[187-91,193,194,203,204,227]. This biological methylation of mercury compounds provides an explanation for the fact that CH_3Hg^+ is found in fish, even if all known sources of mercury in the environment are in the form of inorganic mercury or phenyl mercury. The formation of the volatile CH_3HgCH_3 (b.p. 94 °C) may be a factor in the redistribution of mercury from aqueous industrial wastes. The process of methylation is fundamental to a knowledge of the turnover of mercury. It may be significant in the uptake and distribution of mercury in fish and in the mobilisation of mercury from deposits in bottom sediments into the general environment.

The organic mercury compounds produced, primarily dimethylmercury and methylmercury halides, are potentially more toxic than inorganic forms. There-fore, recent studies of environmental mercury have been concerned with its chemical speciation to determine not only the amounts of mercury present but also the chemical forms. More extensive data in this area will assist in determin-ing the role of organic mercury in the global cycling of the element.

Andren and Harris[228] have reported a methylmercury concentration of $0.02-0.1$ $\mu g\,kg^{-1}$ mercury in unpolluted sediments. In two rockfish samples, the organic mercury concentration was 110 and 190 $\mu g\,kg^{-1}$ (dry weight). This agrees quite well with the reported methylmercury concentration range of $70-200$ $\mu g\,kg^{-1}$ mercury in similar fish[229]. Matsunaga and Takahashi[230] found $0.2-0.4$ $\mu g\,kg^{-1}$ mercury in sediments.

It has been reported that organomercury compounds are significantly concen-trated in fish[16,231-6] predominantly as methylmercury compounds. Fish in contact with water containing 0.01 $\mu g\,l^{-1}$ and sediment containing 30 $\mu g\,kg^{-1}$ of mer-cury have been found with 341 $\mu g\,kg^{-1}$ in their flesh, i.e. a 34 000 bioamplifica-tion in the flesh. At Minamata Bay, Japan, mercury levels in some fish attained 50 $\mu g\,kg^{-1}$ wet weight while levels around 20 $mg\,kg^{-1}$ were common. Experi-ments with brook trout have shown that over a period of 9 months the fish had accumulated in their gonads 900, 2900 and 123 000 $\mu g\,kg^{-1}$ of mercury from water containing 0.09, 0.29 and 0.93 $\mu g\,l^{-1}$ mercury, respectively.

Some of the toxic and adverse effects of organomercury compounds on fish and invertebrates in non-saline waters are discussed, respectively, in Sections 6.2.1 and 6.4.1 and are summarised below:

Teleost	Liver abnormalities
Fish	Carcinogenesis[238]
Codfish	Kidney damage[239]
Microtubes	Severe disruption[240]

Four-day LC_{50} values of 430, 4300 and 507 $\mu g\,l^{-1}$, respectively, were obtained for the catfish *Clarias batrachus L.* exposed to methyl mercuric chloride, hydroxy ethyl mercuric chloride and mercuric chloride in non-saline waters[241].

4.3.4   ORGANOTIN COMPOUNDS

These compounds have been the subject of environmental studies for two obvious reasons. First is the increasing worldwide use of inorganic and organotin compounds in many industrial, chemical and agricultural areas, very little being known about their environmental fate. Second, there is a great difference in toxicity of the various organotin compounds according to the variation of the organic moiety in the molecules.

Organic compounds have found applications in many fields, such as stabilisers for PVC, fungicides and miticides in agriculture, and biocides, algicides, bactericides and molluscicides[242-5], because their properties can be tailored by the variation of the type and the number of substituents to meet widely different requirements. Annual world production was estimated to be 33 000 tons in 1983, most of it dioctyltin maleate[243]. The toxicity and degradation in the environment depend strongly on the number and nature of the substituents[242,248]. Organotin compounds with short alkyl chains or phenyl substituents generally exhibit considerable toxicity towards both aquatic organisms and mammals. Alkyltins with small alkyl chains degrade slowly in the environment[103,184]; phenyltins are less stable and may, under certain conditions, rapidly lose the phenyl substituents. Organotin compounds may accumulate in sediments and aquatic organisms[103].

There is special interest in the biotic and abiotic methylation of tin compounds[249] and the fate of some organotins in the aquatic ecosystems. One possible route is the dealkylation of the trialkyltin species eventually to Sn(IV) and the microbial methylation of Sn(IV) to the various methyltin species. Increasing methyltin concentration with increasing anthropogenic tin influxes has been noted in the Chesapeake Bay[250].

Methyltin species are ubiquitous in natural waters, although their concentration is usually low (less than $1 \, ng \, l^{-1}$) in waters relatively unimpacted by anthropogenic activity[251,252]. Mono- and dimethyltin are the dominant species[251-3], suggesting that methyltins, like methylmercury species, arise via stepwise methylation of the inorganic metal[298]. Not only are sediment slurries capable of methylating added inorganic tin[254] but concentrations of methyltin species increase with estuarine surface-to-volume ratios[251]. Thus, tin methylation in aquatic environments is likely to occur in sediments.

Measurements of sediment methyltin concentrations show monomethyltin to be the dominant species in anoxic sediments while trimethyltin is found in highest concentrations in toxic sediments[255]. This suggests that tin methylation probably occurs in anaerobic sediments, while degradation of higher molecular weight organotins such as tributyltin, an antifouling agent, occurs in oxygenated environments. In recent studies of inorganic tin methylation it has been confirmed that biomethylation occurs preferentially in anaerobic estuarine sediments[256]. Methyltins were produced to a maximum level of about $2 \, ng \, l^{-1}$ (dry weight) of sediment in 21 days[257]. Low concentrations of mono-, di- and trimethyltin compounds found in Baltimore Harbor sediments averaged at 8, 1 and 0.3 $\mu g \, kg^{-1}$ dry weight of sediment while sediment taken in a relatively unpolluted area had much lower organotin content (1.01 and 0.01 $\mu g \, kg^{-1}$).

Rapsonankes and Weber[226] examined the environmental implications of the methylation of tin(III) and methyltin(IV) ions in aqueous samples in the presence of manganese dioxide. Their studies were carried out with particular reference to the mechanisms involved and the role of dimethylcobalt complex carbonion donor, the carbocation donor iodomethane, and the oxidising agent manganese dioxide. The yield of the various methyltin ions were estimated, and some preliminary results were also presented on the further methylation of mono-, di- and trimethyltin, which indicated that the presence of a naturally occurring donor such as methylcobalamin would result in formation of volatile tetramethyltin compounds.

Van Nguyen et al.[299] carried out an investigation of the fate in an aqueous environment of three organotin compounds (triphenyltin acetate, triphenyltin hydroxide and triphenyltin chloride) used in antifoulant paint compositions. The organotin compounds were leached from paint panels by shaking with distilled water for up to 2 weeks at room temperature and the water and undissolved residues were then analysed. The results suggested that the organotin compounds ionised in aqueous media; a simple model was developed to explain the process.

Several investigators have reported ng-$\mu g\,l^{-1}$ concentrations of organotin compounds both in freshwater and marine samples. Inorganic tin, methyltins and butyltins have been detected in marine and freshwater environmental samples[258–61]. The presence of inorganic tin, butyltin and methyltin species has been reported in Canadian lakes, rivers and harbours[262,263]. Both organotins and inorganic tin were reported to be highly concentrated by factors of up to 10^4 in the surface microlayer relative to subsurface water[262,263]. Inorganic tin, mono-, di- and trimethyltins have been detected at $ng\,l^{-1}$ levels in saline, estuarine and freshwater samples[260,264]. Methylation of tin compounds by biotic as well as abiotic processes has been proposed[266].

Possible anthropogenic sources of organotins have recently been suggested. Both polyvinylchloride and chlorinated polyvinylchloride have been shown to

Table 4.13. LC_{50} values obtained for organotin compounds

	Species	Type of water	LC_{50}	Duration of LC_{50} test (days)	Reference
Organotin compounds	Rainbow trout (*Salmo gairdneri*)	Non-saline	1.3 $\mu g\,l^{-1}$	1	294
Organotin compounds	Bivalve mollusc (*Crassostrea virginica*)	Non-saline	1.3 $\mu g\,l^{-1}$ embryo 3.96 $\mu g\,l^{-1}$ larvae	2 2	271
Organotin compounds	Bivalve mollusc (*Mercenaria mercenaria*)	Non-saline	1.13 $\mu g\,l^{-1}$ embryo 1.65 $\mu g\,l^{-1}$ larvae	2 2	271

leach methyltin and dibutyltin compounds respectively, into the environment[267]. Monobutyltin has been measured in marine sediments collected in areas associated with boating and shipping. Butyltin was not detected in areas free of exposure to maritime activity[268]. The use of organotin antifouling coatings in particular has stimulated interest in their environmental impact.

As discussed in greater detail in Sections 6.2.1 and 6.4.4, the LC_{50} values obtained for organotin compounds are extremely low, confirming the high toxicity of these compounds towards water creatures. Reported values for 1 and 2 days LC_{50} is $1-4$ $\mu g\, l^{-1}$ as seen in Table 4.13.

Some adverse toxicity effects of organotin are summarised below (these are discussed further in Sections 6.2.1 and 6.4.1).

	Non-saline waters		*Saline waters*
Fiddler crabs (*Uca pugilatar*)	Retarded limb regeneration[269] Morphological abnormalities	*Gammarus GP* *Brevoortia tyrannus* and larval *Henidia berrylina*	Reduced bodyweight[270] Reduced growth rate
Bivalve molluscs (*Crassostrea virginica mercenaria*)	Acute toxicity to embryos and larvae, delayed clam embryo development[271]	Oyster *Crasspstrea gigas*	Weight, length, width adversely affected[272]

REFERENCES

1. Van Hoof, F. and Van Son, H., *Chemosphere* **10**, 1127 (1981)
2. Mount, D. I. and Stephen, C.E. *J. Wildlife Management* **31**, 168 (1967)
3. Martin, M., Stephenson, M.D. and Martin, J.H., *Californian Fisheries and Game* **63**, 95 (1977)
4. Kariya, T., Haga, Y., Hoga, T. and Tsuda, T., *Bulletin of Japanese Society Scientific Fisheries* **33**, 818 (1967)
5. Brings, W.A., Leonard, E.N. and McKim, J.M., *J. Fisheries Research Board (Canada)* **30**, 583 (1973)
6. Committee on Medical and Biogenic Effects of Environmental Pollution, Selenium, National Academy of Sciences, Washington, DC (1976)
7. Luckey, T.D. and Venegopal, B., *Chemical Engineering News* **54**, 2 (1976)
8. Jaworski, J.F., *Effect of Lead in the Environment 1978, Quantitative Aspects*, National Research Council, Ottawa, NRCC 16736 (1979)
9. Schultz, T.W., Freeman, S.R. and Dumont, J.N., *Archives of Environmental Contamination and Toxicology* **9**, 23 (1980)
10. Johnson, P.A. *Aquatic Toxicology* **10**, 335 (1987)

11. Van der Putte, I., Lubbers, J. and Klar, Z., *Aquatic Toxicology* **1**, 3 (1981)
12. Lima, A.R., Curtis, C., Hammermeister, D.E., Markee, T.E., Northcott, C. and Brocke, L.T., *Archives of Environmental Contamination and Toxicology*, **13**, 595 (1984)
13. Holcombe, G.W., Phipps, G.L. and Fiant, J.T., *Ecotoxicology and Environmental Safety* **7**, 400 (1977)
14. Meisner, J.D. and Hum, W.Q., *Bulletin of Environmental Contamination and Toxicology*, **39**, 898 (1987)
15. Hilmy, A.L., El-Domiaty, N.A., Dabees, A.Y. and Latife, H.A.A., *Comparative Biochemistry and Physiology* **86c**, 263 (1987)
16. Vostal, J., *Mercury in the Environment*, CRL Press, Cleveland, OH (1972)
17. Harrison, F.L., Watness, K., Nelson, D.A., Miller, J.E. and Calabrese, A., *Estuaries*, **10**, 78 (1987)
18. Willis, M., *Archiv für Hydrbiologie* **112**, 299 (1988)
19. Schgal, R. and Saxena, A.B., *International Journal of Environmental Studies* **29**, 157 (1987)
20. Knittel, M.D., Heavy metal stress and increased susceptibility of steelhead trout (*Salino gairneri*) to yersinia poakeri infection. In Eaton, J.G., Parrish, R.R. and Hendricks, A.C. (eds), *Aquatic Toxicology*, pp. 321–7, American Society for Testing Materials, Philadelphia, PA (1980)
21. Snarski, V.M., *Environmental Pollution (Series A)* **28**, 219 (1982)
22. Dixon, D.G. and Hilton, J.W., *J. Fisheries and Biology* **19**, 509 (1981)
23. Hatakeyama, S., *Ecotoxicology and Environmental Safety* **16**, 1 (1988)
24. Nemcsok, J.G. and Hughes, G.M., *Environmental Pollution* **49**, 77 (1988)
25. Shcerban, E.P., *Hydrobiological Journal* **13**, 75 (1979)
26. Ingersoll, C.G. and Winner, R.W., *Environmental Contamination and Toxicology* **1**, 321 (1982)
27. Sinley J.R., Geoltl J.P. and Davies P.H., *Bulletin of Environmental Contamination and Toxicology* **12**, 193 (1974)
28. Nebecker, A.V., Cairns, M.A. and Wize C.M., *Environmental Contamination and Toxicology* **3**, 151 (1984)
29. Ahsanullah, M., Mohley, M.C. and Rankin, P., *Australian Journal of Marine and Freshwater Research* **39**, 33 (1988)
30. Guidici, M., De, N., Migliore, L. and Guirino, S.M., *Hydrobiologia* **146**, 63 (1987)
31. Lee, D. (ed.), *Metallic Contaminants and Human Health*, Academic Press, New York (1972)
32. Friberg, L., Piscator, M. and Norberg, G., *Cadmium in the Environment*, CRC Press, Cleveland, OH (1971)
33. McCaull, J., *Environment* **13**, 3 (1971)
34. Laurie, D.B., Joselow, M.M. and Browder, A.A., *Ann. Intern. Med.* **76**, 307 (1972)
35. De, J.F. and Solbe, L.G., *Water Research* **8**, 389 (1974)
36. Eaton, J.G., *Water Research* **7**, 1723 (1973)
37. Eaton, J.G., McKin, J.M. and Holcombe, G.W., *Bulletin of Environmental Contamination and Toxicology* **19**, 95 (1978)
38. Alabaster, J.S. and Lloyd, R. (eds), *EIFAC Water Quality Criteria for Freshwater Fish*, UN Food and Agriculture Organisation/Butterworths, London (1980)
39. Friberg, L., Kjellstrom, T., Nordberg, G. and Piscator, M., *Cadmium in the Environment*, Environmental Protection Agency EPA 650 (3-75-049) Washington, DC (1975)
40. Reeder, S.W., De Mayo, A. and Taylor, N.C., *Guidelines for Surface Water Quality, Vol. 1, Inorganic Chemical Substances; Cadmium* Environment Canada, Ottawa (1979)

41. Nriagu, J.O. (ed.), *Cadmium in the Environment, Part 1, Ecological Cycling*, Wiley, New York (1980)
42. Nriagu, J.O. (ed.), *Cadmium in the Environment, Part II, Health Effects*, Wiley, New York (1981)
43. Canton, J.H. and Slooff, W., *Ecotoxicology and Environmental Safety* **6**, 113 (1982)
44. Rappe, C., Buser, H.R. and Bosshardt, H.P., in Nicholson, W.J. and Moore, J.A. (eds), *Health Effects of Hologenated Aromatic Hydrocarbons*, pp. 1–18, New York Academy of Sciences, New York (1979)
45. Gill, T.S., Pant, J.C. and Tewari, H., *Ecotoxicology and Environmental Safety* **15**, 153 (1988)
46. Greeve, J.C., Miller, W.E., Debacon, M., Lang, M.A. and Bartels, C.L., *Environmental Toxicology and Chemistry* **7**, 35 (1988)
47. Boge, G., Bussier, D. and Peres, G., *Water Research* **22**, 441 (1988)
48. Carrier, R. and Beitinger, T.L., *Water Research* **22**, 511 (1988)
49. Knudtson, B.K., *Bulletin of Environmental Contamination and Toxicology* **231**, 95 (1975)
50. Attar, E.N. and Maly, E.J., *Archives of Environmental Contamination and Toxicology* **11**, 291 (1982)
51. Chapman, P.M., Farrell, M.A. and Brinkhurst, R.O., *Water Research* **16**, 1405 (1982)
52. Sangalang, G.B. and Freeman, H.C., *Archives of Environmental Contamination and Toxicology* **8**, 77 (1979)
53. Bodlar, C.W.M., Van Leeuewen, C.J., Voogt, P.A. and Zander, P.J., *Aquatic Toxicology* **12**, 301 (1988)
54. Guidici, M., De N. Migliore, S.M., Guarino, S.M. and Gambardella, C., *Marine Pollution Bulletin* **18**, 454 (1987)
55. Thomson, A., Korsgaard, B. and Joensen, J., *Aquatic Toxicology* **12**, 291 (1988)
56. Zischke, J.A. and Arthur, J.W., *Archives of Environmental Contamination and Toxicology* **16**, 225 (1987)
57. Klanda, R.J., Palmer, R.E. and Lenkevich, M.J., *Estuaries* **10**, 44 (1987)
58. Maltby, L., Smart, J.O.H. and Calow, P., *Environmental Pollution* **43**, 271 (1987)
59. Arthur, J.W., West, C.W., Allen, K.N. and Hedtke, S.F., *Bulletin of Environmental Contamination and Toxicology* **38**, 324 (1987)
60. Lemby, A.D. and Smith, R.J.F., *Environmental Toxicology and Chemistry* **6**, 225 (1987)
61. Clark, J.R., Borthwick, P.W., Goodman, L.R., Patrick, J.M., Lores, E.M. and Moore, J.C., *Environmental Toxicology and Chemistry* **6**, 151 (1987)
62. Isensee, A.R., *Ecological Bulletin* **27**, 255 (1978)
63. Decad, G.M., Birnbaum, L.S. and Matthews, H.B., in Hatzinger, O. *et al.* (eds), *Chlorinated Dioxins and Related Compounds: Impact on the Environment*, pp. 307–15, Pergamon Press, New York (1982)
64. His, E. and Robert, R., *Marine Biology* **95**, 83 (1987)
65. Timourian, H. and Watchmaker, G., *Journal of Experimental Zoology* **182**, 379 (1972)
66. Riley, J.P. and Skirrow, R., *Chemical Oceanography*, 2nd edition, Volume I, p. 418, Academic Press, New York (1975)
67. Leatherhead, Y.T.M. and Burton, J.D. *J. Mar. Biol. Assoc. UK* **54**, 457 (1974)
68. Lunde, G., *Environ. Health Perspect.* **19**, 47 (1977)
69. Grieg, R.A., Wenzloff, D.R. and Pearce, J.B., *Marine Pollution Bulletin* **7**, 185 (1976)
70. Dean Luh, M., Baker, R.A. and Henley, D.E. *Science of the Total Environment* **2**, 1 (1973)

71. Coulson, E.J., Remington, R.E. and Lynch, K.M. *J. Nutrition* **19**, 255 (1935)
72. Jacobs, L.W., Syers, J.K. and Keeney, D.R., *Soil. Sci. Soc. Ann. Proc.* **34**, 750 (1970)
73. Andreae, M.O., *Analytical Chemistry* **49**, 820 (1977)
74. Johnson, D.L. and Braman, R.S., *Deep Sea Research* **22**, 503 (1975)
75. Inverson, D.G., Anderson, M.A., Holm, T.R. and Stanforth, R.R., *Environmental Science and Technology* **13**, 1491 (1979)
76. Arsenic in Food Regulations 1959, SI 1959 No. 1831, HMSO, London (1959)
77. Schroeder, H.A. and Balassa, J.J., *J. Chron. Dis.* **19**, 85 (1966)
78. Martin, M., Osborn, K.E., Billig, P. and Glickstein, N., *Marine Pollution Bulletin* **12**, 305 (1981)
79. Bryant, V., McLusky, D.S., Camphill, R. and Newberry, D.M., *Marine Ecology Prog. Series* **24**, 129 (1985)
80. Bryan, G.W., Heavy metal contamination in the sea. In Vernberg, W.B. *et al.* (eds), *Marine Pollution*, Part 3, Academic Press, New York (1976)
81. MacInnes, J.R. and Thuberg, R.P. *Marine Pollution Bulletin* **4**, 185 (1973)
82. Nelson, D.A., Wenzloff, D.R. and Calabrese, A., *Bulletin of Environmental Contamination and Toxicology* **16**, 275 (1976)
83. Walting, H.R. and Watling, R.J., *Bulletin of Environmental Contamination and Toxicology* **29**, 651 (1982)
84. Voyer, R.A., Cardin, J.A., Heltsche, J.F. and Hoffman, G.L., *Aquatic Toxicology* **2**, 223 (1982)
85. Klein-MacPhee, G., Cardin, J.A. and Berry, W.J., *Transactions American Fisheries Society* **113**, 247 (1984)
86. Coglinese, M.P., *Archive of Environmental Contamination and Toxicology* **11**, 297 (1982)
87. Stromgen, T., *Marine Biology* **72**, 69 (1982)
88. Calabrese, A., MacInnes, J.R., Nelson, D.A. and Miller, J.E. *Marine Biology* **41**, 179 (1977)
89. Bryant, V., Newberry, D.M., McLusky, D.S. and Campbell, R., *Marine Ecology Prog. Series* **24**, 139 (1985)
90. Jones, M.B., *Marine Biology* **30**, 13 (1975)
91. Herbert, D.W.M. and Wakeford, A.C., *Inst. Journal of Air and Water Pollution* **8**, 251 (1964)
92. Winlom, H.L. and Smith, R.G., *Marine Chemistry* **7**, 157 (1979)
93. Zirino, A. and Yammamoto, S., *Limnology and Oceanography* **17**, 661 (1972)
94. Turner, D.R., Whitfield, M. and Dickson, A.G., *Geochemica and Cosmochemica Acta* **45**, 855 (1981)
95. Connor, P.M., *Marine Pollution Bulletin* **3**, 190 (1972)
96. Steele, C.W., *Marine Pollution Bulletin* **14**, 168 (1983)
97. Calabrese, A., MacInnes, J.R., Nelson, D.A., Grieg, R.A. and Yevich, P.P., *Marine Environmental Research* **11**, 253 (1984)
98. Phillips, D.J.H., Toxicity and accumulation of cadmium in marine and estuarine biota. In Nriagu, J.O. (ed.), *Cadmium in the Environment Part I. Ecological Cycling*, pp. 426–570, Wiley, New York (1980)
99. Taylor, D., A review of the lethal and sublethal effects of cadmium on aquatic life. In *Proceedings of the Third International Cadmium Conference*, Miami, Florida. pp. 75–81 (1981)
100. Thede, H., Scholtz, N. and Fascher, H., *Marine Ecology Programme Series* **1**, 13 (1979)
101. Pesch, G. and Stewart, G., *Marine Environmental Research* **3**, 145 (1980)
102. Sharp, J.R. and Neff, J.M., *Marine Environmental Research* **3**, 195 (1980)

103. Maguire, R.J., Carey, J.H. and Hale, E.J.J., *Agric. Food Chem.* **31**, 1060 (1983)
104. Jong-Hwa, L., *Bulletin of Fisheries Science Institute* **3**, 11 (1987)
105. Moles, A., Babcock, M.M. and Rice, S.D., *Marine Environmental Research* **21**, 49 (1987)
106. Axiate, V. and George, J., *J. Marine Biology* **94**, 241 (1987)
107. Stromgen, T., *Marine Environmental Pollution* **21**, 239 (1987)
108. *Petroleum in the Marine Environment*, National Academy of Sciences, Washington, DC (1975)
109. Rep. Stud., GESAMP, No. 6 (1977)
110. Neff, J.M., In *Polycyclic Aromatic Hydrocarbons in the Aquatic Environment, Sources, Fates and Biological Effects*, Applied Science Publishers, London (1979)
111. Farrington, I.W., Albaiges, J., Burns, K.A., Dunn, B.P., Eaton, P., Laseter, J.L., Parker, P.L. and Wise, S., in *The International Mussel Watch*, Chapter 2, National Academy of Sciences, Washington, DC (1980)
112. Howard, J.W. and Fazio, T., *Journal of the Association of Official Analytical Chemists* **63**, 1077 (1980)
113. Connel, D.W. and Miller, G.J., *CRC Crit. Rev. Environ. Control* **11**, 37 (1980)
114. Connel, D.W. and Miller, G.J., *CRC Crit. Rev. Environ. Control* **11**, 105 (1980)
115. Davis, K.R., Schultz, T.W. and Dumont, J.N. *Archives of Environmental Contamination and Toxicology* **10**, 371 (1981)
116. Dillon, T.M., Neff, J.M. and Warner, J.S., *Bulletin of Environmental Contamination and Toxicology* **22**, 320 (1978)
117. Dumont, J.N., Schultz, T.W. and Jones, R.D. *Bulletin of Environmental Contamination and Toxicology* **22**, 159 (1979)
118. Giddings, J.M., *Bulletin of Environmental Contamination and Toxicology* **23**, 360 (1979)
119. Parkhurst, B.R., Bradshaw, A.S., Forte, J.L. and Wright, G.P., *Bulletin of Environmental Contamination and Toxicology* **23**, 349 (1979)
120. Southworth, G.R., Keffer, C.C. and Beauchamp, J.J., *Environmental Science and Technology* **14**, 1529 (1980)
121. Wilson, B.W., Pelroy, R.A. and Cresto, J.T., *Mutat. Res.* **79**, 193 (1980)
122. Brule, T., *Journal of Marine Biological Association* **67**, 237 (1987)
123. Holt, G., Froslie, A. and Northeim, G., *Acta Vet. Scand. Suppl.* **70**, 28 (1979)
124. Brevik, E.M., Bjerk, J.E. and Kveseth, N.J., *Bulletin of Environmental Contamination and Toxicology* **20**, 715 (1978)
125. Kveseth, N.J., Bjek, J.E., Fimreite, N. and Stenerson, J., *Archives of Environmental Contamination and Toxicology* **8**, 201 (1979)
126. Lunde, G. and Baumann Ofstad, E., *Fresenius Zeitschrift für Analytische Chemie* **282**, 395 (1976)
127. Baumann Ofstad, E., Lunde, G., Martinsen, K. and Rygg, B., *Science of the Total Environment* **10**, 219 (1978)
128. Norheim, G., *Acta Pharmacol. Toxicol.* **42**, 7 (1978)
129. Koeman, J.H. Ten Noever de Brauw, M.C. and de Vos, R.H., *Nature (London)* **221**, 1126 (1969)
130. Ten Noever de Brauw, M.C. and Koeman, J.H. *Science of the Total Environment* **14**, 27 (1972/73)
131. Tucker, R.E., Young, A.L. and Gray, A.P. (eds), *Human and Environmental Risks of Chlorinated Dioxins and Related Compounds*, Plenum Press, New York (1983)
132. Ferrando, M.D., Andreumoliner, E., Almar, M.M., Cabrion, C. and Nunez, A., *Bulletin of Environmental Contamination and Toxicology* **39**, 365 (1987)

133. Cleland, G.B., McElray, P.N. and Sangegaard, R.A., *Aquatic Toxicology* **12**, 41 (1988)
134. Bridgham, S.D., *Archives of Environmental Contamination and Toxicology* **17**, 731 (1988)
135. Stephenson, M. and MacKie, G.L., *Aquatic Toxicology* **9**, 243 (1986)
136. Weis, J.S. and Ma, A., *Bulletin of Environmental Contamination and Toxicology* **39**, 224 (1987)
137. Kimborough, R.D. (ed.), *Halogenated Biphenyls, Terphenyls, Naphthalenes, Dibenzodioxins and Related Products*, Elsevier/North-Holland, New York (1980)
138. Hutzinger, O., Frei, R.W., Merian, E. and Pocchiari, F. (eds), *Chlorinated Dioxins and Related Compounds: Impact on the Environment*, Pergamon Press, New York (1982)
139. Nicholson, W.J. and Moore, J.A. (eds), *Health Effects of Halogenated Aromatic Hydrocarbons*, New York Academy of Sciences, New York (1979)
140. Lee, D.H.K. and Falk, H.L. (eds), *Environmental Health Perspectives*, Experimental Issue No. 5, US Department of Health, Education and Welfare Publ. No (NTH) 74-218 September (1973)
141. Huff, J.R., Moore, J.A., Saracci, D.R. and Tomatis, L. in Rail, D.P. (ed.), *Environmental Health Perspectives*, pp. 221–40, US Department of Health and Human Services Publ. No. 80–218, No. 36 (1980)
142. Espositio, M.P., Tiernan, T.O. and Dryden, F.E. *Dioxins*, USEPA Report No. EPA-600/2-80-197, November (1980)
143. Olie, K., Vermeullen, P.L. and Hutzinger, O., *Chemosphere* **6**, 455 (1977)
144. Ahling, B., Lindskog, A., Jannson, B. and Sundstrom, G., *Chemosphere* **6**, 455 (1977)
145. Buser, H.R., Bosshardt, H.P., Rappe, C. and Lindahl, R., *Chemosphere* **7**, 419 (1978)
146. Crosby, D.G. and Wong, A.S., *Chemosphere* **5**, 327 (1976)
147. Lamparski, L.L., Stehl, R.H. and Johnson, R.L., *Environmental Science and Technology* **14**, 196 (1980)
148. McConnell, E.E., in Kimbrough, R.D. (ed.), *Halogenated Biphenyls, Terphenyls, Naphthalenes, Dibenzodioxins and Related Products*, pp. 109–50, Elsevier/North-Holland, New York (1980)
149. Golstein, J.A., in Kimbrough, R.D. (ed.), *Halogenated Biphenyls, Terphenyls, Naphthalenes, Dibenzodioxins and Related Products*, pp. 151–90, Elsevier/North-Holland, New York (1980)
150. Bickel, M.H. and Muhelback, S., in Hutzinger, O. *et al.* (eds), *Chlorinated Dioxins and Related Compounds: Impact on the Environment*, pp. 3036, Pergamon Press, New York (1982)
151. DiDomenico, A., Viviano, G. and Zapponi, G., in Kimbrough, R.D. (ed.), *Halogenated Biphenyls, Terphenyls, Naphthalenes, Dibenzodioxins and Related Products*, pp. 105–14, Elsevier/North-Holland, New York (1980)
152. Ward, C.T. and Matsumura, F., *Archives of Environmental Contamination and Toxicology* **7**, 349 (1978)
153. Young, A.L. in Nicholson, W.J. and Moore, J.A. (eds), *Health Effects of Halogenated Aromatic Hydrocarbons*, pp. 173–90, New York Academy of Sciences, New York (1979)
154. Masuda, I. and Kuroki, H., in Kimbrough, R.D. (ed.), *Halogenated Biphenyls, Terphenyls, Naphthalenes, Dibenzodioxins and Related Products*, pp. 561–9, Elsevier/North-Holland, New York (1980)
155. Kuratsune, M., in Kimbrough, R.D. (ed.), *Halogenated Biphenyls, Terphenyls,*

Naphthalenes, Dibenzodioxins and Related Products, pp. 287–302, Elsevier/North-Holland, New York (1980)

156. Vos, J.G., Keoman, J.H., Van der Maas, H.L., Ten Noever de Brauw, M.C. and de Vos R.H., *Food Cosmet. Toxicol.* **8**, 625 (1970)

157. Buser, H.R., Bosshardt, H.P. and Rappe, C., *Chemosphere* **7**, 109 (1978)

158. Smith, R.M., O'Keefe, P.W., Hiker, D.R., Jelus-Tyror, B.L. and Aldous, K.M., *Chemosphere* **11**, 715 (1982)

159. Jansson, B. and Sundstrom, G., in Hutzinger, O. *et al.* (eds), *Chlorinated Toxins and Related Compounds: Impact on the Environment*, pp. 201–8, Pergamon Press, New York (1982)

160. Rappe, C., Markland, S., Bergqvist, P.A. and Hansson, M., *Chem. Scr.* **20**, 56 (1982)

161. Mills, A.I. and Alexander, M., *J. Environmental Quality* **5**, 437 (1976)

162. Firestone, D., Rees, J., Brown, N.L., Barron, R.P. and Damico, J.N.J., *Journal of the Association of Analytical Chemists* **55**, 85 (1972)

163. Rappe, C., Markland, S., Buser, H.R. and Bosshardt, H.P., *Chemosphere* **3**, 269 (1978)

164. Singh, D.K. and Agarwal, R.A., *Science of the Total Environment* **67**, 263 (1987)

165. Upadhyaya, O.V.B. and Shuka, G.S., *Environmental Research* **41**, 591 (1986)

166. Saxena, P.K. and Mani, K., *Environmental Pollution* **5597** (1988)

167. Tripathi, G. and Shulka, S.P., *Ecotoxicology and Environmental Safety* **15**, 277 (1988)

168. Khangarot, B.S. and Ray, P.K., *Archiv. für Hydrobiologie* **113**, 465 (1988)

169. Lall, D.J., Brooke, L.T., Kent, R.J., Knuth, M.L., Polrier, S.H., Huot, J.M. and Lima, A.R., *Archives of Environmental Contamination and Toxicology* **16**, 607 (1987)

170. Rajeswari, K., Kalarani, V., Reddy, D.C. and Ramamurthi, R. *Bulletin of Environmental Contamination and Toxicology* **40**, 212 (1988)

171. Vijayakumari, P., Reddy, D.C. and Ramamurthi, R. *Bulletin of Environmental Contamination and Toxicology* **38**, 742 (1987)

172. Knowles, G.O., McKee, M.J. and Palawski, D.U., *Environmental Toxicology and Chemistry* **6**, 201 (1987)

173. Kocan, R.M., Von Westernhagen, H., Landolt, H.L. and Furstenberg, G., *Marine Environmental Research* **23**, 291 (1987)

174. Russon, C.L., Drummond, R.A. and Hoffmann, A.D., *Bulletin of Environmental Contamination and Toxicology* **41**, 589 (1988)

175. Conti, E., *Aquatic Toxicology* **10**, 325 (1987)

176. Green, D.W., Williams, K.A., Hughes, D.R.L., Shaik, E.A.R. and Pascoe, D., *Water Research* **22**, 225 (1988)

177. Cleland, G.B., McElroy, P.N. and Sangegaard, R.A., *Aquatic Toxicology* **12**, 141 (1988)

178. Dumpert, K., *Ecotoxicology and Environmental Safety* **13**, 324 (1987)

179. Webster, P.W., Canton, J.H. and Darmons, J.A.M.A., *Aquatic Toxicology* **12**, 323 (1988)

180. Bengtsson, B.E., *Water Science and Technology* **20**, 87 (1988)

181. Seeyle, J.G., Marking, L.L., King, E.L., Hanson, L.H. and Bills, T.D., *North American Journal of Fisheries Management* **7**, 598 (1981)

182. Webb, J.L. *Enzyme and Metabolic Inhibitors*, Volume 3, Chapter 6, Academic Press, New York (1966)

183. Staaffer, R.E., Ball, J.W. and Jenne, E.A. Geological Survey Professional Paper 1044F US Government Printing Office, Washington DC (1980)

184. Getzendanner, M.F. and Corbin, H.B., *J. Agric. Food Chem.* **20**, 882 (1972)
185. Wong, P.T.S., Chau, I.K., Luton, L. and Bengert, G.A., Conference Proceedings on Trace Substances in Environmental Health—XI Methylation of Arsenic in the Aquatic Environment, Hemphill University, Missouri (1977)
186. Chapman, A.C., *Analyst (London)* **51**, 548 (1926)
187. Challenger, F., *Chemical Review* **36**, 315 (1945)
188. Vonk, J.W. and Sisperstein, A.K., *Autonie von Leeuwenhoeck* **39**, 505 (1973)
188. Jensen, S. and Jernelov, A., *Nature* **223**, 753 (1968)
190. McBride, B.C., Merilees, H., Cullen, W.R. and Picket, W., *ACS Symposium Series* **82**, 94 (1978)
191. Wood, J.H., Kennedy, F.S. and Rosen, C.G., *Nature (London)* **220**, 173 (1968)
192. Wood, J.H., *Science* **183**, 1049 (1974)
193. Laudner, L., *Nature (London)* **230**, 452 (1971)
194. Bertilsson, L. and Neujahr, H.Y., *Biochemistry* **10**, 2805 (1971)
195. Braman, R.S., in Woolson, E.A. (ed.), *Arsenical Pesticides*, American Chemical Society, Washington, DC, **Ser. 7**, 108 (1975)
196. Cox, D.P., in Woolson, E.A. (ed.), *Arsenical Pesticides*, American Chemical Society, Washington, DC, **Ser. 7**, 81 (1975)
197. Von Ende, D.W., Kearney, P.C. and Kaufman, D.D., *J. Agric. Food Chem.* **16**, 17 (1968)
198. Wong, P.T.S., Silverberg, B.A., Chau, Y.K. and Hodson, P.V. in Nriagu, J.O. (ed.), *Biogeochemistry of Lead*, pp. 270–342, Elsevier, New York (1978)
199. Muddock, B.G. and Taylor, D., The acute toxicity and bioaccumulation of some lead alkyl compounds in marine animals, in Branica, M. and Konrad, Z. (eds), *Proceedings of the International Experts Discussion on Lead—Occurrence, Fate and Pollution in the Marine Environment*, Rovina, Yugoslavia, 1977, Pergamon Press, New York (1980)
200. Grandjean, P. and Neilson, T., *Residue Rev.* **72**, 97 (1979)
201. Grove, J.R., in Branica, M. and Konrad, Z. (eds), *Lead in the Marine Environment*, pp. 45–52, Pergamon Press, Oxford (1980)
202. Wong, P.T.S., Chau, J.K., Kramer, O. and Bengert, G.A., *Water Research* **15**, 621 (1981)
203. Wong, P.T.S, Chau, Y.K. and Luxon, P.L., *Nature (London)* **253**, 263 (1975)
204. Reisinger, K., Stoeppler, M. and Nurnberg, H.W., *Nature (London)* **291**, 228 (1981)
205. Jarvis, A.W.P., Markall, R.N. and Potter, H.R., *Nature (London)* **255**, 217 (1975)
206. Schmidt, U. and Huber, F., *Nature (London)* **259**, 159 (1976)
207. Dumas, J.P., Pazdernik, L., Belloncik, S., Bouchard, D. and Vaillancourt, G., *Proc. 12th Can. Symp. Water Pollut. Res.* **12**, 91 (1977)
208. Sirota, C.R. and Uthe, J.F., *Analytical Chemistry* **49**, 823 (1977)
209. Chau, Y.K. and Wong, P.T.S., Lead in the marine environment, in Branica, M. and Konrad, Z. (eds), *Proceedings of the International Expert Discussion on Lead Occurrence, Fate and Pollution in the Marine Environment*, Rovine, Yugoslavia, 1977, Pergamon Press, New York (1980)
210. Reamer, P.O., Zoller, W.H. and O'Haver, T.C., *Analytical Chemistry* **50**, 1449 (1978)
211. Mor, E.D. and Beccaria, A.M., A dehydration method to avoid loss of trace elements in biological samples, in Branica, M. and Konrad, Z. (eds), *Proceedings of the International Experts Discussion on Lead—Occurrence, Fate and Pollution in the Marine Environment*, Rovina, Yugoslavia, (1977), Pergamon Press, New York (1980)
212. Thayer, J.S., *Occurrence—Biological Methylation of Elements in the Environment*, American Chemical Society, Washington, DC, ACS Advances in Chemistry Series, No. 182, 188 (1978)

213. Uthe, J.F. and Armstrong, F.A.J., *J. Fisheries Research Board, Canada* **27**, 805 (1970)
214. Krenkel, P.A., *International Critical Review of Environmental Control* **3**, 303 (1973)
215. Robinson, S. and Scott, W.B. *A Selected Biogeography on Mercury in the Environment with Subject Listing*, Life Science, Miscellaneous Publication, Royal Ontario Museum, p. 54 (1974)
216. Uthe, J.F. and Armstrong, F.A.J., *Toxicological Environmental Chemistry Review* **2**, 45 (1974)
217. D'Itri, P.A. and D'Itri, F.M., *Environmental Management* **2**, 3 (1978)
218. Furukawa, K. and Tonomura, K., *Agricultural and Biological Chemistry* **35**, 604 (1971)
219. Furukawa, K. and Tonomura, K., *Agricultural and Biological Chemistry* **36**, 217 (1972)
220. Furukawa, K. and Tonomura, K., *Agricultural and Biological Chemistry* **36**, 244 (1972)
221. Brinkmann, F.E. and Iverson, W.P., in Church, T. (ed.), *Marine Chemistry in the Coastal Environment*, American Chemical Society Symposium 18, American Chemical Society, Washington, DC (1975)
222. Fogerstrom, T. and Jernelov, A., *Water Research* **5**, 121 (1971)
223. Yamada, M. and Tonomura, K., *Journal of Fermentation Technology* **50**, 159 (1972)
224. Yamada, M. and Tonomura, K., *Journal of Fermentation Technology* **50**, 893 (1972)
225. Yamada, M. and Tonomura, K., *Journal of Fermentation Technology* **50**, 901 (1972)
226. Rapsonankes, S. and Weber, H., *Environmental Science and Technology* **19**, 352 (1985)
227. Imura, N.E., Sukegawa, S.K. and Pan, K., *Science* **172**, 1248 (1971)
228. Andren, A.W. and Harris, R.C., *Nature (London)* **245**, 256 (1973)
229. Amemiya, T., Takeuchi, M., Ito, K., Ebara, K., Harada, H. and Totani, T., Ann. Rep. Tokyo Metropolitan Res. Lab. Public Health 26-1, 129 (1975)
230. Matsunaga, K. and Takahashi, S., *Analytica Chimica Acta* **87**, 487 (1976)
231. Westoo, G. and Rydalv, M., *Var Foeda* **21**, 20 (1969)
232. Westoo, G. and Rydalv, M., *Var Foeda* **21**, 138 (1969)
233. Westoo, G., *Acta Sc. and Sc. Anal.* **20**, 2131 (1966)
234. Westoo, G., *Acta Sc. and Sc. Anal.* **21**, 1790 (1967)
235. Westoo, G., *Acta Sc. and Sc. Anal.* **22**, 2277 (1968)
236. Jones, P. and Nickless, P., *Analyst (London)* **103**, 1121 (1978)
237. Collett, D.L., Fleming, J.E. and Taylor, G.E., *Analyst (London)* **105**, 897 (1980)
238. Ram, R.N. and Sathyanesan, A.G., *Environmental Pollution* **47**, 135 (1987)
239. Kiruagarum, R. and Joy, P., *Ecotoxicology and Environmental Safety* **15**, 171 (1988)
240. Czuba, M., Seagull, R.W., Tran, H. and Cloutier, L., *Ecotoxicology and Environmental Safety* **14**, 64 (1987)
241. Kiruagarum, R. and Joy, P., *Ecotoxicology and Environmental Safety* **15**, 171 (1988)
242. WHO Task Group, Sharrat, M. (Chairman), Vouk, V.B. (Secretary) *Environ. Health Criter.* **15**, 1 (1980)
243. Zuckerman, J.J., Reisdorf, P.R., Ellis, H.E. III and Wilkinson, R.R., in Brinckman, F.E. and Bellama, J.M. (eds), *Organometals and Organometalloids, Occurrence and Fate in the Environment*, pp. 388–422, ACS Symposium Series No. 28, American Chemical Society Washington, DC (1978)
244. Bock, R., *Residue Rev.* **79**, 216 (1981)
245. Gachter, R. and Muller, H. *Handbuch der Kunststoff-Additive*, Hanser, Munchen (1979)
246. Fishbein, L., *Science of the Total Environment* **2**, 341 (1974)

247. Meinema, H.A., Burger-Weisma, T., Versluis-de-Haau, S. and Gevers, E.C.C., *Environmental Science and Technology* **12**, 288 (1978)
248. Laughlin, R.B., French, W., Johannesen, R.B., Guard, H.E. and Brinckman, F.E., *Chemosphere* **13**, 575 (1984)
249. Guard, H.E., Cobet, A.B. and Coleman, W.H., *Science* **213**, 770 (1981)
250. Jackson, J.A., Brinckman, F.E. and Iverson, W.P., *Environmental Science and Technology* **16**, 110 (1982)
251. Byrd, J.T. and Andreae, M.O., *Science (Washington, DC)* **218**, 565 (1982)
252. Braman, R.S. and Tompkins, M.A., *Analytical Chemistry* **51**, 12 (1979)
253. Hodge, V.F., Seikel, S.L. and Goldberg, E.D., *Analytical Chemistry* **51**, 1256 (1979)
254. Hallas, L.E., Means, J.C. and Cooney, J.J., *Science (Washington, DC)* **215**, 1505 (1982)
255. Turgul, S., Balka, T.I. and Goldberg, E., *Marine Pollution Bulletin* **14**, 297 (1983)
256. Gilmour, C.C., Tuttle, J.H. and Means, J.C., in Sigleo, A.C. and Hattori, A. (eds), *Marine and Estuarine Geochemistry*, pp. 239–58, Lewis Publishers, Chelsea, MI (1985)
257. Gilmour, C.C. and Tuttle, J.H., *Analytical Chemistry* **58**, 1848 (1986)
258. Skopintsev, B.A., *Oceanography* **6**, 361 (1986)
259. Williams, P.J., Determination of organic components, in Riley, J.S. and Skirrow, G. (eds), *Chemical Oceanography* Volume 3, pp. 443–77, Academic Press, New York (1975)
260. Skopinstev, B.A., *Oceanography* **16**, 630 (1976)
261. Wangersky, P.S. and Zika, R.G., The analysis of organic compounds in seawater. Report 3, NRCC 16566, Marine Analytical Chemistry Standards Programme (1978)
262. Van Hall, C.C. and Stenger, V.H., *Water Sewage* **111**, 266 (1964)
263. Van Hall, C.C., Barth, D. and Stenger, V.A., *Analytical Chemistry* **37**, 769 (1965)
264. Van Hall, C.E. and Stenger, V.A., *Analytical Chemistry* **39**, 502 (1967)
265. Van Hall, C.E., Safrenko, J. and Stenger, V.A., *Analytical Chemistry* **35**, 319 (1963)
266. Golterman, H.L., in *Methods for Chemical Analysis of Freshwater*, pp. 133–45, Blackwell, Oxford (1969)
267. Water Pollution Laboratory, Stevenage, Herts, UK, Notes on Water Pollution No. 59, HMSO, London (1972)
268. Slvalla Battori, N., Ribas Solar, F. and Orori Durich, J., *Doc. Inest/ Hidrol.* **17**, 303 (1974)
269. Weis, J.S., Gottlieb, J. and Kwiatkowski, J., *Archives of Environmental Contamination and Toxicology* **16**, 321 (1987)
270. Guidici, M., Migliore, S.M., Guarino, S.M. and Gambardella, C., *Marine Pollution Bulletin* **18**, 454 (1987)
271. Roberts, M.H., *Bulletin of Environmental Contamination and Toxicology*, **39**, 1012 (1987)
272. Noth, K. and Kumar, N., *Chemosphere* **17**, 465 (1988)
273. Knowles, C.D., McKee, H.J. and Palawski, D.U., *Environmental Toxicology and Chemistry* **6**, 201 (1987)
274. McKim, J.M., Schneider, P.K., Carlson, R.W., Hunt, E.P. and Niemi, G.J., *Environmental Toxicology and Chemistry* **6**, 295 (1987)
275. Conti, E., *Aquatic Toxicology* **10**, 325 (1987)
276. Herring, C.O., Adams, J.A., Wilson, B.A. and Pollard, S., *Bulletin of Environmental Contamination and Toxicology* **17**, 731 (1988)
277. Dill, D.C., Murphy, P.G. and Mayes, M.A., *Bulletin of Environmental Contamination and Toxicology* **39**, 869 (1987)

278. Carlson, A.R. and Kosian, P.A., *Archives of Environmental Contamination and Toxicology* **16**, 129 (1987)
279. Smith, P.D., Brockway, D.L. and Stancil, F.E., *Environmental Toxicology and Chemistry* **6**, 891 (1987)
280. Oikawa, A. and Kukkonen, J., *Ecotoxicology and Environmental Safety* **15**, 282 (1988)
281. Cleland, G.B., McElroy, P.N. and Sangegaard, R.A., *Aquatic Toxicology* **12**, 141 (1988)
282. Mallatt, J. and Barron, H.C., *Archives of Environmental Contamination and Toxicology* **17**, 73 (1988)
283. Tripathi, G. and Shulka, G.P., *Ecotoxicology and Environmental Safety* **15**, 277 (1988)
284. Cleland, G.B., McElroy, P.N. and Sangegaard, R.A., *Aquatic Toxicology* **12**, 241 (1988)
285. McKim, J.M., Schmeider, P.K., Niemi, J., Carlson, R.W. and Henry, T.R., *Environmental Toxicology and Chemistry* **6**, 313 (1987)
286. Mitchell, D.G., Chapman, P.M. and Long, T.J., *Bulletin of Environmental Toxicology* **39**, 1028 (1987)
287. Call, D.J., Brooke, L.T., Huot, J.M., Kent, R.J., Knuth, M.L., Poirier, S.H. and Lima, A.R., *Archives of Environmental Contamination and Toxicology* **16**, 607 (1987)
288. Ferrando, M.D., Almar, M.M. and Andreu, E., *Journal of Environmental Science and Health* **B23**, 45 (1988)
289. Wan, M.T., Watts, R.G. and Moul, D.J., *Bulletin of Environmental Contamination and Toxicology* **41**, 609 (1988)
290. Maas Diepeveen, J.L. and Van Leeuwen, C.J., *Bulletin of Environmental Contamination and Toxicology* **40**, 517 (1988)
291. Konouen, D.W., *Bulletin of Environmental Contamination and Toxicology* **41**, 371 (1988)
292. Call, D.J., Poirier, S.H., Knuth, M.L., Harting, S.L. and Lindberg, C.A., *Bulletin of Environmental Contamination and Toxicology* **38**, 352 (1987)
293. Seeyle, J.G., Marking, L.L., King, E.L., Hanson, L.H. and Bills, T.D., *North American Journal of Fisheries Management* **7**, 598 (1987)
294. McGuire, R.J. and Tracz, R.J., *Water Pollution Research Journal of Canada* **22**, 227 (1987)
295. Lasko, J.M. and Peoples, I.A., *J. Agri. Food Chem.* **23**, 674 (1975)
296. Lloyd, R., Effects of lead on marine organisms. Annex B of Technical Report TR208, Water Research Centre, Marlow, Bucks, UK, pp. 41–5 (1984)
297. McKim, J.M., Olson, G.J., Holcombe, G.W. and Hunt, E.P., *Journal of Fisheries Research Board, Canada* **33**, 2726 (1976)
298. Ridley, W.P., Diziker, L.J. and Wood, J.M., *Science* **197**, 329 (1977)
299. Van Nguyen, V., Pasey, J.J. and Eng, G., *Water, Air and Soil Pollution* **23**, 417 (1984)
300. Sadler, K. and Turnpenny, A.W.H., *Water, Air and Soil Pollution* **30**, 593 (1986)

CHAPTER 5

Effect on Creatures of Dissolved Metals in Freshwater and Oceans

Once the concentration of toxic metal has reached a certain level in water the exposure of creatures to that concentration for a certain period of time will produce adverse effects or mortality in them. Adverse effects are commonly identifiable with the occurrence of disease, reduced growth or impaired reproducibility although other adverse effects exist.

Regarding mortalities, characterised by LC_{50}, it is possible to draw up tables correlating for each toxicant and the type of creature the concentration and exposure time to that toxicant above which mortalities occur, i.e. the concentration above which the species is at risk. Such correlations are not always rigidly correct as creatures can develop a reversible immunity to toxic substances when exposed for a period of time. Nevertheless, such correlations do provide a useful indicator to the species at risk.

It is, of course, true to say that even though the concentration of a toxicant is not sufficiently high to cause mortalities in a particular type of fish during exposure for a stipulated period of time, colonies of that creature may still suffer from ill health, reducing colony size due to impaired reproducibility or illness in the young, or because species involved in their food chain are at risk and diminish in numbers.

Toxicity of an element to a particular type of creature can differ appreciably between freshwater (e.g. rivers, ponds, streams), estuarine and bay waters, and open seawater (Chapter 1). For this reason toxicity in freshwater and seawater are discussed separately below.

5.1 FRESHWATER CREATURES

5.1.1 TOXIC EFFECTS ON ORGANISMS

As discussed above, toxic substances can cause either mortalities (characterised by LC_{50}) or adverse effects (characterised by LE_{50}) in exposed creatures. In Table 5.1 are shown such LC exposure time relationships for salmonid and non-salmonid fish for various toxicants. As would be expected, the toxicant concentrations for a particular exposure time that cause adverse effects are lower than

Table 5.1. Effect of trace concentration ($\mu g\,l^{-1}$) of metals on the wellbeing of freshwater non-salmonids and salmonids

Element	Lethal dose (LC_{50})		Reduced growth (LE_{50})		Impaired reproducibility (LE_{50})		
	Days	$\mu g\,l^{-1}$	Days	$\mu g\,l^{-1}$	Days	$\mu g\,l^{-1}$	
Arsenic	100	1400	100	800	–	–	Non-salmonid
	23	5000	23	1000	–	–	Salmonid
Zinc	30	2000	30	210	–	–	Non-salmonid
	140	600	140	210	–	–	Non-salmonid
Cadmium	10	4000	–	–	10	240	Non-salmonid
	100	180	–	–	100	15	Non-salmonid
Chromium	60	2000	60	720	–	–	Non-salmonid
	10	18 300	10	2300	–	–	Salmonid
	100	1150	100	100	100	100	Salmonid
Copper	30	200	30	100	30	–	Salmonid
	40	150	40	–	40	140	Salmonid
	72	80	72	30	72	30	Salmonid
Lead	40	900	–	–	40	70	Non-salmonid
	90	550	–	–	90	400	Non-salmonid
Nickel	4	35 000	–	–	4	100	Non-salmonid
	15	8000	–	–	15	50	Non-salmonid
	100	2200	–	–	100	100	Non-salmonid

those that cause mortalities. Thus, in the case of chromium 100 days' exposure to 1150 $\mu g\,l^{-1}$ chromium causes 50% mortality of salmonid fish, while exposure to 100 $\mu g\,l^{-1}$ chromium for the same time causes reduced growth and impaired reproducibility, but no or few mortalities.

Further information on the effects of metals on freshwater fish and invertebrates is summarised in Table 5.2. The durations of the toxicity tests are not included in this table. They are, however, generally short-term tests, 4 to 14 days' exposure, and the concentrations listed are toxic effect concentrations. This information should be treated with some caution as little experimental work has been conducted on the influence on toxicity of factors such as water hardness, pH, salinity, temperature and acclimatisation of species to metals (Chapter 1). Table 5.2 is, nevertheless, a useful guide in that it highlights those species that are at risk, due to adverse effects or mortality when the concentrations of the element exceed levels quoted. Thus, it can be seen at a glance that when concentrations of mercury, copper and cadmium exceed 1 $\mu g\,l^{-1}$ then crustaceans are at risk, and when concentrations of copper exceed 10 $\mu g\,l^{-1}$ bivalve molluscs are at risk, when concentrations of zinc and nickel exceed 10 $\mu g\,l^{-1}$ crustaceans are at risk and when concentrations of lead and cadmium exceed 10 $\mu g\,l^{-1}$ gastropods and non-salmonid fish are at risk. These observations regarding mercury, copper, zinc, lead and cadmium, in general, confirm those reported in Table 4.6.

There have been some general comparisons of observed water quality and biological status for a wide range of UK rivers for chromium, lead, zinc, nickel and copper[145-149]. These provide information on the range and annual average concentrations of metals, both dissolved and insoluble, and provide information on fishery status, e.g. none, poor, fair, good and whether salmonid or non-salmonid fish are found. Some of these data tabulated in Figure 5.1 show the effect of increasing average metal levels on both yield and type of fish obtained. Zinc does not seem to have an adverse effect, whereas increasing levels of total and dissolved lead, copper, nickel and total chromium do.

It is interesting to compare the data given in Figure 5.1 with the maximum safe concentrations long-term (365 days) toxicity date, i.e. S_x and 95th percentile data (S_{95}) derived by UK authorities as discussed in Table 1.5 (Chapter 1). This information confirms on the whole a high mortality of fish when the yearly average concentration and the maximum metal concentration (Column E, Table 5.3) exceed the maximum safe concentration S_x (Column A, Table 5.3) for copper and zinc. It also confirms the adverse effect of increased metal content on fish yield and type.

Example of river draining rural and urban catchments

The results in panel (A) of Table 5.4 show the effect of animal life of mean toxic metal concentrations found in UK rivers draining rural and urban catchments.

Table 5.2. Toxicity of metals to freshwater fish and invertebrate, species at risk during 4–14 days' exposure to stated concentrations

Concentration above which mortalities can occur		Annelids								Bivalve molluscs								Crustacae							
		Hg	Cu	Cd	Zn	Pb	Cr	As	Ni	Hg	Cu	Cd	Zn	Pb	Cr	As	Ni	Hg	Cu	Cd	Zn	Pb	Cr	As	Ni
$\mu g\,l^{-1}$	$mg\,l^{-1}$																								
0.1–1	0.0001–0.001																	•		•					
1–10	0.001–0.01																		•						
10–100	0.01–0.1									•			•								•	•			•
100–1000	0.1–1	•	•	•																			•		
	1–10				•		•																		
	10–100	↓	↓	↓	↓	•		↓		↓		↓	↓	↓				↓	↓	↓	↓	↓		↓	↓

Reference to panel (A) and Table 5.2 shows that in rivers draining rural catchments concentrations of cadmium, copper, lead and zinc will put crustacae at risk. In addition, high lead concentrations ($10–100\ \mu g\,l^{-1}$) might lead to mortalities in fish and gastropods. Rivers draining urban catchments (panel (B)) have, as would be expected, higher trace metal concentrations and in addition to the above adverse effects concentrations of mercury and nickel might reach levels which have adverse effect on crustacaens. In addition, molluscs might be affected by copper, gastropods by copper and lead, and fish by lead. Thus, many rivers are at risk from the point of view of animal life and it is only in the upper reaches of rivers where industrial activity is small or non-existent, or in clean streams that the risk to such species is consistently low.

Example of polluted rivers

The River Carnon, UK, drains an area of natural mineral enrichment which has a long history of mining for tin and other metals[1-3]. The high concentration of metals in this river (Table 5.4) decimated most forms of animal life, the only forms surviving being those that have developed some tolerance to these metals.

In Table 5.5 are shown the analysis and fish mortality profiles of the River Gwyddan, Wales, both before and after installation of an effluent treatment plant

Table 5.2. (*continued*)

Gastropods								Rotifers								Insects								Non-salmonid fish							
Hg	Cu	Cd	Zn	Pb	Cr	As	Ni	Hg	Cu	Cd	Zn	Pb	Cr	As	Ni	Hg	Cu	Cd	Zn	Pb	Cr	As	Ni	Hg	Cu	Cd	Zn	Pb	Cr	As	Ni

at a steelworks which discharges into the river[4,5]. The beneficial effect of the installation of the effluent treatment on the metal content of the river and on the survival of fish is immediately apparent.

In Tables 5.6 and 5.7, respectively, are shown available information on the toxic effect (LE_{50}) and adverse effects of various metals on fish and other organisms (see also Table 4.1).

5.1.2 ANALYTICAL COMPOSITION

In Table 5.8 is shown a summary of concentrations found in various environmental freshwaters (see Appendix 1, Table A1.1, for a more detailed tabulation). It is seen that concentrations vary over a very wide range the lowest of which would produce no adverse effects on freshwater creatures in short-term exposures, and the highest of which would have severe toxic effects as illustrated in Table 5.9. Thus, in the case of copper at the lower end of the quoted concentrations range ($0.11 \ \mu g \, l^{-1}$) no creatures are at risk during 4–14 days' exposure, while at the higher end of the concentration range quoted ($200 \ \mu g \, l^{-1}$) all creatures are at risk.

It must be emphasised here that the data quoted refer to the wellbeing of freshwater creatures. A further consideration is the wellbeing of humans who eat fish or crustacea. Organisms which survive might well be inedible to humans due

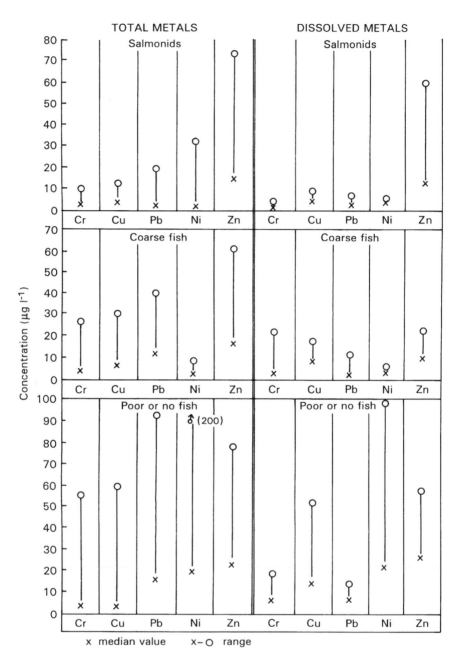

Figure 5.1 Fishery status versus element concentrations. Toxicants: chromium, copper, lead, nickel and zinc

Table 5.3. Comparison of data in Figure 5.1 with toxicity standard data S_x and S_{95} (Table 1.5)

		Toxicity standards (from Table 1.5, Section 1.4)		Concentrations ($\mu g\, l^{-1}$) yearly average of dissolved and total metals found in UK rivers (Table 1.5) supporting:					
		Maximum safe concentration S_x ($\mu g\, l^{-1}$) (365 days)	S_{95} Percentile maximum metal conc. ($\mu g\, l^{-1}$) for 17 days out of 365 days	Salmonids		Coarse Fish		No fish	
				C		D		E	
		A	B	Maximum	Median	Maximum	Median	Maximum	Median
Cr	d	100	800 (s)	2	1	22	2	18	4
	t	100	1000–3000 (ns)	10	2	27	3	55	2
Cu	d	4	17 (s)	9	5	20	8	54	14
	t			12	3	30	5	60	2
Pb	d	20	100 (n/s)	5	2	15	1	12	6
	t			20	2	40	11	95	15
Ni	d	220	900 (n/s)	6	3	5	2	100	20
	t			30	1	5	1	200	18
Zn	d	23	200 (n/s)	60	14	22	10	60	24
	t			75	20	60	18	80	21

s: salmonids
ns: non-salmonids
d: dissolved metals
t: total metals

Table 5.4. Adverse effects of metals on animal life in river waters

	As	Cd	Cr	Cu	Pb	Hg	Ni	Zn
(A) Rivers draining rural catchments								
Composition of water ($\mu g\,l^{-1}$)	1.0	0.19–0.33	1.4–6.0	3.2–8.8	2.4–13.0	0.03–0.09	7.2–9.3	10.0–26.0
Adverse effects expected:								
Crustaceans (cadmium, copper, lead and zinc), fish and gastropods (lead)								
(B) Rivers draining urban catchments								
Composition of water ($\mu g\,l^{-1}$)	3.0	1.22–1.56	7.4–11.0	11.4–13.2	11.0–15.6	0.24–0.7	11.0–45.0	68.0–69.0
Adverse effects expected:								
Crustaceans (cadmium, copper, lead, zinc, mercury and nickel), molluscs (copper), gastropods (copper and lead), and fish (lead)								
(C) River Carnon								
Composition of water ($\mu g\,l^{-1}$)	60	8	–	300	–	–	–	8000
Adverse effects expected:								
Crustaceans (cadmium, copper, zinc), annelids (copper, zinc), molluscs, gastropods, rotifers, insects and non-salmonid fish (copper)								

Table 5.5. Adverse effect of metals on animal life in River Gwyddan, Wales, before and after installation of effluent treatment plant (μg l^{-1})

	Zn	Pb	Ni	Cr	Cu	Fe	Total metals	Fish found
(A) 1969–73: Prior to installation of effluent treatment plant								
Upstream steelworks	0.3	0.1	<0.1	<0.1	<0.05	0.63	1.06	All types, no mortalities
Downstream steelworks	1.26	1.32	0.3	1.0	0.27	325	3.29	No fish survive
Further downstream	0.29	<0.1	0.02	0.34	0.06	50	57	No fish survive
Upstream tidal limit	0.7	0.3	<0.1	0.8	0.09	289	291	No fish survive
(B) 1974–76: After installation of effluent treatment plant								
Upstream steelworks	0.1	0.01	<0.01	<0.1	0.01	3.4	3.5	All types, no mortalities
Downstream steelworks	0.12	0.01	0.02	1.4	0.01	3.0	4.5	No fish survive
Further downstream	—	0.01	–	0.02	0.01	2.5	2.5	*Salmo Vrutta, anguilla, Gasterosteus aculatus, Neomachheilus barbatulus* all found
Upstream of tidal limit	0.08	0.01	0.01	0.01	0.01	1.2	1.3	Above plus flounder all found

Source: Williams, R., Williams, P.F., Benson Evans, K., Hunter, M.D. and Harcup, H.F., *Water Pollution Control* **76**, 428 (1976). Reproduced by permission of the Institution of Water and Environmental Management.

Table 5.6. Concentration of metals in freshwaters and effects on fish

Element	Fish type	Water type	pH	Exposure concentration	Toxicity index	Adverse effects	Ref.
Acidity	Fathead minnows (*Pimephales promelas*)	Natural	5.5–7.0	Hard water 160–200 $\mu g\,l^{-1}$ as $CaCO_3$. Soft water 5–10 $mg\,l^{-1}$ as $CaCO_3$	—	Impairment of feeding behaviour at low pH. This species virtually eliminated at pH 5.8–6.0	6
pH	Brown trout	Loch	4.3	—	—	Reduced growth and survival	7
Aluminium pH and aluminium	Brown trout	Loch	$\leqslant 5.2$	—	—	Growth suppression	7
	Lake trout	Lake	5–0 pulse exposure	200 $\mu g\,l^{-1}$ AC pulse exposure	—	No mortality in 21–32 days' exposure	8
Aluminium and calcium	Rainbow trout (*Salmo gairdneri*)	River	3.7–7.6	150 $mg\,l^{-1}$ Ca	larvae pH, aluminium had LC_{50} of 3.8 $mg\,l^{-1}$ in softwater (1 $mg\,Ca\,l^{-1}$)	pH 3.7: No survival after 5 days. pH 4.6: No survival after 9 days. pH 5.7–6.6: Delayed hatching.	9
Aluminium acidity	Rainbow trout (*Salmo gairdneri*)	Stream	4.83	240 $\mu g\,l^{-1}$ (Al)	—	15% mortality and stress response	10
Aluminium acidity	Blackback herring	Soft freshwater	5.0–7.8	0–400 $\mu g\,l^{-1}$	—	At hardness 23–25 $mg\,l^{-1}$ Ca (as $CaCO_3$) In absence of aluminium, 6.9% embryo mortality at pH 5.0. In aluminium exposed embryos 39–100% mortality at pH 5.0. In absence of aluminium larval mortality ranged from 99% (pH 5.0) to 16% (pH 7.8). 100% mortality within 24 h if larvae exposed to 0.03 $mg\,l^{-1}$ aluminium at pH 5.0.	11

Contaminant	Fish	Water	pH	Concentration	LC₅₀	Comments	Ref.
Aluminium acidity	Brown trout (*Salmo trutta*)	Natural	4.3–6.5	14.8–100 μg l^{-1}	—	In 6 weeks' exposure, only pH 4.3 adversely affected trout growth at aluminium concentrations of > 27 μg l^{-1} and pH 5.5 reduced growth	13
Ammonium	Teleost fish (*Channa punctatus*)	River	—	—	—	Exposure to 100 mg l^{-1} NH₄, the safe concentration (6 months) 500 mg l^{-1}, sublethal concentration (6 months) revealed liver and thyroid degeneration and signs of hyperactivity	14
Ammonia	Rainbow trout (*Salmo gairdneri*)	Natural	—	0.027–0.27 mg l^{-1}	—	0.027 mg l^{-1} ammonium for up to 73 days' 70% mortality when exposure began within 2 h of fertilization of eggs. When exposure did not begin until eggs were 24 days old 40% of eggs, yolk-sac fry and fry died at ammonia concentration of 0.27 mg l^{-1}	15
Aluminium and acidity	Brown trout (*Salmo trutta*)	Natural	5.0	60–230 mg l^{-1}	—	5 days' exposure to acid plus 230 μg l^{-1} aluminium inhibited growth. 8 days' exposure at pH 5.1 of newly hatched fish to 60 mg l^{-1} caused 57% mortality	12
Ammonia (as NH₄Cl)	Rainbow trout (*Salmo gairdneri*) Fathead minnow	River	—	—	LC_{50} 0.53 mg l^{-1}	—	16
Cadmium	*Notropis lutrensis* and (*Pimephales promelas*) (Fathead Minnow)	Natural	—	—	LC_{50} 2.17 mg l^{-1} 96 h LC_{50} < 10 mg l^{-1}	Exposure to sublethal concentration, reduced ability to withstand heat stress	6 17
Cadmium	*Lobistes reticulatus*	Natural	—	—	96 h LC_{50} 350 mg l^{-1} male fish 371 mg l^{-1} female fish	Safe concentration of cadmium 112.9 mg l^{-1} (male) 116.5 mg l^{-1} (female)	18

continued overleaf

Table 5.6. (*continued*)

Element	Fish type	Water type	pH	Exposure concentration	Toxicity index	Adverse effects	Ref.
Cadmium	*Punctius conchonius*	Freshwater	–	630–840 $\mu g\,l^{-1}$	96 h LC_{50} 12.6 $mg\,l^{-1}$	Branched lesions observed. Copious mucus secretion from all over body surface	19
Cadmium	*Tilapia aurea*	Natural	–	6.8–52 $mg\,l^{-1}$	–	In 16 weeks no adverse effects on survival or growth	20
Chromium	*Selenatrum capricornutum*	Ground and surface water	–	–	96 h EC_{50} reported	Inhibition of alkaline phosphatase activity	21
Copper	*Polypedilum nubifer*	Natural	–	10–40 $\mu g\,l^{-1}$	–	Emergence success of eggs decreased from 74% to 2% of the control as the copper content of the water increased from 10 to 40 $mg\,l^{-1}$	22
Copper	Bluegill sunfish (*Lepomis macrochirus*)	Natural	–	40–400 $\mu g\,l^{-1}$	–	40 $\mu g\,l^{-1}$ copper for 8 days; fish 67% as active as in control. 80 $\mu g\,l^{-1}$ copper for 8 days; fish 61% as active as in control, 400 $\mu g\,l^{-1}$ copper for 6 days; fish 44% as active as in control	23
Copper	Rainbow trout (*Salmo gairdneri*)	Natural	–	200 $\mu g\,l^{-1}$	–	24–48 h exposure—tissue necrosis and stress, interference in nervous system functioning	24
Cyanide	Rainbow trout (*Salmo gairdneri*)	Natural	–	–	96 h LC_{50}	96 h LC_{50} varied seasonally and with exercise	25
Lead	*Hebistes reticulatis*	Natural	–	–	96 h LC_{50} 1620 $mg\,l^{-1}$ (male) 1630 $mg\,l^{-1}$ (female)	Safe concentrations of lead 492 $mg\,l^{-1}$ male 487 $mg\,l^{-1}$ female	18

Mercury (as $HgCl_2$)	*Channa punctatus*	Natural	$10\ \mu g\,l^{-1}$	—	6 months' exposure; mercury-induced inhibition of gonadal growth	19, 26
Selenium as selenite, selenomethionine	Bluegill (*Lepomis machochirus*)	Natural	$3\text{–}30\ \mu g\,Se\,g^{-1}$	—	In spawing experiments after 260 days' exposure: cumulative mortality increased, cataract development occurred and larval survival reduced	27
Zinc	Rainbow trout (*Salmo gairdneri*)	Natural	—	96 h LC_{50} 26.0 mg l⁻¹ (juveniles) 24.0 mg l⁻¹ (adults)	—	28
Zinc	*Tilapia zilla, Clarias lazerca*	Natural	—	96 h LC_{50} *Tilapia zilla* 13–33 mg l⁻¹ *Clarias lazerca* 26–52 mg l⁻¹ (depending on season)	—	29

Table 5.7. Concentration of metals in freshwaters and effects on organisms other than fish

Element	Organism	Water type	pH	Exposure	Toxicity index	Adverse effects	Ref.
Ammonia	Fingernail clam (*Musculium transversum*) Crayfish (*Orconectes immunis*)	Natural	–	–	LC_{50} values increased with temperature	–	16, 30
Ammonia	Fingernail clam (*Musculium transversum*)	Stream	–	0.002–0.08 mg l⁻¹ 0.04–0.25 mg l⁻¹ 0.14–0.56 mg l⁻¹	– – –	Survival rate 50–85% 30–55% nil (growth adversely affected)	31
Cadmium	*Idothea baltica* (Crustacea isopoda)	Natural	–	0.01–15 mg l⁻¹	–	In chronic exposure to 0.5 mg l⁻¹ cadmium, did not survive 60 days	32
Cadmium	*Daphnia magna*	Natural	–	0–50 µg l⁻¹	25-day LC_{50} = 10 µg l⁻¹	–	33
Cadmium	Amphipod (*Allorchestes compressa*)	Natural	–	–	96 h LC_{50} 0.78 mg l⁻¹	–	34
Cadmium	*Daphnia magna*	Natural	–	0–72 µg l⁻¹	4, 7 and 21 days, no observed effect concentrations (NOEC) for survival and reproduction, 4 days 2.1 µg l⁻¹ 7 days 0.8 µg l⁻¹ 21 days 0.8 µg l⁻¹	–	35
Chromium	*Cerio daphnia* *L. macrochirus*	Natural	–	–	acute toxicity value, 0.031 mg CrV1 l⁻¹ (*Cerio daphnia*), 182.9 mg CrV1 l⁻¹ (*L. macrochirus*)	–	36

Copper	*Ascellus aquaticus Proasellus coxalis Dollf* (Crustacea isopoda)	River	–	$0.01–0.15$ mg l^{-1}	ST_{50} values recorded	0.005 mg l^{-1} copper did not affect adult survival, juveniles had lower survival at any particular copper level	37
Copper	Amphipod (*Allorchestes compressa*)	Natural	–	–	96 h LC_{50} 0.48 mg l^{-1}	–	34
Cobalt	Freshwater tropical perch (*Colisa fasciatus*)	Freshwaters	–	–	–	In presence of 232.8 mg l^{-1} cobalt decrease in muscle glycogen content	38
Iron (ferrous)	Freshwater isopod (*ascellus aquaticus*)	Pondwater	4.5 or 6.5	$50–1000$ mg l^{-1}	LC_{50} at pH 4.5 in presence of 0.88 mg l^{-1} iron = 256 mg l^{-1}. LC_{50} at pH 6.5 in presence of 0.88 mg l^{-1} iron = 431 mg l^{-1}	–	39
Mercury	Slipper limpet (*Crepidula fornicata*)	Natural	–	$5–50$ μg l^{-1}	2-day LC_{50} 1100 μg l^{-1} at 13.5 °C. No mortality when exposed to 1600 μg l^{-1} at 3 °C	–	40
Selenium (as sodium and selenate) Selenite	*Daphnia magna*	Freshwaters	–	–	48 h LC_{50} (for adults = 0.68 mg l^{-1} (sodium selenite) = 0.75 mg l^{-1} (sodium selenate) For juveniles = 0.55 mg l^{-1} (sodium selenate)	$0.025–0.5$ mg l^{-1} selenium (as sodium selenate) adversely affected growth and reproduction	41
Zinc	Snail (*Ancylus fluviatis*)	Natural	–	$1–18$ μg l^{-1} $100–1000$ μg l^{-1}	60-day LC_{50} 600 μg l^{-1} (large snails) 200 μg l^{-1} (small snails)	Adults exposed to 320 μg l^{-1} zinc survived long enough to breed but levels above 180 μg l^{-1} zinc reduced reproductive capacity	42
Zinc	Amphipod (*Allorchestes compressa*)	Natural	–	–	96 h LC_{50} 2 mg l^{-1}	–	34

Table 5.8. Ranges of metal concentrations found in environmental freshwaters ($\mu g\,l^{-1}$)

Element	River and lake upstream	Surface water	Groundwater	All types
Aluminium total	73–3600	20–1430	–	73–3600
labile	14–520	–	–	14–520
Antimony	0.08–0.42	–	0.77	0.08–0.77
Arsenic	0.42–4.90	–	2.3	0.42–4.90
Barium	10–23	100–103	4.1	10–103
Beryllium	0.4	<0.01–1	–	<0.01–1
Bismuth	0.005	–	<0.00015–0.006	0.005–0.006
Cadmium	0.03–5	4–130	100–2600	0.03–2600
Chromium	0.05–23	0.2–180	1.0	0.05–180
Cobalt	0.2–10	–	0.11	0.11–10
Copper	0.11–200	14–110	3.7	0.11–200
Europium	0.00008–0.018	–	–	0.00008–0.018
Gold	<0.001–0.036	–	–	<0.001–0.036
Iron	1–3925	150–5000	0.15	0.15–5000
Lead	0.13–60	17–42	–	0.13–60
Manganese	0.97–1835	70–500	3.2	0.97–1835
Mercury	0.009–1.3	–	–	0.009–1.3
Molybdenum	0.74–4.08	–	–	0.74–4.08
Nickel	1.5–40	8–40	–	1.5–40
Scandium	–	–	0.009	0.009
Selenium	<0.002–750	–	0.002–0.7	<0.0002–750
Silver	0.3–32	–	–	0.3–32
Titanium	–	3–31	–	3–31
Uranium	0.37–1.36	–	–	0.37–1.36
Vanadium	0.1–24	3.9–24	0.63	0.1–24
Zinc	0.86–630	2.5–250	8.9	0.86–630
Bromine	0.7–7500	40–140	200–28 000	0.7–28 000
Iodine	–	–	10	10
Nitrogen	190–2940	1500–91 000	–	1060–9100
Phosphorus	20–800	–	–	20–800
Silicon	3000–5800	–	–	3000–58
Sulphur	20	–	–	20
Borate	0.12–0.25	–	44	0.12–44
Bromide	–	–	7.8	7.8
Fluoride	100–600	–	–	100–600
Phosphate	160–550	–	–	160–550

Table 5.9. At-risk creatures in typical environmental freshwaters: short-term and long-term exposure to metallic contaminants (from Tables 5.2 and 5.8)

Creatures	Short-term (4–14 days) exposure leading to mortalities. Element concentration (μg l^{-1}) at which creatures do not survive								Long-term (365 days) exposure leading to mortalities. Element concentration (μg l^{-1}) at which salmonids and coarse fish do not survive				
	Hg	Cu	Cd	Zn	Pb	Cr	As	Ni	Cu	Zn	Pb	Cr	Ni
Annelids	>100	>100	>100	>1000	>10000	–	–	>1000	–	–	–	–	–
Molluscs	–	>10	–	–	–	–	–	–	–	–	–	–	–
Crustaceans	>0.1	>1	>0.1	>10	>10	–	>100	>10	–	–	–	–	–
Gastropods	–	>10	>10	–	>10	–	>10000	>10000	–	–	–	–	–
Rotifers	–	>100	–	–	>10000	–	>1000	–	–	–	–	–	–
Insects	>1000	>100	>10	>10000	>10000	–	>10000	>1000	–	–	–	–	–
Non-salmonid fish	>100	>100	>10	>1000	>10	>10	>100	>100	14	24	6	4	20
Range of metal concentrations found in environmental freshwaters (μg l^{-1}) (Table 5.8)	0.009–1.3	0.11–200	0.03–2600	0.86–630	0.13–60	0.05–180	0.42–490	1.5–40	0.11–200	0.86–630	0.13–60	0.05–180	1.5–40
Creatures at risk during short-term exposure in environmental waters	Crustaceans	All types	All types	Crustaceans	Crustaceans, gastropods, fish	Fish	Crustaceans, non-salmonid fish	Crustaceans	Crustaceans				
Creatures at risk during long-term exposure in environmental waters									←——— Non-salmonid fish ———→				

to the presence in their tissues of high levels of metals. Unfortunately, no data are available on the long-term toxic effects of metals on creatures other than non-salmonid fish.

5.1.3   ANALYTICAL METHODS

Various methodologies that have been employed in the past 10 years to determine metals, and some non metals and anions in non-saline natural waters are shown in Table 5.10. The sensitivity of these methods is adequate to meet most requirements in environmental water analysis ranging from the $ng\,l^{-1}$ level (arsenic, chromium, mercury, rhenium and uranium) through the $\mu g\,l^{-1}$ level (alkali earths, alkali metals, iron, nickel, cobalt, copper, manganese, indium, lead, bismuth, boron and silicon) to the 100 $\mu g\,l^{-1}$ level or thereabouts (aluminium. selenium).

Some of the modern techniques employed for the determination of elements and detection limits achievable are listed below:

Flow injection analysis	Detection limit ($\mu g\,l^{-1}$ unless otherwise stated)	
Atomic absorption detector	0.01	Mercury
	150	Selenium
Inductively coupled plasma atomic emission detector	$pg\,l^{-1}$	Gold
Spectrophotometric detector	130	Aluminium
Liquid chromatography and high-performance liquid chromatography	0.01	Uranium
	2.7	Phosphorus
	2.1	Sulphur
As hydride with dc	0.6	Arsenic
plasma atomic emission detector with atomic absorption	0.02	Chromium
Spectrometric detector with spectro-photometric detector	0.01	Uranium
Graphite furnace atomic absorption spectrometry	$\mu g\,l^{-1}$ level	Germanium
Inductively coupled plasma	0.03	Rhenium
Atomic emission spectrometry	30	Mercury
Spectrophotometry	40	Iron
Supercritical fluid chromatography	100–5000	Arsenic, bismuth, cobalt, iron, mercury, nickel, antimony, zinc

continued on page 111

Table 5.10. Methods for the determination of metals, non-metals and anions in natural (non-saline) waters

Element	Technique	Detection limit (μg l^{-1} unless otherwise stated)	Reference
(a) Metals			
Aluminium	Flow injection analysis	130	43
Arsenic	Liquid chromatography with electrochemical detection	–	44
	Reduction to AsH$_3$, analysis by dc plasma atomic emission spectroscopy	0.62	45
Chromium	High-performance flow flame atomic absorption spectrometry	0.03 (CrIII) 0.02 (CrVI)	46
Germanium	Graphite furnace atomic absorption spectrometry		47
Gold	Flow injection inductively coupled plasma quadruple mass spectrometry	10^{-15} ml^{-1}	48
Mercury	Photoacoustic spectrometry of the dithizonate	3ppt	49
	Flow injection atomic absorption spectrometry	0.01	50
Mercury	Inductively coupled plasma mass spectrometry	30	51
	Adsorption on to gold, back extraction with dithiocarbamate and anodic scanning voltammetry		52
Iron	Spectrophotometric	40	53
Rare earths	Preconcentration on silica immobilised 8-hydroxy quinoline	–	54
Rhenium	Inductively coupled plasma atomic emission spectrometry	0.03	5
Selenium	Catalytic flow injection analysis	150	56
Uranium	High-performance liquid chromatography, precolumn derivativisation with Arsenazo 111	0.01	57
Vanadium	Reaction with 2(-8 quinolylazo)-5(dimethylamino)phenol), then reversed phase liquid chromatography–spectrometry	2.6 pg absolute	58
Cd-11, Cu-11, Mg-11, Ca-11, Mn-11, Ni-11, Pb-11, Zn-11	Uptake on 1 immobilised-8-oxine adsorbed on solid sorbent	1000–5000	59

continued overleaf

Table 5.10. (*continued*)

Element	Technique	Detection limit (μg l^{-1} unless (otherwise stated)	Reference
As, Bi, Co, Fe, Hg, Ni, Sb, Zn	Conversion to bis(trifluoroethyl) dithiocarbamate then supercritical fluid chromatography	1–5	60
Cd, Pb, Cu	Stripping on mercury-coated–carbon foam-composite electrode	–	61
Na, Li, Ca, Mg, Ba, Fe, Ni, Cu, Mn, Co, Zn, In, Pb, Bi, B, Si	Laser-induced atomic fluorescence spectroscopy with microwave discharge atomisation	< 1	62
Cu, Fe	High-performance liquid chromatography		63
(b) Inorganic elements			
P, S	Photopyroelectric spectrometry	2.1 (s) 2.7 (P)	64
(c) Anions			
Fluoride	Inductively coupled plasma atomic emissions spectroscopy	30	65
Iodide	Ion selective electrode of Schiff base complexes of cobalt-11	–	66
Chlorite, chlorate	Flow injection analysis–ion chromatography	–	67
Nitrite, nitrate	Differential pulse polarography	–	68
	Reversed-phase ion extraction liquid chromatography	–	69
Sulphide	Ion-selective electrode	–	70
	Differential flow injection analysis using nitroprusside or methylene blue	2	71
	Flow injection–gas diffusion method	0.15	72
	Gas chromatography, flame photometric detector	0.12 pmole^{-1}	73
Borate	Flow injection analysis–UV-visible detection	–	74
Chromium-III and VI	High-performance flow flame atomic absorption spectrometry	0.03 (CrIII) 0.02 (CrVI)	75
Sulphite and dithionate	Ion chromatography	–	76
S″, polysulphide SO$_3$″, S$_2$O$_3$″, SO$_4$″, CN′, SO$_3$″, S″	Photometric flow injection analysis Miscellar analysis (reaction with 5,5-dithiobis(2-nitrobenzoic acid) in cetyltrimethyl ammonium bromide miscelles)	$0.5–1.5 \times 10^{-4}$ M (CN′) $0.2–1 \times 10^{-4}$ M (S″) $0.2–1.5 \times 10^{-4}$ M (SO$_3$″)	78

Table 5.10. (*continued*)

Element	Technique	Detection limit (μg l^{-1} unless otherwise stated)	Reference
Monofluoro-phosphate, orthophosphate, polyphosphates	High-performance liquid chromatography–photo diode array detector	–	79
F$'$, BO$_3''$, Cl$'$, AsO$_4''$, NO$_3'$	Ion chromatography–conductiometric detectors	–	80
I, Br$'$, Cl$'$	Gas chromatography	–	81
NO$_2'$, Br$'$, NO$_3'$, SO$_4''$, I$'$, SCN$'$	Indirect photometric chromatography	0–1–0.4 ng absolute	82
Cl$'$, NO$_2'$, Br$'$, NO$_3'$, PO$_4'''$, PO$_3'''$, SeO$_3''$, SO$_4''$	Ion chromatography	–	83
NO$_3$, Br$'$, NO$_2'$, I$'$, IO$_3$	Micelle exclusion chromatography	–	85
IO$_3'$, BrO$_3'$, Cl$'$, ClO$_3'$, Br$'$, I$'$	High-performance liquid chromatography	–	84
F$'$, ClO$_4'$, Cl$'$, HPO$_2''$, AsO$_4'''$, NO$_3'$, ClO$_3'$, SO$_4''$, CrO$_4''$, BF$_4'$	Liquid chromatography–indirect fluorimetric detection as ruthenium-11 complexes	–	86
SeO$_3''$, SeO$_4''$	Isotope dilution mass spectrometry	200 pg g^{-1}	87

	Detection limit (μg l^{-1} unless otherwise stated)	
Laser-induced atomic fluorescence spectroscopy with microwave discharge atomisation	< 1	Sodium, lithium, calcium, magnesium, barium, iron, nickel, copper manganese, cobalt, zinc, indium, lead, bismuth, boron, silicon

5.2 SEA AND COASTAL WATERS

5.2.1 TOXIC EFFECTS ON ORGANISMS

The natural concentrations of trace metals in relatively unpolluted open seawater where the effect of coastal discharges is minimal are very low and the accurate determination of these concentrations has presented a great challenge to the analytical chemist. As such, low-level contamination of the sample by sampling equipment and neighbouring ships are important factors affecting the accuracy of results and it is only in recent years, in fact, that reliable techniques have evolved. For this reason, in the discussion which follows, only results obtained since 1975 are quoted as these represent the most accurate available. In general, lower values for metals in seawater have been obtained in recent years compared with earlier due to the control of contamination (Table 5.11).

Information on the effects of metals on marine creatures is summarised in Table 5.12. The duration of the toxicity tests are not included in this table as discussed in Section 5.1. They are, however, generally short-term tests and the concentrations quoted are toxic effect concentrations, i.e. concentrations above which mortalities can occur. Table 5.12 is a useful guide in that it highlights those species at risk when concentrations of the stated elements exceed the levels quoted in short-term exposures. If the analytical composition of a marine water is known (Table 5.13) then reference to Table 5.12 shows the adult and larval species at risk. It can be seen, for example, that when concentrations of copper or mercury exceed $1 \mu g l^{-1}$ adult and larval bivalve molluscs are at risk, when concentrations of mercury exceed $1 \mu g l^{-1}$ crustacae larval are at risk and when concentrations of cadmium exceed $1 \mu g l^{-1}$ adult bivalve molluscs are at risk. When concentrations of mercury exceed $10 \mu g l^{-1}$ the following species are at risk: molluscs and crustacae (adult and larval), and adult fish. When concentrations of copper exceed $10 \mu g l^{-1}$ then adult larval annelids, bivalve molluscs, crustacae and fish are at risk, as are adult echinoderms and hyrozoans. With cadmium concentrations above $10 \mu g l^{-1}$ adult bivalve molluscs and crustacae as well as hydrozoans are at risk. Zinc at this concentration causes mortalities in

Table 5.11. Metal determinations ($\mu g l^{-1}$) in seawater (1974–80)

Element	Year	1974	1977	1978/1980
Copper		0.5–6	0.1	0.1–0.2
		(88, 89)	(90)	(91, 92)
Cadmium		–	0.17	15–70
			(93)	(91, 92, 94)
Zinc		–	15	0.03–0.35
			(93)	(91, 92)
Nickel		–	2.0	0.25–0.39
			(95)	(91, 92, 94)
Lead		–	0.4	100
			(95)	(96)
Cobalt		–	0.2	0.02–0.03
			0.2	0.02–0.03
			(95)	(94, 96)

References in parentheses.

adult and larval bivalve molluscs and nickel similarly affects larval echinoderms and fish.

From the reported metal concentrations in open seawaters and coastal waters (Tables 5.13 and 5.14 respectively) it is seen that in each case observed metal concentrations can vary over a wide range. In the case of open seawater (Table 5.12) the only creatures at risk are adult and larval bivalve molluscs due to short-term exposure to copper at the higher end of the concentration range found (i.e. $> 1 \, \mu g \, l^{-1}$) and the same creatures when short-term exposed to zinc at the higher end of the concentration range found (i.e. $> 10 \, \mu g \, l^{-1}$).

A much more serious situation exists in the case of coastal, bay and estuary water. While the maximum concentrations of lead, chromium, arsenic and nickel found in these waters do not present any risk to creatures during short-term exposures, the same cannot be said for mercury, copper, cadmium or zinc (Table 5.12). Some of the creatures at risk in short-term exposures include juvenile and adult bivalve molluscs (mercury, copper, zinc and cadmium), juvenile and adult crustacae (mercury, copper and zinc), juvenile and adult annelids (copper), adult annelids (zinc), juvenile and adult fish (mercury and cadmium), hydrozoans (copper and cadmium), echinoderms and gastropods (copper).

It is interesting at this point to compare the relative toxicities of different elements to freshwater and marine adult creatures when subject to short-term exposure (4–14 days) to these elements. The metal toxicity data in Tables 5.9 and 5.15 are compared in Table 5.16 in those cases where comparable data exists for both types of water. It is seen that in some cases, i.e. where

$$\frac{\text{The concentration of metal } (\mu g \, l^{-1}) \text{ producing mortalities in freshwater fish}}{\text{The concentration of metal } (\mu g \, l^{-1}) \text{ producing mortalities in seawater fish}}$$

$$= (a/b) = \text{unity, creatures are equally sensitive to metals}$$

e.g. annelids to mercury and zinc
    gastropods to copper
    fish to zinc
    crustacae to arsenic
    fish and gastropods to nickel

In other cases when $a/b < 1$ the freshwater creatures are more sensitive than are the marine creatures:

e.g. crustacae to mercury and copper
    annelids, crustacae and fish to cadmium
    crustacae to zinc
    crustacae, fish and gastropods to lead
    fish to chromium

Table 5.12. Toxicity of metals to marine fish and invertebrates, species at risk during 4–14-day exposure to stated concentrations

ADULT SPECIES

Concentration above which mortalities can occur		Annelids								Bivalve molluscs								Crustaceans							
		Hg	Cu	Cd	Zn	Pb	Cr	As	Ni	Hg	Cu	Cd	Zn	Pb	Cr	As	Ni	Hg	Cu	Cd	Zn	Pb	Cr	As	Ni
$\mu g\,l^{-1}$	$mg\,l^{-1}$																								
0.1–1	0.0001–0.001									•															
1–10	0.001–0.01											•	•												
10–100	0.01–0.1	•													•			•	•	•					
100–1000	0.1–1	•				•	•													•			•	•	
	1–10		•	•			•									•		•				•	•		
	10–100	↓	↓	↓	↓	↓	↓	↓	↓	↓	↓	↓	↓	↓	↓	↓	↓	↓	↓	↓	↓	↓	↓	↓	↓

LARVAL SPECIES

		Larval annelids								Larval bivalves							
		Hg	Cu	Cd	Zn	Pb	Cr	As	Ni	Hg	Cu	Cd	Zn	Pb	Cr	As	Ni
0.1–1	0.0001–0.001																
1–10	0.001–0.01									•	•						
10–100	0.01–0.1	•											•				
100–1000	0.1–1			•								•		•		•	•
	1–10														•		
	10–100	↓		↓						↓	↓	↓	↓	↓	↓	↓	↓

Table 5.12. (*continued*)

Echinoderms								Fish								Gastropods								Hydrozoans							
Hg	Cu	Cd	Zn	Pb	Cr	As	Ni	Hg	Cu	Cd	Zn	Pb	Cr	As	Ni	Hg	Cu	Cd	Zn	Pb	Cr	As	Ni	Hg	Cu	Cd	Zn	Pb	Cr	As	Ni

Larval crustaceans								Larval echinoderms								Larval fish							
Hg	Cu	Cd	Zn	Pb	Cr	As	Ni	Hg	Cu	Cd	Zn	Pb	Cr	As	Ni	Hg	Cu	Cd	Zn	Pb	Cr	As	Ni

Table 5.13. Ranges of metal concentrations found in open seawaters (post-1975)

Element	Concentration range found in open surface seawater ($\mu g\ l^{-1}$)	Consensus value ($\mu g\ l^{-1}$)
Aluminium	0.1–0.6	–
Bismuth	< 0.000 003–< 0.000 005	–
Cadmium	0.01–0.126	0.03
Chromium (total)	0.005–1.26	–
Cobalt	0.003–0.16	0.005
Copper	0.0063–2.8	0.05
Iron	0.2–320	0.2
Lead	0.000 041–9.0	–
Manganese	0.018	0.02
Mercury	0.002–0.078[a]	< 0.2
Molybdenum	3.2–12.0	–
Nickel	0.15–0.93	0.17
Rare earths	61.7 (nmole kg^{-1})	–
Rhenium	6–8	–
Selenium	0.000 95–0.029	–
Silver	0.08	–
Thorium	$\leqslant 0.0002$	–
Tin	0.02–0.05	–
Uranium	1.9–2.6	–
Vanadium	0.45–2.0	2.5
Zinc	0.05–10.9	0.49

[a]Generally $< 0.2\ \mu g\,l^{-1}$ except in parts of Mediterranean, where additional contributions due to man-made pollution are found[97–100].

In yet other cases, i.e. $a/b > 1$, the seawater creatures are more sensitive than the freshwater creatures:

e.g. fish to mercury
annelids, bivalve molluscs and fish to copper
annelids to lead

Thus it is dangerous to conclude what the toxic effect of an element on creatures will be in seawater from measurements made on freshwater creatures and vice versa.

While the concentrations of toxic metals in open seawaters might be suffi-ciently low to cause no adverse effects on sea creatures (see below) the same cannot be said for estuary and coastal waters as these might be contaminated by metal originating as industrial discharges (Table 5.17, panels (A) and (B)), or coastal sewage discharges (panel (C)). Reference to Table 5.17 shows that the estuary waters discussed (Severn and Humber, UK) seem to be quite clean and both support fisheries. Only occasional adverse effects on bivalve molluscs, hydrozoans and echinoderms are to be expected, while fish, crustacae, annelids and gastropods survive.

Table 5.14. Ranges of metal concentrations (μg l^{-1}) found in coastal waters compared to concentrations in open seawater

Element	Concentration range found in surface estuary, bay and coastal waters (μg l^{-1})		Concentration range found in open surface seawater (μg l^{-1})		C_{Wmax}/S_{Wmin}
	C_{Wmin}	C_{Wmax}	S_{Wmin}	S_{Wmax}	
Aluminium	6.4	63	0.1	0.6	630
Antimony	0.3	0.82	–	–	–
Arsenic	1.0	1.04	–	–	–
Barium	4.8		–	–	–
Bismuth	0.000 05	0.68	< 0.000 003	0.000 005	> 226 000
Cadmium	0.015	5.0	0.01	0.126	500
Cerium	1.6	16.7	–	–	–
Chromium (total)	0.095	3.3	0.005	1.26	600
Cobalt	< 0.01	0.25	0.003	0.16	83
Copper	0.069	20.0	0.0063	2.8	3200
Iron	1	250	0.2	322	1250
Lanthanum	0.17	0.72	–	–	–
Lead	0.038	7.44	0.000 041	9.0	181 500
Manganese	0.35	250	0.018	–	13 900
Mercury	0.000 02	15.1	0.002	0.078	7550
Molybdenum	2.1	200	3.2	12.0	63
Nickel	0.2	5.33	0.15	0.93	36
Rare earths	–	–	61.7 nmole kg^{-1}	–	–
Rhenium	–	–	6	8	–
Scandium	0.000 95	0.098	–	–	–
Selenium	0.4		0.0095	0.029	–
Thorium	≤ 0.0002			≤ 0.0002	
Tin	–	–	1.9	2.6	–
Uranium	1.36	1.9	–	–	1
Vanadium	0.01	5.1	0.45	2.0	11.3
Zinc	0.007	200	0.05	10.9	5000

Examples of contaminated estuaries

The higher concentrations of metals present in a coastal water adjacent to a sewage outlet (Table 5.15) have more severe adverse effects on a wide range of creatures including adult and larval molluscs, hydrozoans, annelids, echinoderms, gastropods, crustacae and fish. Similar comments can be made for estuary samples taken at the river outfall of a mining area (Table 5.17) and it is not surprising that the waters in either of these areas do not support any animal life.

It must also be recalled that increase in water temperature, say from 10°C to 30°C, causes up to a hundredfold increase in the toxicity of cadmium, chromium, copper, lead, mercury, nickel and zinc. Therefore, more severe adverse effects would be expected in the summer months than in winter.

Table 5.15. At-risk creatures in typical environmental open seawaters and coastal waters, short-term (4–14 days) exposure to metallic contaminants (from Tables 5.12–5.14)

Creature	Element concentration ($\mu g \, l^{-1}$) at which creatures do not survive						
	Hg	Cu	Cd	Zn	Pb	Cr	Ni
Annelids (adult)	>100	>10	>1000	>1000	>100	>100	>10000
Annelid (larval)		>10		>100			
Bivalve mollusc (adult)	>0.1	>1	>1	>10	>1000	>10000	>10000
Bivalve mollusc (larval)	>1	>1	>100	>10	>100	>1000	>100
Crustacae (adult)	>10	>10	>10	>100	>1000	>1000	>100
Crustacae (larval)	>1	>10	>100	>100	>100	>1000	>1000
Echinoderm (adult)		>10		>1000	>1000	>10000	>10
Gastropods		>10			>100		>10000
Hydrozoans		>10					
Fish (adult)	>10	>10	>1	>1000	>1000	>10000	>100
Fish (larval)		>10	>100	>1000			
Range of metal concentrations found in open seawaters ($\mu g \, l^{-1}$) (Table 5.13)	0.002–0.078	0.0063–2.8	0.01–0.126	0.05–10.9	0.000041–9.0	0.005–1.26	0.15–0.93
Creatures at risk during short-term exposure in open seawater	None	Bivalve molluscs (adult and larval) at high end of concentration range	None	Bivalve molluscs (adult and larval) at high end of concentration range	None	None	None
Range of concentrations found in bay, coastal and estuary waters ($\mu g \, l^{-1}$) (Table 5.14)	0.00002–15.1	0.069–20.0	0.015–5.0	0.007–200	0.038–7.44	0.095–3.3	0.2–5.33
Creatures at risk during short-term exposure in above waters when metal concentrations are at higher end of range quoted	Bivalve molluscs (adult and larval) Crustacae (adult and larval) Fish	Annelids (adult and larval) Bivalve molluscs (larval) Crustacae (adult and larval) Echinoderm gastropods Hydrozoans Fish (adult and larval)	Bivalve molluscs (adult) Hydrozoans	Bivalve molluscs (adult and larval) Annelid (larval) Crustacae (adult and larval)	None	None	None

Table 5.16. Comparison of short-term (4–14 days) concentrations of metals (μg l^{-1}) producing mortalities in adult creatures in (a) freshwater and (b) seawater

Creature	Hg			Cu			Cd			Zn		
	a	b	a/b	a	b	a/b	a	b	a/b	a	b	a/b
Annelid	>100	>100	1	>100	>10	10	>100	>1000	0.1	>1000	>1000	1
Bivalve mollusc	–	–	–	>10	>1	10	–	–	–	–	–	–
Crustacae	>0.1	>10	0.01	>1	>10	0.1	>0.1	>10	0.01	>10	>100	0.1
Fish	–	>10	10	>100	>10	10	>10	>100	0.1	>1000	>1000	1
Gastropod	–	–	–	>10	>10	1	–	–	–	–	–	–

Creature	Pb			Cr			As			Ni		
	a	b	a/b	a	b	a/b	a	b	a/b	a	b	a/b
Annelid	>10000	>100	100	–	–	–	–	–	–	>1000	>10000	0.1
Bivalve mollusc	–	–	–	–	–	–	–	–	–	–	–	–
Crustacae	>10	>1000	001	–	–	–	>100	>100	1	>10	>100	0.1
Fish	>10	>1000	0.01	>1000	>10000	0.1	>100	>1000	0.1	>100	>100	1
Gastropod	>10	>100	0.1	–	–	–	–	–	–	>10000	>10000	1

Table 5.17. Adverse effects of metals on marine life in estuary, coastal and seawaters

Composition of water ($\mu g\,l^{-1}$)	As	Cd	Cr	Cu	Pb	Hg	Ni	Zn
(A) Severn Estuary water (1975–80)	nd	0.31–1.48	nd	2.2–4.2	1.5–4.1	nd	1.9–3.6	11–22

Adverse effects reported: bivalve molluscs and hydrozoan adults (cadmium), bivalve molluscs, adults and larvae (copper and zinc). The water quality of this estuary is having no adverse effect on harvested creatures and it is in fact sustaining salmon, eel and shellfisheries

Composition of water ($\mu g\,l^{-1}$)	As	Cd	Cr	Cu	Pb	Hg	Ni	Zn
(B) Humber Estuary (1980–82)	–	0.2–0.7	–	0.1–8.0	0.5–1.0	1.0–10	1.0–15	2.0–50

Adverse effects expected: bivalve molluscs and their larvae (zinc and copper), bivalve molluscs (mercury) and echinoderm larvae (nickel). The water in this estuary at times contains sufficiently high concentrations of nickel to adversely affect echinoderm larvae and of zinc and copper to affect adult bivalve molluscs. Nevertheless, these waters support fisheries of salmon, eels, sole, flounders, sprat, shrimp, cockles, whelks, crabs and lobster, bass, whiting, pouting, weever, coley, mullet, mackerel, dab and plaice

Composition of water ($\mu g\,l^{-1}$)	As	Cd	Cr	Cu	Pb	Hg	Ni	Zn
(C) Coastal water adjacent to sewage discharge	1.0	1.5–2.5	13.0–16.5	48.6–49.5	30–31	0.03–0.09	17–18	113–115

Adverse effects expected: Adult bivalve molluscs (cadmium, copper, zinc), mollusc larvae (copper, zinc), adult hydrozoans (cadmium, copper), adult annelids (copper), annelid larvae (copper, zinc), adult echinoderms (copper), echinoderm larvae (nickel), adult gastropods (copper), adult crustacae and crustacae larvae (copper and zinc), and adult fish and fish larvae (copper). Little or no fishery would be expected in this area

Composition of water ($\mu g\,l^{-1}$)	As	Cd	Cr	Cu	Pb	Hg	Ni	Zn
(D) Restronguet Creek (estuary of River Carnon, UK)	4–8	–	2–8	–	2–30	–	–	500–700

Adverse effects expected: Adult bivalve molluscs (cadmium, copper, zinc) and their larvae (copper, zinc), hydrozoans (cadmium, copper), adult annelids (copper), and their larvae (copper, zinc) adult echinoderms (copper), adult gastropods (copper), adult and larvae crustacae (copper, zinc), and adult and larvae fish (copper). The high concentration of metals has decimated most forms of adult and larval life in these waters, the only forms surviving being those which have developed some tolerance to metals

A limited amount of work has been carried out on the effect of metals and cyanide on organisms that live in the sea.

Aluminium

Atlantic salmon smolts were exposed to acidic, aluminium-rich salt water, pH 5.1, (calcium 1 mg l^{-1}, labile aluminium (60 μg l^{-1}). Mortalities occurred when the pH was 6.05 or below and did not occur at pH 6.45 or above[35].

Striped bass (*Morone saxatilis*) of ages between 4 and 195 days upon exposure to 25–400 μg l^{-1} aluminium at a pH between 5.0 and 7.2 undergo mortalities at lower pH values[101].

Cadmium

Exposure of the estuarine mysid (*Mysidopsis bahia*) for 96 hours to cadmium chloride at salinities between 6 and 38 per thousand, show increased toxic effects as indicated by 96-hour LC_{50} values, with increasing salinity.

Chromium

The effect has been studied by using sodium chromate (0.001–2 mole l^{-1}) on the survival and development of young adult and larval crustaceans (*Palaeomonetes varians*, *Palaemon elegans*, *Neomsis integer*, *Praunus flexosus* and *Daphnia magna*) during 40 days' exposure. Minimal concentrations affect adult and larval mortality and larval development (MEC). No observed effect concentration (NOEC) and 4- and 10-day LC_{50} values were obtained[103]. Uptake of chromium by juvenile plaice (*Pleuronectes platessal*) has been shown to reduce average fish size and produce other histological changes[104].

Cyanide

Cyanide concentrations of 0.005–0.03 mg l^{-1} have been reported to be safe levels for aquatic organisms. Ruby et al.[105] have shown, however, that 0.005 mg l^{-1} have adverse effects on yolk synthesis in Atlantic salmon (*Salmo solar*).

Lead

It has been shown that there is significant relationship between lead concentration in mussel (*Mytilus edulis*) and lead concentrations in seawater. The lead level in seawater should not exceed 1.27 μg l^{-1} to avoid adverse effects on the mussels and on humans who eat them.

Calcium oxide

Calcium oxide is used to treat mussel (*Mytilus edulis*) beds. It has been found that whereas quicklime has no adverse effect on mussels, it does adversely affect starfish (*Asterias vulgaris*) in coastal waters. It has no adverse effect on other creatures likely to be present in mussel beds, i.e. bloodworms (*Glycra dibranchiata*), sandworms (*Littorina littorea*) and juvenile lobsters (*Homarus americanus*)[107].

Heavy metals

The acute toxicity has been determined of copper, mercury, cadmium and zinc in fiddler crabs (*Uca annuliges* and *Uca triangularis*) collected in Visakhapatan Harbour. The 96-hour LC_{50} values for the two creatures were respectively 2.75 and 2.83 Hg l^{-1}, 12.82 and 14.81 mg Cu l^{-1}, 48.21 and 43.23 mg Cd l^{-1} and 76.95 and 66.42 mg Zn l^{-1}. Verriopoulos *et al.*[109] determined LC_{50} values for copper and chromium in *Artemia salina*.

5.2.2 ANALYTICAL COMPOSITION

In Table 5.13 is presented a summary of the best available values for trace metals in open surface seawater (see Appendix 1, Table A1.2 for more details). With the exception of iron, manganese, zinc and aluminium, metal concentrations are usually below 1 μg l^{-1} and except molybdenum, uranium, arsenic and barium this applies to all the toxic metals. In general, minimum metal concentrations reported in numerous surveys agree with the consensus values reported in 1986 by Paulson[110].

It would be expected, and is indeed found, that the concentrations of metals in coastal waters and estuaries would be higher than in open seawater due to pollution from rivers and coastal discharges. That this is so is shown in Table 5.14, which compares ranges of metal contents in coastal waters with those of open seawater (further information is given in Appendix 1, Table A1.3). Some idea of the relative concentration of metals in coastal waters compared to that in open seawater can be obtained by dividing the maximum concentration found in coastal water (i.e. C_{Wmax}) by the minimum concentration found in open seawater (i.e. S_{Wmin}). It is seen in Table 5.14 that values of C_{Wmax}/S_{Wmin} from about 200 000 (bismuth, lead) through intermediate values of 1000 to 8000 (iron, copper, mercury, zinc) to relatively low values below 1000 (aluminium, cadmium, chromium, cobalt, molybdenum, nickel, uranium and vanadium). For relatively unpolluted coastal waters the quotient C_{Wmin}/S_{Wmin} is, as would be expected, close to unity.

Table 5.18 compares metal concentrations in coastal waters with metal concentrations of rivers discharging into the sea. It is seen that, as would be

Table 5.18. Comparison of metal contents (μg l^{-1}) of coastal waters and rivers

	Coastal water		River water		
	C_{Wmin}	C_{Wmax}	R_{Wmin}	R_{Wmax}	C_{Wmax}/R_{Wmax}
Aluminium (total)	6.4	63	73	6300	0.01
Antimony	0.3	0.82	0.08	0.42	2
Arsenic	1.0	1.04	0.42	490	0.002
Barium	0.48		10	23.0	0.021
Bismuth	0.000 05	0.68	0.005		1.43
Cadmium	0.015	5.0	0.03	5.0	1
Chromium (total)	0.095	3.3	0.05	23.0	1.14
Cobalt	< 0.01	0.25	0.2	10.0	0.025
Copper	0.069	20.0	0.11	200	0.1
Iron	1	250	1	3925	0.062
Lead	0.038	7.44	0.13	60	0.12
Manganese	0.35	250	0.97	1835	0.14
Mercury	0.000 02	15.1	0.009	1.3	11
Molybdenum	2.1	200	0.74	4.1	50
Nickel	0.2	5.3	1.5	4.40	0.13
Selenium	0.4		< 0.0002	> 50	0.008
Uranium	1.3	1.9	0.37	1.36	0.14
Vanadium	0.1	5.1	0.1	24	0.20
Zinc	0.007	200	0.86	630	0.33

Maximum concentration in coastal water less than maximum concentration in rivers, i.e. $C_{Wmax}/R_{Wmax} \leqslant 1$,

Al, As, Ba, Cd, Co, Cu, Fe, Pb, Mn, Ni, Se, V, U, Zn

Maximum concentration in coastal water up to ten times greater than maximum concentration in river water, i.e. $C_{Wmax}/R_{Wmax} > 1$ to < 10, Sb, Hg, Cr

Maximum concentration in coastal water more than ten times greater than maximum concentration in river water, i.e. $C_{Wmax}/R_{Wmax} > 10$, Bi, Mo

expected, due to the diluting effect of seawater with the possible exception of antimony, mercury, bismuth and molybdenum the maximum metal contents of coastal waters are considerably lower than the maximum in river waters. Consequently, some creatures which do not survive in rivers might do so in coastal waters.

5.2.3 ANALYTICAL METHODS

Various methodologies for determining trace metals, non-metals and some anions in seawater are reviewed in Table 5.19. Concentrations of elements in seawaters are considerably lower than in non-saline inland waters and hence greater

Table 5.19. Methods for the determination of metals and anion in seawaters

Element	Technique	Detection limit	Ref.
(a) Metals			
Ammonium	High-performance liquid chromatography	–	111
Barium	Direct injection graphite furnace absorption spectrometry	100 pg l^{-1}	112
Copper	Ion-selective electrodes	–	113
	Anodic scanning voltammetry	5×10^{-10} M	114
Gold	Flow injection inductively coupled plasma quadruple mass spectrometry	10^{-15} ML^{-1} (0.2 pg l^{-1})	115
Iron-II	Flow injection analysis with chemiluminescent detector	0.45 mM l^{-1}	116
Iron-III	Chelating resin colelution— chemiluminescent detection	0.05 nM l^{-1} (3 pg l^{-1})	117
Lanthanides yttrium	Inductively coupled plasma atomic emission spectrometry	$600–3000 \text{ pg l}^{-1}$	118
Radium	Thermal ionisation mass spectrometry	fg level	119
	α-liquid scintillation counting	0.02 dpm kg^{-1}	120
Rare earths	Resin complexation—inductively coupled mass spectrometry	–	121
Vanadium	Chelating resin preconcentration inductively coupled plasma atomic emission spectrometry	–	122
V, Ni, As	Inductively coupled plasma mass spectrometry	10^4 pg l^{-1} (V) 30 pg l^{-1} (Ni) $4 \times 10^4 \text{ pg l}^{-1}$ (As)	123
Al, Fe, Mn	High-performance liquid chromatography with spectrophotometric and electrochemical detection	–	124
Al, Tl, Cr, Mn, Fe, Co, Ni, Cu, Zn, Y	Inductively coupled plasma atomic emission spectroscopy	$5 \times 10^5 \text{ pg l}^{-1}$ (Pb) to 10^4 pg l^{-1} (Cu, Zn)	125
Cd, Co, Cu, Fe, Mn, Ni, Pb, Zn	Preconcentration technique using acid-ammonium hydroxide buffer system	–	126
Misc. trace metals	Preconcentration techniques using bis(2-hydroxyethyl) dithiocarbamate	–	127
(b) Anions			
Iodide	Cathodic stripping square wave voltrammetry	$1.2 \times 10^3 \text{ pg l}^{-1}$	128
	High-performance liquid chromatography	$2 \times 10^3 \text{ pg l}^{-1}$	129

Table 5.19 (*continued*)

Element	Technique	Detection limit	Ref.
Nitrate, Nitrite	Anion chromatography on octedecyl silane reverse phase column	2×10^6 pg l^{-1} (NO$_2$') 8×10^6 pg l^{-1} (NO$_3$')	130
Silicate	Conversion to silane then chemiluminescence analysis	0.5μg absolute	131
Sulphide	Gas chromatography with flame ionisation detection	1000 pg l^{-1}	132
Sulphate	Suppressed ion chromatography	–	133

sensitivity in the analytical method is usually required. It will be noted that detection limits are reported in pg l^{-1}, not μg l^{-1} as is the case in non-saline waters (Table 5.10). Some of the techniques that have been employed are reviewed. It is worth noting that sample preconcentration techniques are often employed in seawater to improve analytical sensitivity.

		Detection limit (pg l^{-1}) *unless otherwise stated*
High-performance liquid chromatography with spectrometric and electrochemical detection	Ammonium, iodide aluminium, iron, manganese	2×10^3
Inductively coupled plasma atomic emission spectrometry		
	Lanthanides and yttrium	600–3000
	Rare earths	
	Vanadium	10^5
	Nickel	30
	Arsenic	4×10^4
	Aluminium, lead, thallium, chromium, manganese, iron, cobalt, nickel, copper, zinc yttrium	5×10^5 (lead) 10^4 (copper, lead)
Graphite furnace atomic absorption spectrometry		
	Barium	100

		Detection limit (pg l^{-1}) *unless otherwise stated*
Flow injection analysis		
With inductively coupled plasma atomic emission spectrometric detector	Gold	0.2
With chemiluminescent detector	Iron II	0.4 mM l^{-1}
Anodic scanning voltammetry		
	Copper	5×10^{-10} mole
Cathodic scanning voltammetry		
	Iodide	1.2×10^3
Gas chromatography with flame ionisation detector	Sulphide	1000
Thermal ionisation mass spectrometry with α-liquid scintillation counting	Radium	fg level

5.3 METAL LOAD ON THE NORTH SEA

In Table 5.20 the inputs of metals to the North Sea from all land-based European sources are compared with the median estimates of the direct atmosphere fallout of metal from European sources onto the surface of the North Sea. It is seen that atmosphere fallout is a major proportion of total metal contamination from all sources, ranging from less than 20% (chromium and zinc) to more than 30% (copper, arsenic, lead, zinc, mercury, nickel and cadmium). Such fallout will be the consequence of smokestack emissions from metal industries, and a large proportion of these metals are swept from the atmosphere into the oceans by rain (Table 5.21).

When total annual land-based plus atmospheric emissions of metals are considered as percentages of the estimated quantities of metals in the North Sea water column, it is seen that the input of arsenic is small (annual input 24% of column load), copper and nickel are intermediate (about 50% of column load), while cadmium, chromium, mercury and zinc are major contributors, being comparable with those present in the water column. The annual land-based plus atmospheric input of lead is four times the column load, this mainly being contributed by land-based sources.

Table 5.20. Total quantities of metals entering North Sea from all land-based sources and from direct atmospheric deposition contrasted with the estimated mass of metals in the water column of the North Sea[134]

	Total land-based, tonne annum⁻¹ A	Total atmospheric, tonne annum⁻¹ B	Grand total land-based plus atmospheric, tonne annum⁻¹ $C = A + B$	North Sea water column tonne D	Total load based as % of mass in water column $E = A \times 100/D$	Total atmospheric as % of mass in water column $F = B \times 100/D$	Grand total land-based plus atmospheric as % of mass in water column in North Sea $E + F$	Concentration of soluble metal in North Sea ($\mu g\,l^{-1}$)
As	800	230	1030	4300	18.6	5.3	24.0	1.0
Cd	245	569	814	860	28.4	66.1	94.6	0.02
Cr	666	667	7328	8600	77.4	7.75	85.2	0.4
Cu	4705	3942	8647	17 200	27.3	22.9	50.3	0.2
Pb	6521	2920	9441	2150	0.303	135.8	439	0.05
Hg	66.5	51	117.5	86	77.3	59.3	136.6	0.002
Ni	4052	1569	5621	10 750	37.7	14.6	52.3	0.25
Zn	31 508	7008	38 516	43 000	73.3	16.3	89.5	1.0

Source: Hill, J.M., O'Donnell, A.R. and Mance, G., The Quantities of Some Heavy Metals Entering the North Sea, Tech. Report TR205, Water Research Centre, Stevenage, Herts, UK (1984). Reproduced by permission of the Water Research Centre.

Table 5.21. Metal contents ($\mu g\,l^{-1}$) of rain and snow aqueous precipitation into oceans

Element	Sample	Concentration ($\mu g\,l^{-1}$)	Range of metal content in open ocean (Table 5.13) ($\mu g\,l^{-1}$)	Atmospheric fallout of metals as % of metal contamination of North Sea from all sources (Table 5.20)	Reference
Mercury	Rain	0.0017–0.0023	0.002–0.078	43.6	135
Mercury inorganic	Rain	0.014	0.002–0.078	43.6	135
Mercury total	Rain	0.015	0.002–0.078	43.6	135
Cadmium	Snow	0.034	0.010–0.126	69.9	136
Cadmium	Snow	0.005	0.010–0.126	69.9	137
Copper	Snow	0.097	0.0063–2.8	45.5	136
Copper	Snow	0.02	0.0063–2.8	45.5	137
Lead	Snow	2.48	0.000 041–9.0	30.9	136
Lead	Snow	0.05	0.000 041–9.0	30.9	137
Lead	Rain	2–40	0.000 041–9.0	30.9	138
Lead	Rain	4.7	0.000 041–9.0	30.9	138
Nickel	Rain	5.0	0.15–0.93	27.9	139
Selenium	Snow	0.005–0.025	0.000 95–0.029	–	140
Silver	Snow	3–300	0.08	–	141
Tin	Rain	0.025	0.02–0.05	–	142
Bismuth dissolved	Rain	0.0006	–	–	143
Bismuth total	Rain	0.003	< 0.000 003–< 0.000 005	–	143
Antimony	Rain	0.002–0.089	–	–	144

Also included in Table 5.20 are the average soluble metal contents found in open waters in the North Sea. As the data show, the average annual concentration of metal from land-based and atmospheric sources ranges from 24% (arsenic) to 439% (lead) of the weight of metals in the water column. If all these annual additions of metals were to distribute themselves evenly through the water column and remain in solution then the soluble metal contents of the North Sea would be expected to undergo quite dramatic annual increases, ranging from 24% to 439%.

In these circumstances a lead level of 0.02 μg l^{-1} in 1990 would increase to 0.1 μg l^{-1} in 1991; 0.5 μg l^{-1} in 1992: 2.5 μg l^{-1} in 1993; 12.5 μg l^{-1} in 1994 and 62.5 μg l^{-1} in 1995. Careful monitoring of the composition of North Sea levels reveals that such increases in dissolved or suspended metal content with time do not occur. This is presumably because metals absorbed onto bottom sediments are accumulated by animal life, and are converted into chemically insoluble forms which settle on the sea bed. Also, of course, soluble and total metals will be swept to areas outside the North Sea water column. The fact that soluble metal contents in the North Sea water column are not increasing perceptibly with time is thus not as reassuring as at first may seem—the total metal load on the North Sea and surrounding areas will be increasing each year.

The results in Table 5.22 present a more detailed breakdown of the sources of metal pollution of the North Sea water column. Examination of the percentage data shows that rivers and atmospheric pollution are major contributors of pollution in the North Sea and that sewage and sea dumping are the lowest contributors with dredging spoils occupying an intermediate position:

	Metal load as % of total metal load
Rivers	19.3% (cadmium) to 56.6% (arsenic)
Atmospheric	9.1% (chromium) to 69.9% (cadmium)
Dredging spoil	0.5% (arsenic) to 53.9% (chromium)
Sewage	1.6% (lead) to 20% (arsenic)
Sea dumping	0.5% (arsenic) to 8.1% (chromium)

Sea dumping of sewage and industrial waste, i.e. direct discharge, and also sea dumping from ships are clearly minor contributors to pollution of the North Sea, amounting in total to no more than 3.4% (cadmium) to 20.5% (arsenic) of pollution from all sources. Having said this, a framework for regulating these sources of pollution is essential. Stopping direct discharge would necessitate finding alternative disposal routes for some 10.7 million tonnes per annum of sewage sludge and industrial waste by methods such as landfill, incineration, and farm application, which would themselves pose threats to the environment.

By far the largest reductions in pollution load on the North Sea would be achieved by controlling pollution from rivers and atmospheric pollution, and

Table 5.22. Total quantities of metals entering the North Sea from all land-based sources and from direct atmospheric deposition contrasted with estimated mass in the water column of the North Sea[134]

	Land-based, tonne annum^{-1}					Atmospheric deposition, tonne annum^{-1}	Grand total (land-based plus atmospheric, tonne annum^{-1})	As % of grand total				
	River	Direct discharges (incl. input from industrial discharges)	Sea dumping	Dredging spoils	Total			River	Direct discharge	Sea dumping	Dredging spoils	Atmospheric
As	584	206	5	5	800	230	1030	56.6	20.0	0.48	0.48	22.3
Cd	157	22	6	60	245	569	814	19.3	2.7	0.7	7.4	69.9
Cr	1761	355	596	3949	6661	667	7328	24.0	4.8	8.1	53.9	9.1
Cu	2600	276	360	1469	4705	3942	8647	30.0	3.2	4.2	17.0	45.5
Pb	2554	150	377	3440	6521	2920	9441	27.0	1.6	4.0	36.4	30.9
Hg	27	7	2.5	30	66.5	51	117.5	22.9	6.0	2.1	25.6	43.6
Ni	2466	500	97	989	4052	1569	5621	43.9	8.9	1.7	17.6	27.9
Zn	14017	1160	950	15381	31508	7008	38516	36.4	3.0	2.5	40.0	18.2

Source: Hill, J.M., O'Donnell, A.R. and Mance, G., *The Quantities of Some Heavy Metals Entering the North Sea*, Tech. Report TR205, Water Research Centre, Stevenage, Herts, UK (1984). Reproduced by permission of the Water Research Centre.

dredging spoils which together represent between 70% (arsenic) and 96.6% (cadmium) of the metals entering the North Sea. Control of these will be essential to the future wellbeing of this sea and the surrounding oceans.

REFERENCES

1. Klumpp, P.W. and Peterson, P.J., *Environmental Pollution* **19**, 11 (1979)
2. Luoma, S.N. and Bryan, G.W., *Nereis Diverseolor Estuarine Coastal and Shelf Science* **15**, 95 (1982)
3. Bryan, G.W. and Gibbs, P.E., Heavy metals in the Fal Estuary, Cornwall. A study of longterm contamination by mining water and its Effects on Estuarine Organisms. Occasional Publication No. 2, Plymouth Marine Biology Association of the UK (1983)
4. Williams, R., Williams, P.F., Benson-Evans, K., Hunter, M.D. and Harcup, H.F., *Water Pollution Control* **76**, 428 (1976)
5. Turnpenny, A.W.H. and Williams, P., *Environmental Pollution (Series A)* **26**, 39 (1981)
6. Lemly, A.D. and Smith, R.J.F., *Environmental Toxicology and Chemistry* **6**, 225 (1987)
7. Sadler, K. and Turnpenny, A.W.H., *Water, Air and Soil Pollution* **30**, 593 (1986)
8. Gunn, J.M. and Noakes, D.L.G., *Canadian Journal of Fisheries and Aquatic Sciences* **44**, 1418 (1987)
9. Thomsen, A., Korsgaard, B. and Joensen, J., *Aquatic Toxicology* **12**, 291 (1988)
10. Muniz, I.P., Andersen, R. and Sullivan, J., *Water, Air and Soil Pollution* **36**, 371, (1987)
11. Klauda, R.J., Palmer, R.E. and Lenkevich, M.J., *Estuaries* **10**, 44 (1987)
12. Segner, H., Marthaler, R. and Linnenbach, M., *Environmental Biology and Fishes* **21**, 153 (1988)
13. Sadler, K. and Lynam, S., *Journal of Fish Biology* **31**, 209 (1987)
14. Ram, R.N. and Sathyanesan, A.G., *Ecotoxicology and Environmental Safety* **13**, 185 (1987)
15. De, L.G. Solbe, J.F. and Shurhen, D.G., *Water Research* **23**, 127 (1989)
16. Arthur, J.W., West, C.W., Allen, K.N. and Hedtke, S.F., *Bulletin of Environmental Contamination and Toxicology* **38**, 324 (1987)
17. Carrier, R. and Beitinger, T.L., *Water Research* **22**, 511 (1988)
18. Sehgal, R. and Saxena, A.B., *International Journal of Environmental Studies* **29**, 157 (1987)
19. Gill, T.S., Pant, J.C. and Tewari, H., *Ecotoxicology and Environmental Safety* **15**, 153 (1988)
20. Greene, J.C., Miller, W.E., Debacon, M., Long, M.A. and Bartels, C.L., *Environmental Toxicology and Chemistry* **7**, 35 (1988)
21. Boge, G., Bussière, D. and Perez, G., *Water Research* **22**, 441 (1988)
22. Hatakeyama, S., *Ecotoxicology and Environmental Safety* **16**, 1 (1988)
23. Ellgaard, E.J. and Guillot, J.L., *Journal of Fish Biology* **33**, 601 (1988)
24. Nemcsok, J.G. and Hughes, G.M., *Environmental Pollution* **49**, 77 (1988)
25. McGeachy, S.M. and Leduc, G., *Archives of Environmental Contamination and Toxicology* **17**, 313 (1988)
26. Ram, R.N. and Joy, K.P., *Bulletin of Environmental Contamination and Toxicology* **41**, 329 (1988)

27. Woock, S.E., Garrett, W.R., Partin, W.E. and Bryson, W.T., *Bulletin of Environmental Contamination and Toxicology* **39**, 998 (1987)

28. Meisner, J.D. and Hum, W.Q., *Bulletin of Environmental Contamination and Toxicology* **39**, 898 (1987)

29. Hilmy, A.L., El-Domiaty, N.A., Daabees, A. and Latife, H.A.A., *Comparative Biochemistry and Physiology* **86C**, 263 (1987)

30. Arthur, J.W., West, C.W., Allen, K.N. and Hedtke, S.F., *Bulletin of Environmental Contamination and Toxicology* **38**, 324 (1987)

31. Zischke, J.A. and Arthur, J.W., *Archives of Environmental Contamination and Toxicology* **16**, 225 (1987)

32. Giudici, M. de N., Migliore, S.M., Guarino, S.M. and Gambardella, C., *Marine Pollution Bulletin* **18**, 454 (1987)

33. Bodar, C.N.M., Van Leeuwen, C.J., Voogt, P.A. and Zander, D.J., *Aquatic Technology* **12**, 301 (1988)

34. Absanullah, M., Mobley, M.C. and Rankin, P., *Australian J. of Marine and Freshwater Research* **39**, 33 (1988)

35. Skogheim, O.K., Rosseland, B.O., Hoell, E. and Kroglund, F., *Water, Air and Soil Pollution* **30**, 587 (1986)

36. Dorn, P.B., Rodgers, J.H., Kop, K.M., Raia, J.C. and Dickson, K.L., *Environmental Toxicology and Chemistry* **6**, 435 (1987)

37. Giudici, M. de N., Migliore, L. and Guariano, S.M., *Hydrobiologia* **146**, 63 (1987)

38. Nath, K. and Kumar, N., *Chemosphere* **17**, 465 (1988)

39. Maltby, L., Snart, J.O.H. and Calow, P., *Environmental Pollution* **43**, 271 (1987)

40. Harrison, F.L., Watness, K., Nelson, D.A., Miller, J.E. and Calabrese, A., *Estuaries* **10**, 78 (1987)

41. Johnson, P.A., *Aquatic Toxicology* **10**, 335 (1987)

42. Willis, M., *Archiv. für Hydrobiologie* **112**, 299 (1988)

43. Chung, H.K. and Ingle, J.D., *Analytical Chemistry* **62**, 2547 (1990)

44. Stojavovic, R.S., Bond, A.M. and Butler, E.C.V., *Analytical Chemistry* **62**, 2692 (1990)

45. Chan, H., Brindle, I.D. and Le, X.C., *Analytical Chemistry* **64**, 667 (1992)

46. Posta, J., Berndt, H., Luo, S.K. and Schaldach, G., *Analytical Chemistry* **65**, 2590 (1993)

47. Zheng, Y. and Zhang, D., *Analytical Chemistry* **64**, 1656 (1992)

48. Falkner, K.K. and Edmond, J.M., *Analytical Chemistry* **62**, 1477 (1990)

49. Chen, N., Guo, R. and Lac, E.P.C., *Analytical Chemistry* **60**, 2435 (1988)

50. Hanna, C.D. and Tyson, J.F., *Analytical Chemistry* **65**, 653 (1993)

51. Powell, M.J., Quan, E.S.K., Boomer, D.W. and Wiederin, D.R., *Analytical Chemistry* **64**, 2253 (1992)

52. Lu, Jem-Man and Lee, Jun, D.B., *Analytical Chemistry* **66**, 1242 (1994)

53. Hirayama, K. and Unohara, N., *Analytical Chemistry* **60**, 2573 (1988)

54. Jones, K., *Analytical Chemistry* **66**, 1786 (1994)

55. Ketterer, M.E., *Analytical Chemistry* **62**, 2522 (1990)

56. Jones, S.K., *Analytical Chemistry* **63**, 692 (1991)

57. Kerr, A., Kupferschmidt, W. and Alta, S.M., *Analytical Chemistry* **60**, 2729 (1988)

58. Miura, J., *Analytical Chemistry* **62**, 1424 (1990)

59. Abbolino, O., Mentasti, E., Porta, V. and Sarzanini, C., *Analytical Chemistry* **62**, 21 (1990)

60. Laintz, K.E., Yu, J.J. and Wai, C.M., *Analytical Chemistry* **64**, 311 (1992)

61. Wang, J., Breun-Steiner, A., Agnes, L., Sylwester, A., La Grasse, R.R. and Bitsch, N., *Analytical Chemistry* **64**, 151 (1992)

62. Oki, Y., Tashiro, E., Maeda, M., Handa, C., Izumi, S. and Matsuda, K., *Analytical Chemistry* **65**, 2096 (1993)
63. Ichinaka, S., Hongo, N. and Yamazaki, M., *Analytical Chemistry* **60**, 2099 (1988)
64. Hinoue, T., Kaji, J., Yakoyama, Y. and Murata, M., *Analytical Chemistry* **63**, 2086 (1991)
65. Manzoori, J.L. and Miyazaki, A., *Analytical Chemistry* **62**, 2457 (1990)
66. Yuan, R., Chai, Ya Qin, Liu, D., Gao, D., Lij, Z. and Ya, R.Q., *Analytical Chemistry* **65**, 2572 (1993)
67. Dietrich, A.M., Ledder, J.D., Gallagher, D.L., Grabeel, M.N. and Hochn, R.C., *Analytical Chemistry* **64**, 496 (1992)
68. Holak, W. and Specchio, J.J., *Analytical Chemistry* **64**, 1313 (1992)
69. Marengo, E., Gennaro, M.C. and Abrigo, C., *Analytical Chemistry* **64**, 1885 (1992)
70. Tse, Y.H., Janda, P. and Lever, A.B.P., *Analytical Chemistry* **66**, 384 (1994)
71. Kuban, V., Dasgupta, P.K. and Marx, J.N., *Analytical Chemistry* **64**, 36 (1992)
72. Milosalvjevic, E.B., Solujie, L., Hendrix, J.L. and Nelson, J.H., *Analytical Chemistry* **60**, 2791 (1988)
73. Radford-Knoery, J. and Cutter, G.A., *Analytical Chemistry* **65**, 976 (1993)
74. Lussier, T., Gilbert, R. and Hubert, J., *Analytical Chemistry* **64**, 2201 (1992)
75. Posta, J., Berndt, H., Luo, S.K. and Scholdach, G., *Analytical Chemistry* **65**, 2590 (1993)
76. Petrie, L.M., *Analytical Chemistry* **65**, 952 (1993)
77. Sonne, K. and Dasgupta, P.K., *Analytical Chemistry* **63**, 427 (1991)
78. Gonzalez, V., Moreno, B., Sicilia, D., Rubio, S. and Perez-Bendito, P., *Analytical Chemistry* **65**, 1897 (1993)
79. Yaza, N., Nakashima, S., Nakazato, V.T., Uida, N., Kodama, H. and Tateda, A., *Analytical Chemistry* **64**, 1499 (1992)
80. Berglund, I., Dasgupta, P.K., Lopez, J.L. and Nara, O., *Analytical Chemistry* **65**, 1192 (1993)
81. Mack, R.S. and Grimsrud, E.P., *Analytical Chemistry* **60**, 1684 (1988)
82. Maki, S.A. and Danielson, N.D., *Analytical Chemistry* **63**, 699 (1991)
83. Saari Nordhaus, R. and Anderson, J.M., *Analytical Chemistry* **64**, 2283 (1992)
84. Salov, V.V., Yoshinaga, J., Shibata, Y. and Morita, M., *Analytical Chemistry* **64**, 2425 (1992)
85. Okada, T., *Analytical Chemistry* **60**, 1511 (1988)
86. Rigas, P.G. and Pietrzyk, D.J., *Analytical Chemistry* **60**, 1650 (1988)
87. Tanzer, D. and Henmann, K.G., *Analytical Chemistry* **63**, 1984 (1991)
88. Muzzareli, R.A.A. and Rocchetti, R., *Analytica Chimica Acta* **69**, 35 (1974)
89. Ediger, R.D., Peterson, G.E. and Kerber, J.D., *Atomic Absorption Newsletter* **13**, 61 (1974)
90. Sperling, K.R.Z., *Analytical Chemistry* **287**, 23 (1977)
91. Bruland, K.W., Franks, R.P., Knaner, G.A. and Martin, J.H., *Analytica Chimica Acta* **105**, 233 (1979)
92. Smith, R.G. and Windom, H.L., *Analytica Chimica Acta* **113**, 39 (1980)
93. Campbell, W.C. and Ottaway, J., *Analyst (London)* **102**, 495 (1977)
94. Yeats, P.A., Bowers, J.M. and Walton, A., *Marine Pollution Bulletin* **9**, 264 (1978)
95. Batley, G.E. and Matorsek, J.P., *Analytical Chemistry* **49**, 2031 (1977)
96. Sturgeon, R.E., Berman, S.J., Desauliniers, A., Mytytiuk, A., McLaren, J.W. and Russell; private communication
97. Thibaud, Y., *Science et Pêche, Bulletin International Pêche Maritime* **209**, 1 (1971)
98. Cumont, G., Viallex, G., Lebevre, H. and Bobenrieth, P., *Revue International Oceanographie Mediterranean* **26**, 956 (1972)

 99. Renzoni, A. and Baldi, F., *Accua e Aria* 597 (1975)
100. Stopple, M., Backhaus, R. and Matthes, W., *Proc. Verb. XXVth Congress and Plenary Assembly M ICSEM*, Split, Yugoslavia (1976)
101. Buckler, D.R., Mehrle, P.H., Cleveland, L. and Dwyer, F.J., *Water, Air and Soil Pollution* **35**, 97 (1987)
102. De Lisle, P.F. and Roberts, M.H., *Aquatic Toxicology* **12**, 357 (1988)
103. Van der Meer, C., Tenunissen, C. and Boog, T.F.M., *Bulletin of Environmental Contamination and Toxicology* **40**, 204 (1988)
104. Kranz, H. and Gercken, J., *Journal of Fish Biology* **31**, 75 (1987)
105. Ruby, S.M., Idler, D.R. and Jo, Y.D., *Archives of Environmental Contamination and Toxicology* **16**, 507 (1987)
106. Talbot, V., *Marine Biology* **94**, 557 (1987)
107. Shummay, S.E., Card, D., Getchell, R. and Newell, C., *Bulletin of Environmental Contamination and Toxicology* **40**, 503 (1988)
108. Devi, V.U., *Bulletin of Environmental Contamination and Toxicology* **39**, 1020 (1987)
109. Verriopoulos, G., Morantou-Apostolopoulou, M. and Milliou, E., *Bulletin of Environmental Contamination and Toxicology* **38**, 483 (1987)
110. Pauson, A.J., *Analytical Chemistry* **58**, 183 (1986)
111. Gardner, W.S., Herch, L.R., St Johns, P.A. and Seitzuger, S.P., *Analytical Chemistry* **63**, 1838 (1991)
112. Bishop, J.K.B., *Analytical Chemistry* **62**, 553 (1990)
113. Belli, S.L. and Zirino, A., *Analytical Chemistry* **65**, 2583 (1993)
114. Scarano, G., Morelli, E., Seritti, A. and Zirino, A., *Analytical Chemistry* **62**, 943 (1990)
115. Falkner, K.K. and Edmond, J.M., *Analytical Chemistry* **62**, 1477 (1990)
116. Eirod, U.A., Johnson, K.S. and Coale, K.H., *Analytical Chemistry* **63**, 893 (1991)
117. Obata, H., Faratani, H. and Nakayama, E., *Analytical Chemistry* **65**, 1524 (1993)
118. Shahani, M.B.S., Akagi, T., Shimizu, H. and Masuda, A., *Analytical Chemistry* **62**, 2709 (1990)
119. Cohen, A.S. and O'Nions, R.K., *Analytical Chemistry* **63**, 2705 (1991)
120. Buenelt, W. and Tai Weichich, *Analytical Chemistry* **64**, 1691 (1992)
121. Shabani, M.B., Akagi, T. and Masuda, A., *Analytical Chemistry* **64**, 737 (1992)
122. Dupont, V., Auger, Y., Jeandel, C. and Warter, M., *Analytical Chemistry* **63**, 520 (1991)
123. Alves, L.C., Allen, L.A. and Houk, R.S., *Analytical Chemistry* **65**, 2468 (1993)
124. Nagaosa, Y., Kawabe, H. and Bond, A.M., *Analytical Chemistry* **63**, 28 (1991)
125. Akagi, T. and Haraguchi, H., *Analytical Chemistry* **62**, 81 (1990)
126. Pai, S.C., Chen, T.C. and Wong, G.T.F., *Analytical Chemistry* **62**, 774 (1990)
127. Von Geen, A. and Boyle, E., *Analytical Chemistry* **62**, 1705 (1990)
128. Luther, G.W., Swartz, C.B. and Ullman, W.J., *Analytical Chemistry* **60**, 1721 (1988)
129. Verma, K.K., Jain, A. and Verma, A., *Analytical Chemistry* **64**, 1484 (1992)
130. Ito, K., Ariyoshi, Y., Tanabiki, F. and Sunabara, H., *Analytical Chemistry* **63**, 273 (1991)
131. Fujiwara, K., Uchida, M., Chen, M., Kuwamoto, Y. and Kumamaru, T., *Analytical Chemistry* **65**, 1814 (1993)
132. Leck, C. and Bagander, L.E., *Analytical Chemistry* **60**, 1680 (1988)
133. Singh, R.P., Pambid, E.R. and Abbas, N.M., *Analytical Chemistry* **63**, 1897 (1991)
134. Hill, J.M., O'Donnell, A.R. and Mance, G., The quantities of some heavy metals entering the North Sea. Technical Report TR205, Water Research Centre, Stevenage, Herts, UK (1984)
135. Fitzgerald, W.F., Research Institute for Metals, Sicily, Italy (1978)

136. Mart, L., Nurnberg, H.W. and Dryssen, D., in Wong, C.S. *et al.* (eds), *Trace Metals in Seawater*, Proceedings of a NATO Advanced Research Institute on Trace Metals in Seawater, 30 March to 3 April 1981, Sicily, Italy, p. 113, Wiley, New York (1981)
137. Landy, M.P., *Analytica Chimica Acta* **121**, 39 (1980)
138. Dhabashwar, R.G. and Zarapskar, L.R., *Analyst (London)* **105**, 386 (1980)
139. Braun, H. and Metzger, M., *Fresenius Zeitschrift für Analytische Chemie* **318**, 321 (1984)
140. Weiss, H.V., *Analytica Chimica Acta* **56**, 136 (1971)
141. Morozumi, M., Nakamura, S. and Patterson, C.C., *Japan Analyst* **19**, 1057 (1970)
142. Braman, R.S. and Tompkins, M.A., *Analytical Chemistry* **51**, 12 (1979)
143. Pillay, K.K.S., Thomas, C.C., Sondel, J.A. and Hyde, C.M., *Analytical Chemistry* **43**, 1419 (1971)
144. Bentine, K.K. and Dong Soohee Du, in Wong, C.S. *et al.* (eds), *Trace Metals in Seawater*, Proceedings of a Nato Advanced Research Institute on Trace Metals in Seawater, 30 March to 3 April 1981, Sicily, Italy, p. 25, Wiley, New York (1981)
145. Mance, G., Brown, V.M. and Yates, J., Proposed Environmental Quality Standards for List III Substances in Water; Copper. Water Research Centre, Marlow, Bucks, UK (1984)
146. Mance, G., Brown, G.M., Gardiner, J. and Yates, J., Proposed Environmental Quality Standards for List III Substances in Water; Chromium. Technical Report TR207, Water Research Centre, Marlow, Bucks, UK (1984)
147. Mance, G., Brown, G.M., Gardiner, J. and Yates, J., Proposed Environmental Quality Standards for List II Substances in Water; Inorganic Lead. Technical Report TR208, Water Research Centre, Marlow, Bucks, UK (1984)
148. Mance, G. and Yates, J., Proposed Environmental Quality Standards for List II Substances in Water; Zinc. Technical Report TR209, Water Research Centre, Marlow, Bucks, UK (1984)
149. Mance, G. and Yates, J., Proposed Environmental Quality Standards for List II Substances in Water; Nickel. Water Research Centre, Marlow, Bucks, UK (1984)

CHAPTER **6**

Effects on Creatures of Organic and Organometallic Compounds in Water

A wide variety of organic compounds can occur in fresh and marine waters. Also, naturally occurring organic compounds such as amino acids and fatty acids involved in food chains are present. The majority of organic compounds found occur as a result of human activities. Possible causes of water, land or atmospheric pollution by organics and organometallics are industrial and other discharges, whether accidental or deliberate, land use of chemicals, substances produced as a result of fires and industrial smoke emissions and domestic waste and discharges.

Organometallic compounds can originate in one of two ways, either by direct contamination of the water with organometallic compounds, or by the production of organometallic compounds in the water or sedimentary matter or living creatures by a biomethylation of inorganic metals caused by various types of organisms (see Chapter 4 for details).

6.1 ORGANIC COMPOUNDS: FRESHWATER

6.1.1 TOXIC EFFECTS ON ORGANISMS

LC_{50} values and adverse effect data on fish and creatures other than fish for a range of organic compounds in freshwaters are reviewed in Tables 6.1 and 6.2 (see also Section 4.2).

6.1.2 ANALYTICAL COMPOSITION

Information on the types and concentrations of organic compounds that have been found in freshwaters, including rivers, lakes and surface waters, is presented in Table 6.3. A summary of these results (Table 6.4) shows that up to 13.4 μg l^{-1} total haloforms have been found in river waters (up to 70.5 μg l^{-1} in lakes) and up to 4.3 μg l^{-1} polychlorinated biphenyls in rivers. At these maximum levels there is cause for ecological concern. Thus, the World Health Organisation quotes a maximum permitted level of 0.2 μg l^{-1} for six carcinogenic

Table 6.1. Concentration of substances in freshwaters and effect on fish

Substance 1	Fish type 2	Water type 3	Exposure concentration 4	Toxicity index 5	Adverse effects 6	Ref. 7
Methylene dichloride	Juvenile fathead minnow (*Pimphales promelas*)	Natural	–	48 h LC_{50} = 502 mg kg^{-1} 192 h LC_{50} = 471 mg kg^{-1}	–	1
Methyl bromide and sodium bromide	Guppy (*Poecilia reticulata*)	Natural	Methyl bromide 0.032–3.2 mg l^{-1} Sodium bromide 0–32 000 mg l^{-1}	–	NaBr concentrations of ≥ 100 mg l^{-1} caused paralysis in 1–3 months. No observed lethal concentration in 1 month at 10 mg l^{-1}. Exposure to 1.8 mg l^{-1} methyl bromide for 4 days caused degenerative changes in gills and oral mucosa.	
1,3-dichlorobenzene 1,4-dichlorobenzene 1,2,3,4-tetrachloro-benzene Pentachlorobenzene Hexachlorobenzene	Fathead minnows (*Pimphales promelas*)	Natural	–	96 h LC_{50} 1,3-dichlorobenzene, 7800 μg l^{-1} 1,2,3,4-tetrachlorobenzene 1100 μg l^{-1} 1,4-dichlorobenzene, 4200 μg l^{-1} No effect concentrations (NOEC) (highest) 1,3-dichlorobenzene, 1000 μg l^{-1} 1,4-dichlorobenzene, 570 μg l^{-1} 1,2,3,4-tetrachlorobenzene, 250 μg l^{-1} lowest effect concentrations (LOEC) 1,3-dichlorobenzene, 2300 μg l^{-1} 1,4-dichlorobenzene, 1000 μg l^{-1} 1,2,3,4-tetrachlorobenzene, 400 μg l^{-1} Pentachlorobenzene and hexachlorobenzene non-toxic at 5.5 and 4.8 μg l^{-1} respectively.	–	2

Diethylhexyl phthalate	Daphnia and fish	Natural	–	Oral 4-day LD_{50} 10–100 kg g^{-1}	Chronic sublethal exposure causes deterioration of reproductive capacity and immune system and carcinogenic activity.	3
Acrylates and methacrylates	Juvenile fathead minnows (*Pimphales promelas*)	Natural	–	6 h LC_{50} reported to be lower than predicted values.	Toxicity ratio 4–6: respiratory and metabolic inhibition toxicity ratio: 42–56 Neurotoxicity (toxicity ratio = predicted LC_{50} divided by observed LC_{50})	4
1,2,4-trichloro-benzene	Fathead minnows (*Pimphales promelas*)	Natural	–	96 h LC_{50} 2.76 mg l^{-1}	–	5
Pentachlorophenol	*Selenastrum capricornutium*	Natural	–	96 h LC_{50} 0.11 mg l^{-1} (soft water) 0.15 mg l^{-1} (soft water) 0.76 mg l^{-1} (hard water) (pH 7.6–8.4)	–	6
Bleached Kraft mill effluent containing 2,4,6-trichlorophenol pentachlorophenol 2,3,4,6-tetrachlorophenol and resin acid	Roach (*Rutilus rutilus*)	Kraft mill effluent diluted × 2000	–	2,4,6-trichlorophenol, 0.05 mg l^{-1} 2,3,4,6-trichlorophenol, 0.071 mg l^{-1} pentachlorophenol, 0.028 mg l^{-1}	Exposure of roach to 0.035 LC_{50} for 38 days, then to 0.07 LC_{50} for 14 days produced no evident adverse effects	7
Pentachlorophenol 2,4-dichlorophenol Tricaine methane-sulphonate, 1-octanol	Rainbow trout (*Salmo gairdneri*)	Natural	–	96 h LC_{50} (1) Pentachlorophenol, 0.09 mg l^{-1} (2) 2,4-dichlorophenol, 4.64 mg l^{-1} (3) Tricaine methane sulphonate, 50.2 mg l^{-1} (4) 1-octanol, 15.8 mg l^{-1}	Survival time (h) (1) 31.6 (2) 15.2 (3) 47 (4) 5.6	8
Arochlor 1254	Rainbow trout (*Salmo gairdneri*)	Natural	3–300 mg l^{-1}	–	Severe weight reduction in 12 months' exposure (also liver weight reduction)	9

continued overleaf

Table 6.1. (continued)

Substance 1	Fish type 2	Water type 3	Exposure concentration 4	Toxicity index 5	Adverse effects 6	Ref. 7
Arochlor 1254 plus Mirex	Rainbow trout (*Salmo gairdneri*)	Natural	30 mg l^{-1} Arochlor plus 5 mg l^{-1} Mirex	–	Severe weight reduction in 12 months' exposure (also liver weight reduction)	9
3,4-dichloroaniline	Fathead minnows (*Pimephales promelas*)	Natural	–	24 h LC_{50} = 9.03–12 mg l^{-1} 48 h LC_{50} = 8.8–10.0 mg l^{-1} 96 h LC_{50} = 6.99–8.06 mg l^{-1}	Exposure to 5.1–15.7 μg l^{-1} 3,4-dichloroaniline had no effect on egg hatchability or egg survival in 5 days fry survival reduced on exposure to 23 μg l^{-1} 3,4-dichloroaniline for 28 days	10
Insecticides Carbaryl	Catfish (*Clarias batrachus*)	Freshwater	–	24 h LC_{50} = 6.1–16.1 mg l^{-1} 48 h LC_{50} = 53.7–134 mg l^{-1} 72 h LC_{50} = 48.6–123 mg l^{-1} 96 h LC_{50} = 46.9–107.7 mg l^{-1}	–	11
Carbaryl	Rainbow trout (*Salmo gairdneri*)	Natural	5.17 mg l^{-1}	–	Survival time 13 hours	12
Malathion	Rainbow trout (*Salmo gairdneri*)	Natural	0.3 m gl^{-1}		Survival time 40.1 hours	12
Malathion	Freshwater teleosts (*Channa punctatus*)	Freshwater	–	96 h LC_{50} = 1.73 mg l^{-1}	–	13
Trichlorofon	*Cichalsoma urophthalmus* fry	Natural	0–80 mg l^{-1}	100% mortality in 24 hours in presence of 60–80 mg l^{-1} trichorofon	Erratic swimming and hyperventilation prior to death seen at 30–80 mg l^{-1}. At lower concentrations fish showed loss of reflexes but survived	14

Compound	Species	Environment	Dose	LC_{50}	Effects	Ref
Roundup and Rodeo herbicides	Rainbow trout Chinock, Coho Salmon	Lake and stream	—	Roundup 96 h LC_{50} = 7.2–12 mg l^{-1} Rodeo-x-77 surfactant 96 h LC_{50} = 120–290 mg l^{-1} Rodeo 96 h LC_{50} = 580 mg l^{-1}	—	15
Bromacil and Diuron	Fathead minnows (*Pimphales promelas*)	Natural	—	Bromacil 24 h LC_{50} = 185 mg l^{-1} 48 h LC_{50} = 183 mg l^{-1} 96 h LC_{50} = 182 mg l^{-1} 168 h LC_{50} = 167 mg l^{-1} Diuron 24 h LC_{50} = 23.3 mg l^{-1} 48 h LC_{50} = 19.9 mg l^{-1} 96 h LC_{50} = 14.2 mg l^{-1} 168 h LC_{50} = 7.7 mg l^{-1}	1.0–29.0 mg l^{-1} Bromacil had no effect on fish hatch % mortality or juvenile fish survival. These concentrations did affect fish growth. 2.6–78 mg l^{-1} Diuron did not affect fish hatch or growth. 78 μg l^{-1} Diuron decreased fish survival and increased numbers of dead fish and deformed fry	16
Lindane	Carp (*Cyprinus carpio*)	Natural	0–1000 mg l^{-1} in food pellets	—	No adverse effects	17
Lindane	Teleost fish (*Anguilla anguilla*)	Lake	—	96 h LC_{50} = 0.32–0.54 mg l^{-1} (15 °C) 0.67–0.68 mg l^{-1} (22 °C) 0.45–0.55 mg l^{-1} (29 °C)	—	18
Methoxychlor	Rainbow trout (*Salmo gairdneri*)	Natural	0–580 μg l^{-1} for 2 days and 0–30 μg l^{-1} for 68 days	—	Methoxychlor had no adverse effect on survival, growth and development parameters. Long-term effects on reproduction not clarified	19
Benomyl	Rainbow trout Channel catfish Bluegills	Natural	—	96 h LC_{50} values reported	—	20

Table 6.2. Concentration of organic substances in fresh waters and effect on organisms other than fish

Substance 1	Fish type 2	Water type 3	pH 4	Exposure concentration 5	Toxicity index 6	Adverse effects 7	Ref. 8
Polycyclic aromatic hydrocarbons	Daphnid (*Daphnia magna*)	Natural	–	–	LC_{50} values reported for various PAHs	–	21
Sodium dodecyl sulphate, Triton-X-100 Sodium dodecylbenzene sulphonate	Lugworm (*Arenicola marina*)	Natural	–	–	48 h LC_{50} sodium dodecyl sulphate 15.2 mg l^{-1} Triton X-100, 15.2 mg l^{-1} sodium dodecylbenzene sulphonate 12.5 mg l^{-1}	Lugworm gills and epidermic receptors are the most sensitive to detergents	22
di-2-ethyl hexylphthalate	Daphnid (*Daphnia magna*)	Natural	–	0–811 μg l^{-1}	21-day maximum allowable toxicant concentration (MATC) between 158 and 811 μg l^{-1}	Surfacing behaviour increased on day 1 of test on daphnids exposed to 158 or 811 μg l^{-1} di-2-ethyl hexylphthalate	23
Phenol	*Ascellus aquaticus*	Natural	–	–	–	Immobilisation, paralysis and mortality reported	24
Phenol, *o/m/p* Trimethylphenol Cresol, Xylenols	Daphnid (*Daphnia magna*)	Natural	–	–	–	24 h LC_{50} reported	25
2,4-dichlorophenol and aniline	Daphnid (*Daphnia magna*)	Natural	–	–	Maximum allowable toxicant concentration (MATC) 24.6–46.7 μg l^{-1} (aniline) 0.7–1.48 mg l^{-1} (2,4 dichlorophenol)	–	26
Ethylene dibromide	*Hydra oligactis*	Natural	–	–	72 h LC_{50} = 50 mg l^{-1}	–	27
2,2-dichloro-biphenyl	*Daphnia pulicarria*	Lake	–	50 ng l^{-1} to 10 μg l^{-1}	–	Significant mortality in inhibition of reproduction at concentrations down to 50–100 ng l^{-1}	28

Insecticides

Carbaryl	Lugworm (*Arenicola marina*)	Natural	–	–	48 h $LC_{50} = 7.2$ mg l^{-1}	–	22
Parathion ethyl	Lugworm (*Arenicola marina*)	Natural	–	–	48 h $LC_{50} = 2.7$ mg l^{-1}	–	22
Malathion	Toad embryos (*Bufo arenarum*)	Natural	–	0–70 mg l^{-1}	–	0–30 mg l^{-1} embryonic development normal 44 mg l^{-1}, 67% mortality in 5 days	29
Methyl parathion	Penaeid prawn (*Metapenaeus monoceros*)	Freshwater	–	0.04–1.2 mg l^{-1}	Sublethal concentration 0.04 mg l^{-1} Lethal concentration 1.2 mg l^{-1}	–	–
Fenitrothion Carbofuran	Freshwater mullet (*Channa punctatus*)	Freshwaters	–	1.5 mg l^{-1} (fenitrothion) 5 mg l^{-1} (carbofuran)	–	Growth abnormalities of follicle and epithelium reported	30
Permethrin	Snail (*Lymnaea acuminata*)	Natural	–	–	–	At 40 and 80% of 24 h LC_{50} dose evidence for nerve poisoning blocking aerobic and anaerobic metabolism of snail	31
Lindane, endosulphan	European eel (*Anguilla anguilla*)	Natural	–	–	LC_{50} values reported	Toxicity at 15 °C and 29 °C greater than at 22 °C	32
Phosphamidon	Freshwater prawn (*Macrobrachium lamarrei*)	Freshwater	–	–	Extremely toxic to *Macrobrachum lamarrei* 40% of 96 h LC_{50} caused glycogen depletion in muscle	–	33
Phosphamidon	Penaeid prawn (*Metapenaeus monoceros*)	Freshwater	–	0.04–1.2 mg l^{-1}	Sublethal concentration 0.04 mg l^{-1} Lethal concentration 1.2 mg l^{-1}	–	34

continued overleaf

Table 6.2. (continued)

Substance 1	Fish type 2	Water type 3	pH 4	Exposure concentration 5	Toxicity index 6	Adverse effects 7	Ref. 8
2.4-dinitrophenoxyl acetic acid	Macro invertebrates	Pond	—	—	—	No adverse short-term effect on macro invertebrate communities. Death of macrophytes over one year	35
P-chloroaniline	South African clawed toad (*Xenopus laevis*)	Natural	—	—	—	*p*-chloroaniline at 100 mg l^{-1} kills embryos	36
Aniline			—	—		Aniline at $\geqslant 1$ mg inhibited embryo development	
Sodium dodecyl benzene sulphonate			—	—		Sodium dodecyl benzene sulphonate at 50 mg l^{-1} kills embryos	
Phenol						Phenol at 50 mg l^{-1} was lethal to tadpoles at early stages of larval development	
Methylene bis thiocyanate	(1) *Nitromonas nitrobacter*	Freshwater	6–8	—	(1) 3 h MIC = 3200 μg l^{-1}	—	37
	(2) *Photobacterium phosphoreum*	Freshwater	6–8	—	(2) 15 min $EC_{50} = 54$ μg l^{-1}		37
	(3) *Chlorella pyrenoidosa*	Freshwater	6–8	—	(3) 96 h $EC_{50} = 42$ μg l^{-1}		37
	(4) *Daphnia magna*	Freshwater	6–8	—	(4) 48 h $EC_{50} = 25$ (pH6) 48 h $EC_{50} = 73$ (pH 8)		37
	(5) *Poecilia reticulata*	Freshwater	6–8	—	(5) 96 h $LC_{50} = 390$ μg l^{-1}		37
	(6) *Salmo gairdneri*	Freshwater	6–8	—	(6) 14 day $LC_{50} = 84$ μg l^{-1} 60 day LC_{50} 65 μg l^{-1}		37

Table 6.3. Organics in rivers, lakes and surface waters

		Concentration ($\mu g\,l^{-1}$)	Ref.
	Haloforms		
$CHCH_3$	River	0.56–0.75	39
		0.20–0.67	40
	Lakewater	54.6–59.1	41
$BrCl_2CH$	River	< 0.1	42
		0.06	39
		0.1–7.6	40
Br_2ClCH	River	< 0.1	42
		0.08	39
		4.66	40
Br_3CH	River	< 0.1	42
		0.15–0.21	39
		0.51	40
CCl_4	River	0.02–0.12	39
	Lakewater	11.8–14.3	41
$Cl_2CH\,CH_2Cl$	River	0.05–0.09	39
	Lakewater	7.8–11.4	41
$Cl_2CH\,CH_2Cl$	Lakewater	8–20	43
$Cl_2CHCHCl_2$	Lakewater	2–5	43
	Total haloforms		
	River	0.92–1.31	39
	River	5.47–13.44	40
	River	11.8–14.3	42
	Lake	62.4–70.5	41
	Polyaromatic hydrocarbons		
Benzo(a)pyrene	River	0.032–0.038	44
Fluoranthrene		0.02–1.1	45
Benzo(k)fluoranthrene		0.03–0.49	
Benzo(a)pyrene		0.10–0.65	
Perylene		0.03–0.20	
Indene (1, 2, 3-ed) pyrene		0.4–0.32	
Benzo(ghi)perylene		0.04–0.12	
Pyrene		0.05–0.43	
Benzo(a)anthracene/ Chrysene		0.14–0.53	
Benzo(b)fluoranthrene		0.13–0.57	
Total PAH		< 0.1–4.3	
	Chlorinated insecticides		
α-BHC	River	0.003	46
α-BHC	River	0.002	47
α- and γ-BHC plus hexachlorobenzene	River	0.018	48
β-BHC	River	0.0004	46

continued overleaf

Table 6.3. (*continued*)

		Concentration ($\mu g\,l^{-1}$)	Ref.
	River	0.013	47
	River	0.023	48
	Surface water	0.006–0.078	49
γ-BHC	River	0.69	48
	River	0.006	47
	Surface water	0.006–0.078	49
γ-BHC	River	0.69	48
	River	0.006	47
	Surface water	0.004–0.02	49
δ-BHC	River	0.016	47
DDT	River	0.042	48
p, p^1-DDT	River	0.051	50
	Surface water	0.009–0.037	49
o, p^1-DDT	Surface water	0.005–0.025	49
DDE	River	0.022	48
p, p^1 DDE	Surface water	0.002–0.010	49
Lindane	River	0.01	50
	River	0.001	46
Dieldrin	River	0.031	50
Aldrin	River	0.02	50
Endrin	River	0.038	50
γ-Chlodane	River	0.03	50
Methoxychlor	River	0.12	50
Endosulphan	River	0.028–0.28	51
Heptachlor	Surface water	0.001–0.007	49
Hexachlorobenzene	Surface water	0.002–0.008	49
Total chlorinated insecticides			
	River	0.0034	46
	River	0.037	47
	River	0.761	48
	River	0.300	50
	Surface water	0.029–0.185	49
Other types of insecticides			
Ronnel		0.002–0.022	52
Dursban		0.030–0.043	
Diazinon		0.020–0.037	
Malathion		0.027–0.032	
Parathion		0.037–0.039	
Parathion-methyl		0.021–0.038	
Polychlorinated biphenyls	Ground water	0.0001–0.0002 (as Arochlor 1016)	53
Pentachlorophenol	River	10–250	54
	Well water	0.1	55
Dibutyl phosphate	River	< 0.1–1.0	56, 57
	River	45	58
	River Meuse	< 0.1–0.9	56, 60

Table 6.3. (*continued*)

		Concentration (μg l^{-1})	Ref.
di-2-ethylhexyl	River	0.4–4.2	56, 57
phthalate	River	10	56, 59, 60
	River Meuse	0.1–1.1	56, 60
Non-ionic detergents (phenol polyoxyalkylene condensates)	River	8–70	61
Alkyl benzene	River	270–600	62
Sulphonates	River	10–600	62
	2C10	4400–4800	63
	3-C10	3700–6200	
	4,5-C10	1200–16 200	
	2-C11	3900–12 200	
	3-C11	8200–8300	
	4,5,6C11	42 700–43 700	
	2-C12	< 200–4200	
	3-C12	1200–3400	
	4,5,6-C12	13 700–17 100	
	2-C13	< 200–800	
	3-C13	< 200–800	
	4,5,6-C13	< 200–6200	
Methylene blue active substances	Ground water	20–22	64
Fluorescent whitening agents	River	5–7	62
	River	1–7	70
Fatty acids	River (total C10–C19 FA)	4.13–527	65
Nitriloacetic acid	River	0.4	66
Dissolved organic carbon	River	6000–10 000	67
	Lakewater	1500–3080	68
	Ground water	300–6300	64
Dissolved inorganic carbon	Lakewater	1060–6190	69
	Ground water	1000	64
Gaseous organic carbon	Lakewater	1900–2310	71
Particulate organic carbon	Lakewater	100–300	71

polyaromatic hydrocarbons (fluoranthrene, benzo(d)-fluoranthrene, benzo(k)-fluoranthrene, benzo(a)-pyrene, benzo(ghi)perylene and indeno-1,2,3(ed)-pyrene), while the Germany specification for total polyaromatic hydrocarbons is 0.25 μg l^{-1}.

Table 6.4. Summary of organics in natural waters (rivers, lakes and surface waters) (concentration $\mu g\,l^{-1}$)

Haloforms	Rivers	Lakes	Surface waters
$CHCl_3$	0.02–0.75	54.6–59.1	–
$BrCl_2CH$	< 0.1–7.6	–	–
Br_2ClCH	< 0.1–4.66	–	–
Br_3CH	< 0.1–0.51	–	–
CCl_4	0.02–0.12	11.8–14.3	–
$CH_2CH\ CH_2Cl$	0.05–0.09	7.8–11.4	–
$Cl_2CH\ CH_2Cl$	–	8–20	–
$Cl_2CH\ CHCl_2$	–	2–5	–
Total haloforms	0.92–13.4	62.4–70.5	–
Total polyaromatic hydrocarbons	< 0.1–4.3	–	–
Chlorinated insecticides			
α-BHC	0.002–0.003	–	–
β-BHC	0.0004–0.023	–	0.006–0.078
γ-BHC	0.006–0.69	–	0.004–0.02
δ-BHC	0.16	–	–
DDT	0.042	–	–
p, p^{1}DDT	0.051	–	0.009–0.037
o, p^{1}DDT	–	–	0.005–0.025
DDE	0.022	–	–
p, p^{1}DDE	–	–	0.002–0.010
Lindane	0.001–0.01	–	–
Dieldrin	0.031	–	–
Aldrin	0.02	–	–
Endrin	0.035	–	–
γ-chlordane	0.03	–	–
Heptachlor	–	–	0.001–0.007
Methoxychlor	0.12	–	–
Endosulphan	0.028–0.28	–	–
Hexachlorobenzene	–	–	0.002–0.008
Total chlorinated insecticides	0.003–0.76	–	0.029–0.185
Other insecticides			
Ronnel	0.002	–	–
Dursban	0.030–0.043	–	–
Diazinon	0.020–0.037	–	–
Malathion	0.027–0.032	–	–
Parathion	0.037–0.039	–	–
Parathion methyl	0.021–0.038	–	–
Total organophosphorous insecticides	0.14–0.18		
Polychlorinated biphenyls	0.0001–0.0002 (as Arochlor 1016)	–	–
Pentachlorophenol	0.1–250	–	–
Dibutylphosphate	< 0.1–45	–	–
di-2-ethylhexylphthalate	01–4.2	–	–
Non-ionic detergents	8–70	–	–

Table 6.4. (*continued*)

Haloforms	Rivers	Lakes	Surface waters
Alkyl benzene sulphonates	10–600	–	–
Fatty acids	4.1–527	–	–
Nitriloacetic acid	0.4	–	–
Dissolved organic carbon	1500–10 000	1500–3080	300–6300
Dissolved inorganic carbon	–	1060–6190	1000
Gaseous organic carbon	–	1900–2310	–
Particulate organic carbon	–	100–300	–

Cross-checking of the toxicity data (Tables 6.1 and 6.2) and actual concentrations occurring in water samples (Tables 6.3 and 6.4) makes it possible to evaluate creatures that will be subject to adverse effects or mortalities for any particular water.

Thus in the case of pentachlorophenol up to 250 μg l^{-1} of this compound has been found in river waters (Table 6.4). The 96-hour EC_{50} value of this compound lies in the range 90–760 μg l^{-1} (Table 6.1). Thus concentrations of 250 μg l^{-1} could cause adverse effects or even fatalities, water hardness being an important parameter in this respect. Di-2-ethyl hexylphthalate can occur in rivers at concentrations up to 4.2 μg l^{-1} (Table 6.3). As the reported 21-day maximum allowable concentration (MATC) for this compound lies between 158 and 811 μg l^{-1} (Table 6.2) no abnormal effects such as abnormal surfacing behaviour would be expected at the maximum concentration of this compound likely to occur in rivers.

6.1.3   ANALYTICAL METHODS

A wide variety of analytical techniques have been used to determine organic compounds in non-saline natural waters. Detection limits achievable range from the μg l^{-1} and ng l^{-1} ranges to the pg l^{-1} level (Table 6.5). The most important types of organic compounds can be determined at these levels. A summary of the methods, types of compounds determinable and detection limits achievable is given below. Gas chromatography is generally used to determine the more volatile contaminants while high-performance liquid chromatography is reserved for less volatile contaminants. Coupling of either of these techniques with a mass spectrometric detector enables definitive identifications of unknown contaminants to be made. Spray and trap or purge and trap methods when coupled with gas chromatography enable a wide range of volatile contaminants to be determined with great sensitivity.

Table 6.5. Methods for the determination of organic compounds in natural (non-saline) waters

Compound	Technique	Detection limit	Found	Ref.
Halogenated aliphatic and aromatic compounds				
Trichloroethane	Extraction kinetics	–	–	72
	Gas chromatography	$0.4 \ \mu g \, l^{-1}$		73
Chlorinated hydrocarbons	Gas sensor/permeation apparatus	$1000-5000$ $\mu g \, l^{-1}$	–	74
p-dibromobenzene and chloroform	Gas chromatography	3 pg ($CHCl_3$) 29 pg ($pC_6H_4Br_2$) absolute	–	75
Chloroaromatics	Membrane mass spectroscopy	–	–	76
Haloforms	Membrane mass spectroscopy	–	–	77
CH_3I, C_2H_5I	Gas chromatography, electron capture detector with negative ion hydration and photodetachment	$0.01 \ \mu g \, l^{-1}$	–	78
$CHCl_3$, CCl_4, $CHBrCl_2$, $CHBr_2Cl$, $CHBr_3$	Gas chromatography	$0.9-2.6 \ \mu g \, l^{-1}$	–	79
$CHCl_3$, $BrCl_2CH_2$ $Br_2ClCH_2CHBr_3$	Purge and trap gas chromatography with electron capture detector	$0.000 \, 01 -$ $0.000 \, 05 \ \mu g \, l^{-1}$	–	80
CF_2Br_2, CH_2Br_2 $CFCl_3$, $CHCl_3$, CCl_4	Gas chromatography with photodetachment modulated electron capture detector (PDM)	–	–	81
CHI_3, $CHCl_3$, CCl_4, CF_2Br_2, $CFCl_3$ $CH_2 = CClF$, C_2H_5I, CH_2ClI, CCl_3Br CH_3CCl_3, $CHCl_3$	Gas chromatography with photochemical modulated pulsed electron capture detector	–	–	82
Aromatic hydrocarbons, benzene, toluene	Dynamics of extraction from water using liquid coated fused silica films	–	–	83
Toluene	Membrane mass spectroscopy	–	–	76

Table 6.5. (*continued*)

Compound	Technique	Detection limit	Found	Ref.
Anthracene, chloronaphthalene	Spray and trap method— thermal desorption gas chromatography—mass spectroscopy	–	–	84
Polynuclear aromatic hydrocarbons	Cyclodextran modified solvent extraction	–	–	85
	Supercritical fluid extraction	–	–	86
Surfactants (alkoxy) polyethoxy carboxyates	Mass spectrometry	–	–	87
	Mass spectrometry and fast ion atom bombardment	–	–	87
Linear primary alcohol ethoxylates	Thermospray liquid chromatography—mass spectrometry	–	–	8
Polyoxyethylenes	Liquid chromatography with indirect conductiometric detector	–	–	89
Linear alkylbenzene sulphonates and dialkyltertalin sulphonates	Gas chromatography—mass spectrometry	$0.1\ \mu g\,l^{-1}$	–	90
Alkyl sulphates, alkyl ethoxy sulphates	Liquid chromatography— ion mass spectrometry	–	–	90
Alkylbenzene sulphonates	Continuous flow fast atom bombardment spectrometry	$0.5\ \mu g\,l^{-1}$	–	92
Cationic detergents	Fast atom bombardment— mass spectrometry	$\mu g\,l^{-1}$ level	–	93
	Ion pairing with fluorospar	$0-1$ M level	–	94
Linear alkylbenzene sulphonate	Liquid chromatography	–	–	95
Polychlorinated biphenyls	Gas chromatography	–	–	96, 97
	Enzyme-linked immunosorbent assay	–	–	98
Polychlorinated dibenzo-*p*-dioxins	High-performance liquid chromatography	–	–	99
	Negative ion mass spectrometry	–	–	100
	High-resolution mass spectrometry, selected ion monitoring	–	–	101
	Column chromatography	–	–	102

continued overleaf

Table 6.5. (*continued*)

Compound	Technique	Detection limit	Found	Ref.
52 pesticides	High-performance liquid chromatography—mass spectrometry	–	–	103
Carbamate type (carbofuran, propoxur, carbaryl)	Flow-through sensor—high-performance liquid chromatography	ng l^{-1} to pg l^{-1}	–	104
Acidic, basic and neutral pesticides	Liquid chromatography—mass spectrometry	–	–	105
Chlorinated phenoxy acid and ester herbicides	Liquid chromatography, particle beam mass spectrometry and UV absorption spectrometry	10^3 μg l^{-1}	–	106
35 pesticides (oxamyl, methomyl, phoxan, 2,4,5,T, 2,4 DB, MCPB, etc.)	Preconcentration on graphitised carbon black cartridge—liquid chromatography	0.003–0.07 μg l^{-1}	–	107
Propoxur, carbofuran, propham, captan, chloropropham, barban, butyrate	Automated high-performance liquid chromatography	–	–	108
Carbofuran	On-line immunoaffinity chromatography with coupled column liquid chromatography mass spectrometry	–	–	109
	Mass spectrometry			
Terbutryn, aldicarb sulphone, propoxur, carbofuran	Ion-trap mass spectrometry	10–30 pg absolute	–	110
Atrazine and others	High-resolution fast atom bombardment—mass spectrometry	0.001 μg l^{-1} level	6–94 μg l^{-1}	111
	Gas chromatography			
Atrazine	Gas chromatography—high-resolution mass spectrometry	200–500 pp quadrillion	–	112
Chlorinated pesticides and polychlorinated biphenyls	Gas chromatography (dual column)	–	–	113
Triazine herbicides, (atrazine, ametryne, propazine)	Gas chromatography—mass spectrometry	0.2–2 μg l^{-1}	–	114

Table 6.5. (*continued*)

Compound	Technique	Detection limit	Found	Ref.
Diazepham	Gas chromatography—chemical reaction interface mass spectrometry	–	–	115
	Miscellaneous methods			
Dichlorvos, methoate, oxamyl, methomyl	Preconcentration carbon black—high-performance liquid chromatography	$0.01~\mu g\,l^{-1}$	–	116
Trifluralin, simazine, atrazine, propazur, diazinon, parathion methyl, alachlor, malathion, parathion, chlorpyrifos	Solid phase extraction	–	0	117
Organosulphur compounds	Surface acoustic wave sensors (SAWS)—pattern recognition	$0.5~mg\,m^3$	–	118
CH_3S, Me_2S, dimethyl disulphide, $(Me-S-S-Me)$	Gas chromatography with flame ionisation detection	0.6×10^{-3} $\mu g\,l^{-1}$ (as S), (CH_3SH)	–	119
Thiabendazole	Solid phase spectro-fluorimetry	$0.1~\mu g\,l^{-1}$	–	120
Organophosphorus compounds	Surface acoustic wave sensors (SAWS)—pattern recognition	$0.01~mg\,m^3$	–	118
	Selective adsorption on chemiresistant sensors	–	–	121
	Detection with supported copper and cuprous oxide island film ac and dc methods	$\mu g\,l^{-1}$ level	–	122, 123
Carboxylic acids, Palmitate, 1-pyrenyl palmitate, lactate, propionate, formate	High-performance liquid chromatography after fluorescent labelling with 1-pyrenyl diazomethane	20–30 fmole	–	124
ascorbate	Derivativisation—high performance liquid chromatography	–	–	125
formate	Coupled enzymatic high-performance liquid chromatography	$0.5~\mu M$ absolute	0.4–10 μm	126

continued overleaf

Table 6.5. (*continued*)

Compound	Technique	Detection limit	Found	Ref.
glycollate, lactate, propionate, acetoacetate, α-hydroxy butryate, chloroacetate, isobutyrate, butyrate	High-performance liquid chromatography of ruthenium-111 complexes	–	–	127
	Ion exclusion chromatography			
Tartrate, maleate, malonate, citrate, glycollate, fumarate, formate	Ion exclusion chromatography	–	–	128
	Ion chromatography			
Glyoxylate, malonate	Ion chromatography with conductiometric detection	–	–	129
Acetate, lactate, propionate, butyrate, isobutyrate	Ion chromatography on mixed-mode stationary phase	–	–	130
Benzoate, chlorobenzoate, anisates, dimethylbenzoate, salicylate, hydroxybenzoate, resorcylate, photocatechuate, gallate	Ion chromatography	–	–	131
Malonate, hexabenzene carboxylate, octanoate, octanedioate	Isotope dilution gas chromatography—Fourier transform infrared spectroscopy of methyl esters	–	–	132
Chloroligno-sulphamic acid	Pysolysis-gas chromatography—mass spectrometry	16–32 μg l^{-1}	18–310 μg l^{-1} (Rivers) 60 μg l^{-1} (potable water)	133
Acetate	Potentiometric gas sensing probe	–	–	134
Organics	Spray extraction then gas chromatography—mass spectrometry	0.01–0.03 μg l^{-1}	–	135
Dissolved organic carbon	Flow method	–	–	136
	Nitrogen compounds			
Chloroaniline	Liquid chromatography	0.1 μg l^{-1}	–	137

Table 6.5. (*continued*)

Compound	Technique	Detection limit	Found	Ref.
Nitriloacetate, ethylene diaminetetra acetate	Liquid chromatography with amperometric detector	100 μg l^{-1} (NTA) 150 μg l^{-1} (EDTA)	–	138
N-nitrosodimethyl amine	Proton NMR spectroscopy	510 μg l^{-1}	–	139
Nitrogen compounds	Purging kinetics	200 μg l^{-1}	–	140
Pentachlorophenol	Membrane liquid chromatography	3 μg l^{-1}	–	141
Phthalate esters	Micellar electrokinetic chromatography	–	0	142
Trace organics	Membrane gas chromatography	–	–	143
Vinyl chloride, benzene, carbon tetrachloride	Helium-purged hollow fibre membrane mass spectrometry interface	sub-μg l^{-1}	–	144
Benzothiazole, naphtholnols, 2-chloroethanol phosphate, 3-methyl 3-octanol, diazinon, chlorfenvinphos	Gas chromatography—mass spectrometry	–	–	145
Miscellaneous organics	Gas extraction with single hollow fibre membrane and gas chromatographic cryotrapping	–	–	146

Liquid chromatography and high-performance liquid chromatography	*Detection limit* (μg l^{-1} *unless otherwise stated*)
Polychlorinated dibenzo-*p*-dioxins	0.003–0.7 ng l^{-1}
Carbamate insecticides, dichlorvos, methoate, oxamyl, methomyl	ng l^{-1}
Carboxylic acids	0.01
Linear alkyl benzene sulphonates Chloroaniline	20–30f mole 0.1

Liquid chromatography and high-performance liquid chromatography		Detection limit ($\mu g\,l^{-1}$ unless otherwise stated)
With mass spectrometric detector	Linear primary alcohol etholxylates	
	Linear alkyl benzene sulphonates	0.1
	Dialkyl tetralin sulphonates	0.1
	Acidic, basic and neutral pesticides	
	Chlorinated phenoxyl and ester herbicides	10^3
With ion spray mass spectrometric detector	Alkyl sulphates	
	Alkyl ethoxy sulphates	
With conductiometric detector	Polyoxyethylenes	
With amperometric detector	Nitrilo acetic acid	
	Ethylenediamine tetracetate	100
Ion exclusion chromatography		
	Carboxylic acids	
Ion chromatography		
	Carboxylic acids	
Supercritical fluid chromatography		
	Polynuclear hydrocarbons	
Gas chromatography		
	Polychlorinated biphenyls	
	Trichloroethane	0.4
	p-dibromobenzene	20 pg absolute
	haloforms, chloroform	3 pg absolute
	$CHCl_3$, CCl_4, $CHBrCl_2$, $CHBr_2\,Cl$, $CHBr_3$, CF_2Br_2, CH_2Br_2, $CFCl_3$, CHI_3, CF_2Br_2, $CFCl_3$, CH_2FCClF_2, C_2H_5I, CH_3CCl_3, C_2HCl_3, CH_2ClI, CCl_3Br	0.9–2.6
With electron capture detector	CH_3I, C_2H_5I,	0.01

Liquid chromatography and high-performance liquid chromatography		*Detection limit ($\mu g \, l^{-1}$ unless otherwise stated)*
With photodetachment modulated electron capture detector	Chlorinated pesticides and polychlorinated biphenyls	
With flame ionisation detector	Methyl sulphides, dimethyl disulphide	0.6×10^3
With mass spectrometric detector	Atrazine,	200–500 ppquadrillion
	triazine herbicides, diazepham,	0.2–2
	Misc. organics Benzothiazole Naphtholnols 2-chloroethyl phosphate 3-methyl-3 octanol Diazinon chlorfenvinphos	0.01–0.03

Pyrolysis gas chromatography

	Carboxylic acids	16–32

Isotope detection gas chromatography

	Carboxylic acids	

Purge and trap gas chromatography

With electron capture detector	$CHCl_3$, $BrCl_2CH_2BrCH_2Cl$, $CHBr_3$	0.000 01– 0.000 05
	Organonitrogen compounds	200

Spray and trap gas chromatography

With mass spectrometric detector	Anthracene Chloronaphthalene	

Mass spectrometry

Membrane mass spectrometry	Chloroaromatic compounds Haloforms Toluene Pentachlorophenol Trace organics	3
Fast atom bombardment mass spectrometry	Cationic detergents Atrazine	0.001

Liquid chromatography and high-performance liquid chromatography		*Detection limit (μg l^{-1} unless otherwise stated)*
Negative ion mass spectrometry	Polychlorinated dibenzo-p-dioxins	
High-resolution mass spectrometry	Polychlorinated dibenzo-p-dioxins	
Ion trap mass spectrometry	Insecticides	10–30 pg absolute
Surface acoustic wave sensors (SAWS)		
	Organosulphur compounds	0.5 mg m^3
	Organophosphorus compounds	0.01 mg m^3
Proton nuclear magnetic resonance spectroscopy		
	N-nitrosodimethylamine	500 μg L^{-1}

6.2   ORGANOMETALLIC COMPOUNDS: FRESHWATERS

6.2.1   TOXIC EFFECTS ON ORGANISMS

A limited amount of work has been carried out on the adverse effect of various types of organometallic compounds in non-saline waters on fish and on organisms other than fish (see also Section 4.3).

Fish

Organomercury compounds

Adult and 6-month-old teleost fish (*Channa punctatus*) were exposed to the organomercury fungicide Emison (methoxy ethyl mercuric chloride). Examination of the fish after 6 months exposure revealed liver abnormalities including hyperplasir and fatty necrosis indicative of carcinogenesis. Severe physio-metabolic dysfunction would lead to mortalities in teleost fish[147].

The toxic effect has been examined of methylmercuric chloride, methoxyethyl mercuric chloride and mercuric chloride on the survival of the catfish, *Clarias batrachus L. LC$_{50}$* of 0.43, 4.3 and 0.507 mg l^{-1} were obtained. Kidney damage was evident in exposed specimens[148].

Organotin compounds

A 24-hour *LC$_{50}$* value of 1.3 μg l^{-1} has been reported for adult rainbow trout[149]. The concentrations of tributyl tin found in surface micro layers of natural waters were in the range 1.9–473 μg l^{-1}. Consequently surface swimming rainbow trout could be at risk.

Organisms other than fish

Organomercury compounds

Microtubes were unaffected upon exposure to 1 mg l^{-1} of methylmercury for 1–24 hours, and severely disrupted upon exposure to 6 mg l^{-1} of methylmercury for 1–24 hours[150].

Organotin compounds

Exposure of adult fiddler crabs (*Uca pugilator*) to tributyltin concentrations as low as 0.5 μg l^{-1} retarded limb regeneration and of ecolysis and produced morphological abnormalities in regenerated limbs[151].

Roberts[152] has reported on acute toxicity of tributyltin to embryos and larvae of bivalve molluscs (*Crassostrea virginica* and *Mercenaria mercenaria*). Forty-eight-hour LC_{50} values of 1.30 and 3.96 μg l^{-1} were obtained in *C. virginica* embryos and straight-hinge stage larvae, respectively, and 1.13 and 1.65 μg l^{-1} in *M. mercenaria* embryos and larvae respectively. The 24-hour LC_{50} values for both species were greater than 1.3 μg l^{-1} in embryos and 4.2 μg l^{-1} in larvae. Evidence suggested that below the LC_{50} value tributyltin causes delayed clam embryo development. Tributyltin concentrations above 0.77 μg l^{-1} in the water caused abnormal shell development.

6.2.2 ANALYTICAL COMPOSITION

In Table 6.6 are quoted some typical levels of organometallic compounds of tin, lead and mercury found in river and surface water and rain. The high levels of

Table 6.6. Organometallic compounds in freshwaters and rain

Compound	Type of sample	Concentration (μg l^{-1})	Ref.
Sn1V$_{3+}$	River	0.005–0.57	245
MeSn		0.001–0.04	
Me$_2$Sn^{2+}		0.007–0.005	
Me$_3$Sn$^+$		0.0006–0.004	
Total Sn		0.005–0.58	
Me$_3$Sn$^+$	Rain	0.006	246
Bu$_2$Sn^{2+} }	Rain	< 0.001	247
Bu$_3$Sn$^+$ }	Switzerland	< 0.001	
BuSn^{3+} }	Rivers	0.035–0.050	247
BuSn^{2+} }	Switzerland	0.010–0.040	
BuSn^{3+} }		0.005–0.015	
MeHg	River Waal	0.31–1.15	248
MeHg	River	0.0059–0.012	249
MeHg	Rain	0.009	249
PbEt$_4$	Surface water	50–530	250

methylmercury in the polluted River Waal and of tetraethyl lead originating from
gasoline are notable.

6.2.3   ANALYTICAL METHODS

Methods used for the determination of various types of organometallic compounds in non-saline waters and the detection limits achievable are summarized
in Table 6.7. These methods are extremely sensitive and are capable of determining organometallic compounds in water at levels of environmental concern.

High-performance liquid chromatography		*Detection limit*
With graphite furnace atomic absorption detector	Organotin compounds	0.5 μg Sn absolute
Using thermochemical hydride generation and atomic absorption spectrometric detector	Arsenobetaine Arsenocholine Trimethylarsonium cation	13.3 ng absolute 14.5 ng absolute 7.8 ng absolute
With electrochemical detection	Selenols Diselinides Selenyl sulphides	
Supercritical fluid chromatography With inductively coupled plasma mass spectrometric detector	Organotin compounds	0.04 pg as SnBu$_4$, 0.047 pg as SnPh$_4$
Gas chromatography With atomic absorption spectro- metric detector	Et$_2$Pb^{2+} Et$_3$Pb^{3+}	fg g^{-1} range
With microwave-induced plasma emission spectrometric detector	Bu$_3$SnCl Bu$_2$SnCl$_2$ BuSnCl$_3$	0.05 pg as Sn absolute
With flame photometric detector	Methyl-cyclo-pentadienyl manganese tricarbonyl	
Flow injection analysis With atomic absorption spectrometric detector	Organomercury compounds	0.01 μg l^{-1}

*High-performance liquid
chromatography*   *Detection limit*

Isotope dilution mass spectrometry

Trimethyl selenonium   200 pg g^{-1} to
15 ng g^{-1}

Nuclear magnetic resonance spectroscopy of Hg^{199}

Methylmercury
compounds

6.3 ORGANIC COMPOUNDS: SEAWATER

6.3.1 TOXIC EFFECTS ON ORGANISMS

Available toxicity data for organisms in seawater are reviewed in Tables 6.8 (fish) and Table 6.9 (creatures other than fish) (see also Section 4.2).

6.3.2 ANALYTICAL COMPOSITION

Due to the diluting effect, much lower concentrations of organics are to be expected in seawater. This is borne out by comparing the total haloform content of river water (up to 13.4 μg l^{-1}) (Table 6.4) with that in seawater (0.119 μg l^{-1}, Table 6.10). An exception is, of course, the naturally occurring amino acids found in seawater where concentrations for total combined amino acids of up to 1350 μg l^{-1} have been found in the North Sea and up to 120 μg l^{-1} in the open ocean.

Total organic carbon levels in seawater are a reflection of the total amount of carbon present originating from natural and polluting sources. Levels range from about 500 to 3000 μg l^{-1}, of which only a negligible proportion is particulate or volatile (Table 6.11).

6.3.3 ANALYTICAL METHODS

The number of organic compounds for which methods of analysis in seawater are available are far fewer than in the case of non-saline waters. Classes of environmentally important compounds for which methods are available for the analysis at the ng l^{-1} level or less include polychlorinated biphenyls, poly-chlorinated-benzo-*p*-dioxins, polychlorinated dibenzofurans, organosulphur compounds and organic amines and carboxylic acids (Table 6.12).

Table 6.7. Methods for the determination of organometallic compounds in non-saline waters

Compound	Technique	Detection limit	Found	Ref.
Organomercury compounds				
Organomercury	Flow injection atomic absorption spectrometry	0.01 μg l^{-1}	—	153
Methylmercury complexes, e.g. CH$_3$Hg-11-thiol complexes	NMR of Hg-199	—	—	154
Organolead compounds				
Et$_3$Pb$^+$	Gas chromatography (capillary column)— atomic absorption spectrometry	fg g^{-1}	0.02–0.25 pg g^{-1} EtPb$^+$, 0.02–0.5 pg g^{-1}	155
Et$_2$Pb^{2+}			Et$_2$Pb^{2+} found in Greenland snow	
Organotin compounds	Supercritical fluid chromatography with indirectly coupled plasma mass spectrometric detector	0.04 pg (SnBu$_4$) 0.047 pg (SnPh$_4$) absolute	—	156
	High-performance liquid chromatography with graphite furnace atomic absorption detector	0.5 μg absolute as Sn	—	157
Tributytin	Capillary gas chromatography coupled to helium microwave-induced plasma emission spectrometric detector	0.05 pg absolute (as Sn)	—	158
Dibutyltin				
Monobutyltin				
Organoarsenic compounds				
Arsenobetaine	Liquid chromatography – thermochemical hydride generation atomic absorption spectrometry	13.3 ng (AB) 14.5 ng (AC) 7.8 ng (TMA)	—	159
Arsenocholine				
Tetramethyl arsonium cation				
Organoselenium compounds				
Selenols, Diselenides, selenyl sulphides	Reverse phase high-performance liquid chromatography with electrochemical detection	—	—	160
Trimethyl-selenonium	Isotope dilution mass spectrometry	200 pg g^{-1} to 15 ng g^{-1}	—	161
Organogermanium compounds				
Methylated germanium	Separation from inorganic germanium by solvent extraction	—	—	162
Organomanganese compounds				
methylcyclopenta-dienyl manganese-tricarbonyl	Gas chromatography with flame photometric detector	—	—	163

Table 6.8. Concentration of organic substances in seawater and effect on fish

Substance	Fish type	Water type	Exposure pH	Toxicity concentration	index	Adverse effects	Ref.
Hydrocarbon oils	*Myagropsis myagroides* Fendsholt	Sea	–	$10-10\,000$ $mg\,l^{-1}$	–	Allometric growth inhibited by $10\,000\,mg\,l^{-1}$ oil or $10-10\,000\,mg\,l^{-1}$ dispersant and $10-10\,000\,mg\,l^{-1}$ oil plus dispersant	164
	Pink salmon (*Oncorhychus gorbuscha*)	Sea	–	$0.7-2.4\,mg\,l^{-1}$	–	Exposure to $0.7-2.4$ $mg\,l^{-1}$ oil for 30 days reduced yolk reserves	165
	Plaice	Sea in vicinity of *Amoco Cadiz* spill	–	–	–	Polyaromatics in sediments $100\,mg\,kg^{-1}$; adverse effect on reproduction	166
	Baltic herring (*Clupea horengus*)	Sea	–	$3-200\ \mu g\,l^{-1}$	–	$180-200\ \mu g\,l^{-1}$ hydrocarbon toxic to developing fish	167
Phthalate esters	Atlantic cod (*Gadus morhua*)	Sea	–	$7-25\ \mu g\,l^{-1}$	–	$7-25\ \mu g\,l^{-1}$ phthalate esters toxic to developing fish	167
Hexazinone	Juvenile Pacific salmonids	Sea	–	–	96 h LC_{50} hexazinone = $276\,mg\,l^{-1}$ Pronone 109 = $904\,mg\,l^{-1}$ Velpar L = $1686\,mg\,l^{-1}$ Pronone carrier = $4330\,mg\,l^{-1}$ Velpar carrier = $20\,000\,mg\,l^{-1}$	–	168

continued overleaf

Table 6.8. (*continued*)

Substance	Fish type	Water type	pH	Exposure concentration	Toxicity index	Adverse effects	Ref.
3-trifluoromethyl-4-nitrophenol lampricide	Walleye (*Stizostedium vitreum*)	Sea	–	–	12 h LC_{25} 4.1 mg l^{-1} (gametes) 2.6 mg l^{-1} (eggs)	Adverse effects on eggs and fry	169
	Larval sea lampreys (*Petromyzon marinus*)	Sea	–	–	8 h $LC_{99.9}$ 1 mg l^{-1}		
Roundup herbicide	Coho salmon smolts (*Oncorhynchus kisutch*)	Sea	–	0.03–2.78 mg l^{-1}	–	10 days' exposure to 0.03–2.78 mg l^{-1} Roundup had no effect on survival or growth	170
Garlon 3A, Garlon 4A, Trichloropyrester, 3,5,6-trichloro-2-pyridinol, 2-methoxy 3,5,6-trichlorophyridine	Juvenile Pacific salmonids	Sea	–		24 h to 96 h LC_{50} reported	Garlon 4 and 3,5,6 trichloro-2-pyridinol highly toxic	171

Table 6.9. Concentration of organic substances in seawater and effect on organisms other than fish

Substance	Organism	Water type	pH	Exposure	Toxicity index	Adverse effects	Ref.
Hydrocarbon oil	Mussel (*Mytilus edulis*) Periwinkle (*Littorina littoreah*)	Sea	–	30–129 $\mu g\,l^{-1}$	–	*Mytilis edulis*, 4–5 months' exposure, investigation of mixed function oxidase system and its recovery during subsequent 2–9 months in absence of oil	172
Diesel oil	Mussel (*Mytilus edulis*)	Sea	–	27.4–127.7 $\mu g\,l^{-1}$	–	6- and 9-month tests exposure storage, comparison of survival rates in high and low nutrient seawater	173
Oil dispersants	Mussel (*Mytilus edulis*)	Sea	–	2–2.5 $\mu g\,l^{-1}$ oil ctg 5% dispersants	–	Growth rate reduced 80–90% in 170 h exposure	174
Hydrocarbon oil	Marine bivalve (*Venus verrucosa*)	Sea	7.2–7.8	–	–	Reduced pumping activity of lateral cilia. Mucus production increased	175
Phenol Fuel oil Kerosene Solar oil	*Crustaceans* (1) Shrimps (*Palaemon elegans*) (*Palaemon adsperus*) (2) Crab (*Rhithropano-peusharrisi tridentatus*) (3) Amphipod (*Pontogammarus maeoticus*)	Sea	–	–	–	Mean critical concentrations Phenol, 0.001 mg l^{-1} Fuel oil, 0.4 mg l^{-1} Kerosene, 0.01 mg l^{-1} Solar oil, 0.01 mg l^{-1}	176

continued overleaf

Table 6.9. (*continued*)

Substance	Organism	Water type	pH	Exposure	Toxicity index	Adverse effects	Ref.
	(4) Crayfish (*Balanus improvisus*)						
Polychloro-biphenyls	Common seal	Sea	–	–	–	Study of effect of PCB on seal reproduction	177
Diflubenzuron	Larval horseshoe crabs (*Limulus polyphemus*)	Sea	–	–	–	50 μg l^{-1} severe mortalities 5 μg l^{-1} slight delay in moulting	178
Endosulphan	Crab (*Oziotelphusa senex senex*)	Sea	–	0.2–18.62 mg l^{-1}	96 h LC_{50} (sublethal) = 6.2 mg l^{-1} (lethal) 18.62 mg l^{-1}	Increase in body weight, haemolymph volume and hydration level decrease in above	179 180
Kepone	Lamprey (*Petromyzon marinus*)	Sea	–	–	96 h LC_{50} 444 μg l^{-1} (12 °C) 414 μg l^{-1} (20 °C) incipient lethal concentrations = 145 μg l^{-1}		181
Tetrachloro-1,2-benzoquinone (paper mill effluent constituent)	Fourhorn sculpin (*Myoxocephalus quadricornis*) also bleak (*Alburnus alburnus*) also Perch (*Perca fluviatilis*)	Gulf	–	0.1–0.5 mg l^{-1} (sublethal concentrations)	–	Skeletal abnormalities	182

Table 6.10. Organics in seawater

Substance	Sample	Concentration (μg l^{-1})	Ref.
CHCl$_3$	Seawater	0.026	186, 187
CH$_3$C Cl$_3$		0.046	186, 187
CCl$_4$		< 0.005	186, 187
CHClCCl$_2$		0.015	186, 187
CCl$_2$CCl$_2$		0.005	186, 187
CHBr$_3$		0.027	186,187
Total haloforms		0.119	
2.6-Dinitro toluene	Dobkai Bay, Japan	< 0.02–2.1	188, 189
2.4-Dinitro toluene	Dobkai Bay, Japan	0.13–28.3	188, 189
Azarenes	Dobkai Bay, Japan		190
Quinoline		0.022	
Iso-quinoline		0.013	
2-methyl quinoline		0.046	
1-methylisoquinoline		0.043	
6-methylquinoline		0.004	
4-methylquinoline		0.003	
2,6-dimethylquinoline		0.016	
2,4-dimethylquinoline		0.055	
4-azafluorene		0.006	
Benzo (b) quinoline		< 0.0001	
Acridine		0.009	
Phenanthridine or benzo (b) quinoline		0.002	
10-azabenzo (a) pyrene		< 0.0001	
Dibenz (c,b) acridine		0.0007	
Dibenz (a,b) acridine		0.003	
Dibenz (a,z) acridine		0.004	
Non-ionic detergents exposed as C$_{12}$H$_{25}$(C$_6$H$_4$O(C$_2$H$_4$O)$_6$H)	Trieste Harbour	39–216	191
Total free amino acids	Open ocean	0–180	192–197
	North Sea	20–180	
	Baltic Ocean	4.8–84.5	198
	Mediterranean	5–92	199
Total combined amino acids	Open ocean	3–130	200
	Open ocean	10.5–87.5	201
	Open ocean	10–120	202
	Mediterranean	28–200	200
	Baltic Ocean	500	199
	North Sea	35–1350	198

Table 6.11. Organic carbon in seawater

	(μg l^{-1})	Ref.
(A) *Total organic carbon* *filtered seawater*	0.035–1.22	263
	0.57–1.74	204
	1.49–3.08	68
	0.74–2.44	205
	0.13–1.63	206
Scotian Shelf (filtered)	*Depth (m)* 0 0.75–1.07	206
	10 0.81–0.93	
	25 0.73–1.14	
	50 0.62–0.97	
	100 0.62–0.85	
	150 0.71–0.81	
	200 0.56–0.76	
	400 0.51–0.72	
	500 0.53–1.64	
	0–500 0.77–0.88	
Halifax Harbour (Filtered)	1 1.04–1.29	206
	10 1.03–1.32	
(Unfiltered)	1 1.27–1.90	
	10 1.18–1.41	
Coastal area (Filtered)	1.12–1.31	206
(Unfiltered)	1.38–1.68	
Sargossa Sea	0.85	59
Vineyard Sound	1.07	
Santa Cruz	0.99	
Norwegian fiord	1.00	
(B) *Particulate organic carbon*		
Surface seawater	0.025–0.2	
Deep seawater	0.003–0.015	
(C) *Volatile organic carbon* Gulf of St Lawrence	*Depth (m)*	
	0–10 0.032	70
	10–50 0.036	
	50–100 0.030	
	100–250 0.030	
Scotian Shelf	0–10 0.041	70
	10–25 0.038	
	100–250 0.035	
	250–750 0.033	
	750–1500 0.026	

Table 6.11. (*continued*)

	$(\mu g\, l^{-1})$		Ref.
Central and north-west Atlantic	0–10	0.030	70
	10–25	0.028	
	25–100	0.032	
	100–250	0.028	
	250–750	0.025	
	750–1500	0.026	
	1500–3000	0.024	
	3000–5000	0.026	
St Margaret's Bay, Nova Scotia	0–40	0.031	70
Halifax Harbour	0–10	0.033	70

Table 6.12. Methods for the determination of organic compounds in seawater

Compound	Technique	Detection limit	Found	Ref.
Polychlorinated biphenyls, polychlorinated dibenzo-*p*-dioxins, dibenzofurans	Mass spectrometry	–	–	207
Methyl mercaptan, dimethyl disulphide	Gas chromatography with flame ionisation detector	$0.6\ ng\, l^{-1}$ (CH_3SH)	–	208
Amines (methylamine), dimethylamine trimethylamine, and organic acids (acetate, propionate, butyrate, valerate, pyruvate, acrylate, benzoate)	Diffusion across hydrophilic membrane to concentrate and remove inorganic salts	–	–	209
Formate	Coupled enzymatic high-performance liquid chromatography	$0.5\ \mu M$ absolute	$0.2–0.8\ \mu m$	210

6.4 ORGANOMETALLIC COMPOUNDS: SEAWATER

6.4.1 TOXIC EFFECTS ON ORGANISMS

Zischke and Arthur[183] have determined 96-hour LC_{50} values of tributyltin compounds to mysids (*Mysidopsis bahia*). The age of the fish was an important factor in determining the sensitivity of juveniles to tributyltin compounds.

In chronic toxicity tests[184] carried out in the Chesapeake bay area on biota exposed to tributyltin the survival of *Gammarus SP* was unaffected by 24-hour exposure to concentrations up to 0.58 $\mu g\, l^{-1}$, although body weight was reduced by 64% relative to controls. Survival of *Brevoortia tyrannus* and larval *Menidia beryllina* was unaffected by 28 days' exposure to concentrations of tributyltin up to 0.49 $\mu g\,l^{-1}$. Growth was reduced by 20–22% following exposure to 0.09 or 0.49 $\mu g\,l^{-1}$ tributyltin Noth and Kumar[185] have studied the effect of tributyltin containing paints over 13 months' exposure to the oyster *Crassostrea gigas*. Oyster weight, length and width were adversely affected. Embryonic and larval viability were unaffected. The toxicity of organometallic compounds on seawater is also discussed in Section 4.3.

6.4.2   ANALYTICAL COMPOSITION

In Table 6.13 is listed the available information of the presence of organometallic compounds in seawater. Traces of organically bound arsenic are ubiquitous. Organotin compounds are found only in certain coastal areas where these compounds are used as antifoulants on boats and harbour works. Several governments (the UK, France and several states in the USA) have banned the use of organotin compounds in recreational craft, while the EU and Scandinavian countries are debating the issue[211].

6.4.3   ANALYTICAL METHODS

Methods for the determination of organometallic compounds in seawater are reviewed in Table 6.14. Detection limits range from $0.4 \times 10^{-8}\,\mu g\,l^{-1}$ (i.e. 0.004 $pg\,l^{-1}$) (organotin compounds) to 0.02–20 $\mu g\,l^{-1}$ (organomercury compounds) and 0.02–0.25 $\mu g\,l^{-1}$ (organoarsenic compounds). The major techniques employed are summarised below:

Atomic absorption spectrometry		*Detection limit ($\mu g\,l^{-1}$ unless otherwise stated)*
Cold vapour	Organo mercury	0.02
	Methylmercury	20
Conversion to hydride—flame atomic absorption spectrometry	Butyltin compounds	0.000 01
	Dimethyl arsinate	
	Monomethyl arsinate	
	Trimethyl arsine oxide	
Graphite furnace atomic absorption spectrometry	Butyltin compounds	0.000 01
	Dimethyl arsinate	0.02 (asAs)
Gas chromatography		
With flame photometric detector	Organomercury compounds	10 $\mu g\,kg^{-1}$
	Volatile organotin compounds	

Atomic absorption spectrometry		Detection limit (μg l^{-1} unless otherwise stated)
Reduction to hydride—gas chromatography with atomic absorption detector	Mono-, di- and trimethyl tin compounds	0.000 01 μg Sn absolute
Methylation then glc	mono-, di- and tributyl tin chlorides	
With mass spectrometric detector	Butyltin compounds	0.000 01
	Methyl arsenic compounds	0.25
Column chromatography Reduce to arsine, atomic absorption spectrometric detector	Organoarsenic compounds	
Anodic scanning voltammetry	Trimethyl lead compounds	
Spectrofluorimetric methods	Triphenyltin compounds	0.004×10^3
Nuclear magnetic resonance spectroscopy	Tributyl tin chlorides	0.4×10^{-8} to 2×10^{-8}

6.5   ORGANIC COMPOUNDS: COASTAL WATERS

6.5.1   TOXIC EFFECTS ON ORGANISMS

Trim[242] has discussed the results obtained in static 96-hour toxicity tests with malathion, endosulphan and fenvalerate on estuarine waters on the Mummichog, *Fundulas heteroclitus*. All three insecticides were highly toxic to estuarine and coastal water fish.

Eggs (late blastula) or 406-day larvae of striped bass (*Morone saxatilus*) or sheepshead minnow (*Cyprinodon variegatus*) when exposed to cyclophosphamide or N-methyl-N-nitro-N nitroguanidine of concentrations of 1–1000 μm for 1–4 days show a close dependent relationship between aberration frequency of chromosomes and the concentration of toxicant in eggs and larvae of both species[243]. Ram and Sathanesan[244] determined LC_{50} values obtained when mysid (*Mysidopsis bahia*), grass shrimps (*Palaemonetes pugio*), pink shrimps (*Penaeus duorarum*) and sheepshead minnow (*Cyprinodon variegatus*) were exposed to spray applications (336 g per hectare) of fenthion to water on an estuarine shore line. Mortalities and non-lethal effects occurred in these species[244]. The toxic effects on organisms in saline waters are also discussed in Section 4.2.

Table 6.13. Organometallic compounds in seawater

	Concentration (μg l^{-1})	Ref.
Mercury		
Me Hg in seawater	0.06	212
Arsenic		
Irish Sea	2.49–2.65	213
Tin		
Gulf of Mexico		147
SnIV	0.0022–0.062	
MeSn	< 0.00001–0.015	
Me$_2$Sn	0.00074–0.007	
Me$_3$Sn	< 0.00001–0.00098	
Total Sn	0.0036–0.085	
Old Tampa Bay		147
SnIV	< 0.0003–0.0027	
MeSn	0.00086–0.0011	
Me$_2$Sn	0.0006–0.002	
Me$_3$Sn	< 0.00001–0.00095	
Total Sn	0.0025–0.005	
Estuary		147
SnIV	0.0003–0.020	
MeSn	< 0.00001–0.008	
Me$_2$Sn	0.00079–0.0022	
Me$_3$Sn	< 0.00001–0.0011	
Total Sn	0.0025–0.023	
Harbour water		214
Me$_2$Sn	< 0.01–0.02	
Me$_3$Sn	< 0.01–0.02	
SnH$_4$	0.2–20	
Me$_4$Sn	< 0.01–0.3	
BuSnH$_3$	0.05–0.3	
Surface water		215
SnIV	0.001–0.009	
BuSnH$_3$	0.01–0.06	
Bu$_2$SnH$_2$	0.13–0.46	
Bu$_3$SnH	0.06–0.78	
Bottom water		215
SnIV	0.003–0.005	
BuSnH$_3$	0.03–0.04	
Bu$_2$SnH$_2$	0.13	
Bu$_3$SnH	0.01–0.10	
Estuary water		216
Bu$_3$Sn	0.08–0.19	
Bay samples		147
SnIV	0.003–0.02	
MeSn	0.0007–0.008	
Me$_2$Sn	0.0008–0.002	
Me$_3$Sn	0.0003–0.001	
Total Sn	0.0002–0.023	

Table 6.13. (*continued*)

	Concentration (μg l^{-1})	Ref.
Lake Michigan, adjacent to coast		147
SnIV	0.08–0.49	
MeSnCl$_3$	0.006–0.0013	
Me$_2$SnCl$_2$	< 0.0001–0.063	
BuSnCl$_3$	0.002–1.22	
Bu$_2$SnCl$_2$	0.01–1.6	
San Diego Bay, surface water		147
SnIV	0.006–0.038	
MeSnCl$_3$	0.0002–0.0008	
Me$_2$SnCl$_2$	0.015–0.045	
BuSnCl$_3$	< 0.0001	
Bu$_2$SnCl$_2$	< 0.0001	
San Francisco Bay		147
SnIV	0.0002–0.0003	
MeSnCl$_3$	< 0.0001	
Me$_2$SnCl$_2$	< 0.0001	
BuSnCl$_3$	< 0.0001	
Bu$_2$SnCl$_2$	< 0.0001	
Coast adjacent to San Francisco		147
SnIV	0.0003–0.0008	
MeSnCl$_3$	< 0.0001	
Me$_2$SnCl$_2$	< 0.0001	
BnSnCl$_3$	< 0.0001	
Bu$_2$SnCl$_2$	< 0.0001	

6.5.2   ANALYTICAL COMPOSITION

See Section 6.3.2. and Table 6.10.

6.5.3   ANALYTICAL METHODS

See Section 6.3.3 and Table 6.12.

6.6   ORGANOMETALLIC COMPOUNDS: COASTAL WATERS

6.6.1   TOXIC EFFECTS ON ORGANISMS

See Sections 6.4.1 and 4.3.

Table 6.14. Methods for the determination of organometallic compounds in seawater

Compound	Technique	Detection limit (μg l^{-1} unless otherwise stated)	Ref.
Organomercury compounds			
Methylmercury	Cold vapour atomic absorption spectrometry	0.02	217
	Ultraviolet digestion—cold vapour atomic absorption spectrometry	–	218
Methylmercury	Benzene extraction gold amalgamation—atomic absorption spectrometry	20	220
Methylmercury	Dithizone-carbon tetrachloride extraction, reduction to mercury then atomic absorption spectrometry	–	221, 222
Methylmercury	Collect mercury on mussels, then solvent extraction gas chromatography	10	219
Organolead compounds			
Trimethyl lead	Anodic stripping voltrammetry	–	222
Organic compounds			
Mono-, di- and trimethyltin compounds	Reduce to CH_3SnH_3 $(CH_3)SnH_2$ and $(CH_3)_3SnH$ with sodium borohydride, separation by gas chromatography with atomic absorption detector	0.000 01 μg Sn absolute	223
Butyltin chlorides	Atomic absorption spectrometry	0.4 ng absolute	224
Butyltin compounds			
Incl. tri-*n*-butyltin, bis(tri-*n*-butyltin) oxide	Methyl isobutyl ketone extraction—graphite furnace atomic absorption spectrometry and gas chromatography with mass spectrometric detection or hydride reduction—flame atomic absorption spectrometry	0.000 01	225, 226
Tributyltin trichloride, dibutyltin dichloride, tributyltin chloride	Methylation then gas chromatography or organic solvent tropalone extract of water	–	227
Volatile organotin compounds	Collection organotin on Tenax GV Purge column on to gas chromatograph with flame photometric detection	–	228

Table 6.14. (*continued*)

Compound	Technique	Detection limit (μg l^{-1} unless otherwise stated)	Ref.
Tributyltin	Chloroform extraction, nuclear magnetic resonance spectroscopy	$0.4 \times 10^8 - 2 \times 10^{-8}$	229
Triphenyltin	Toluene extraction spectrofluorometric method with 3-hydroxy flavone	0.004×10^3	230
Organoarsenic compounds			
Dimethyl arsenate, monomethyl arsenate, trimethyl arsenate oxide	Convert to hydrides (NaBH$_4$)— cold trap vaporise slowly by warming, absorption spectrometry	4	231–238
Dimethyl arsenate	Preconcentration on cation exchange resin (Dowex AG50.W-XB) then graphite furnace atomic absorption spectrometry	0.02 (as As)	239
Organoarsenic compounds	Sample photolysed by ultraviolet light, organic arsenic extracted with dipyrolidone dithio carbonate in chloroform then atomic absorption spectrometry	0.14	240
Methyl arsenic compounds	Extracted from sample with cold toluene, then gas chromatography with mass spectrometric detector	0.25	232, 234
Organoarsenic compounds	Organoarsenic compounds separated from water by column chromatography, reduced to arsines and analysed by atomic absorption spectrometry	–	241

6.6.2 ANALYTICAL COMPOSITION

See Sections 6.4.2 (Table 6.13) and 4.3.

6.6.3 ANALYTICAL METHODS

See Section 6.4.3 and Table 6.14.

REFERENCES

1. Dill, D.C., Murphy, P.G. and Mayes, M.A., *Bulletin of Environmental Contamination and Toxicology* **39**, 869 (1987)
2. Webster, P.W., Cunton, J.H. and Dormans, J.A.M.A., *Aquatic Toxicology* **13**, 323 (1988)
3. Wams, T.J., *Science of the Total Environment* **66**, 1 (1987)
4. Russom, C.L., Drummond, R.A. and Hoffman, A.D., *Bulletin of Environmental Contamination and Toxicology* **41**, 589 (1988)
5. Carlson, A.R., *Bulletin of Environmental Contamination and Toxicology* **38**, 667 (1987)
6. Smith, P.D., Brockway, D.L. and Stancil, F.E., *Environmental Toxicology and Chemistry* **6**, 891 (1987)
7. Oikara, A. and Kulkonen, J., *Ecotoxicology and Environmental Safety* **15**, 282 (1988)
8. McKim, J.M., Schmeider, P.K., Carlson, R.W., Hunt, E.P. and Niemi, G.J., *Environmental Toxicology and Chemistry* **6**, 295 (1987)
9. Cleland, G.B., McElroy, P.N. and Sangegard, R.A., *Aquatic Toxicology* **12**, 141 (1988)
10. Call, D.J., Poiriee, S.H., Knuth, M.L., Harting, S.L. and Lindberg, C.A., *Bulletin of Environmental Contamination and Toxicology* **38**, 352 (1987)
11. Tripathi, G. and Shukla, S.P., *Ecotoxicology and Environmental Safety* **15**, 277 (1988)
12. McKim, J. M., Schmeider, P.K., Niemi, J., Carlson, R.W. and Henry, T.R., *Environmental Toxicology and Chemistry* **6**, 313 (1987)
13. Khangarrot, B.S. and Ray, P.K., *Archiv für Hydrobiologie* **113**, 465 (1988)
14. Flores-Nava, A. and Vizcarra-Quiroz, J.J., *Aquaculture and Fisheries Management* **19**, 341 (1988)
15. Mitchell, D.G., Chapman, P.M. and Long, T.J., *Bulletin of Environmental Contamination and Toxicology* **39**, 1028 (1987)
16. Call, D.J., Brooke, L.T., Kent, P.J., Knuth, M.L., Poirier, S.H., Huot, J.M. and Lima, A.R., *Archives of Environmental Contamination and Toxicology* **16**, 607 (1987)
17. Lossarni-Dunier, M., Monod, G., Demael, A. and Lepot, D., *Ecotoxicology and Environmental Safety* **13**, 339 (1987)
18. Ferrando, M.D., Almar, M.M. and Andreu, E., *Journal of Environmental Science and Health* **B23**, 45 (1988)
19. Henning, T.A., McGuiness, E.J., George, L.M. and Blumhagen, K.A., *Bulletin of Environmental Contamination and Toxicology* **40**, 764 (1988)
20. Palawski, D.U. and Knowles, C.O., *Environmental Toxicology and Chemistry* **5**, 1039 (1986)
21. Newsted, J.L. and Giesy, P., *Environmental Toxicology and Chemistry* **6**, 445 (1987)
22. Conti, E., *Aquatic Toxicology* **10**, 325 (1987)
23. Knowles, C.O., McKee, M.J. and Palawski, D.U., *Environmental Toxicology and Chemistry* **6**, 201 (1987)
24. Green, D.W., Williams, K.A., Hughes, D.R.L., Shaik, G.A.R. and Pascoe, D., *Water Research* **22**, 225 (1988)
25. Devilliers, J., *Science of the Total Environment* **76**, 79 (1988)
26. Gersich, F.M. and Milazzo, D.P., *Bulletin of Environmental Contamination and Toxicology* **40**, 1 (1988)
27. Herring, C.O., Adams, J.A., Wilson, B.A. and Pollard, S., *Bulletin of Environmental Contamination and Toxicology* **40**, 35 (1988)

28. Bridgham, S.D., *Archives of Environmental Contamination and Toxicology* **17**, 731 (1988)
29. Rosenbaum, E.A., Caballero de Castro, A., Guana, L. and Pelhen de D'Angelo, A.M., *Archives of Environmental Contamination and Toxicology* **17**, 831 (1988)
30. Saxena, P.K. and Mani, K., *Environmental Pollution* **55**, 97 (1988)
31. Singh, D.K. and Agarwal. R.A., *Science of the Total Environment* **67**, 263 (1987)
32. Ferrando, M.D., Andreumoliner, E., Almar, M.M., Cabrian, C. and Nunez, A., *Bulletin of Environmental Contamination and Toxicology* **39**, 365 (1987)
33. Upadhyay, O.V.B. and Shuka, G.S., *Environmental Research* **41**, 591 (1986)
34. Reddy, M.S. and Rae, K.V.R., *Bulletin of Environmental Contamination and Toxicology* **40**, 752 (1988)
35. Stephenson, M. and MacKie, G.L., *Aquatic Toxicology* **9**, 243 (1986)
36. Dumpert, K., *Ecotoxicology and Environmental Safety* **13**, 324 (1987)
37. Maas-Diepeveen, J.L. and Van Leeuwen, C.J., *Bulletin of Environmental Contamination and Toxicology* **40**, 517 (1988)
38. Kononen, D.W., *Bulletin of Environmental Contamination and Toxicology* **41**, 371 (1988)
39. Von Rensburg, V.F.F., Van Huyssteen, J.T. and Hassat, A.J., *Water Research* **12**, 127 (1978)
40. Kirschen, N.A., *Varian Instrument Applications* **14**, 10 (1980)
41. Dietz, E.A. and Singley, K.F., *Analytical Chemistry* **51**, 1809 (1979)
42. Fielding, M., McLoughlin, K. and Steel, C., Water Research Centre Enquiry Report ER532 August 1977, Water Research Centre, Stevenage Laboratory, Elder Way, Stevenage, Herts, UK (1977)
43. Hagenmaier, H., Werner, G. and Jager, W., *Zeitschrift für Wasser und Abwasser Forschung* **15**, 195 (1982)
44. Monarco, S., Causey, B.S. and Kirkbright, G.C., *Water Research* **13**, 503 (1979)
45. Acheson, M.A., Harrison, R.M., Perry, R. and Wellings, R.A., *Water Research* **10**, 207 (1976)
46. Stachel, B., Bactjer, K., Cetinskaja, M., Deuszelu, J., Lahl, U., Liese, K., Thiemann, W., Gabiel, W. and Kozicki, R., *Analytical Chemistry* **53**, 1469 (1981)
47. Suzuki, M., Yamoto, Y. and Wanaltabe, T., *Environmental Science and Technology* **11**, 1109 (1977)
48. Sackmauereva, M., Pal'usova, O. and Szokalay, A., *Water Research* **11**, 551 (1977)
49. Leoni, V., Pucetti, G. and Grella, A., *Journal of Chromatography* **106**, 119 (1975)
50. Aspila, I., Carron, J.M. and Chau, A.S.I., *J. Ass. Analytical Chemistry* **60**, 1097 (1977)
51. Bergemann, H. and Hellman, H., *Deutsch Gewasserkundliche Mitteilungen* **24**, 31 (1981)
52. McIntyre, A.E., Perry, R. and Lester, J.N., *Environmental Technology Letters* **1**, 157 (1980)
53. Le'Bel, G.L. and Williams, D.T., *Bulletin of Environmental Contamination and Toxicology* **132**, 277 (1977)
54. Ervin, H.E. and McGinnis, C.D., *Journal of Chromatography* **190**, 203 (1980)
55. Morgade, C., Barquit, H. and Ptaffenberger, C.D., *Bulletin of Environmental Contamination and Toxicology* **24**, 257 (1980)
56. Schouten, M.J., Copius Peereboom, J.W. and Brinkman, U.A.J., *International Journal of Environmental Analytical Chemistry* **7**, 13 (1979)
57. Schouten, M.J., Copius Peereboom, J.W., Brinkman, U.A.J., Schwart, H., Anzion, C.J.M. and Van Vleit, H.M., *International Journal of Environmental and Analytical Chemistry* **6**, 133 (1979)

58. Mori, S.J., *Journal of Chromatography* **129**, 53 (1976)
59. Ton, N. and Takahashi, Y., *International Laboratory*, September, 49 (1985)
60. Schouten, M.J., Copius Peereboom, J.W., Brinkman, U.A.J., Schwart, H., Anzion, C.J.M. and Van Vleit, H.M., *International Journal of Environmental Analytical Chemistry* **6**, 133 (1979)
61. Jones, P. and Nickless, G., *Journal of Chromatography* **156**, 99 (1978)
62. Uchiyama, M., *Water Research* **13**, 847 (1979)
63. Hun Nami, H. and Hanya, T., *Journal of Chromatography* **161**, 205 (1978)
64. Hughes, J.L., Eccles, L.A. and Malcolm, R.L., *Ground Water* **12**, 283 (1974)
65. Hullett, D.A. and Eisenreich, S.J., *Analytical Chemistry* **51**, 1953 (1979)
66. Aue, W.A., Hastings, C.R., Gerkardt, K.O., Pierce, J.D., Hill, H.M. and Marsemian, R.F., *Journal of Chromatography* **72**, 259 (1972)
67. Baker, C.D., Bartlett, P.D., Farr, I.S. and Williams, G.J., *Freshwater Biology* **4**, 467 (1974)
68. Goulden, P.D. and Brooksbank, P., *Analytical Chemistry* **47**, 1943 (1975)
69. Games, L.M. and Hayes, J.H., *Analytical Chemistry* **48**, 130 (1976)
70. Schwarzenbach, R.P., Bromund, R.H., Gschwend, P.M. and Zafiron, O.C., *Organic Geo. Chemistry* **1**, 93 (1978)
71. Kraubeck, H.J., Lampert, W. and Bredie, H., *Fachzeitschrift für das Laboratorium* **25**, 2009 (1981)
72. Pratt, K.F. and Pawliszyn, J., *Analytical Chemistry* **64**, 2101 (1992)
73. Yong, H.J. and Pawliszyn, J., *Analytical Chemistry* **65**, 1758 (1993)
74. Stetter, J.R. and Cao, Z., *Analytical Chemistry* **62**, 182 (1990)
75. Ryan, D.A., Argentine, S.M. and Rice, G.W., *Analytical Chemistry* **62**, 853 (1990)
76. La Pack, M.A, Ton, J.C. and Enke, C.G., *Analytical Chemistry* **62**, 1265 (1990)
77. Arbon, R.E. and Grimsudrud, E.P., *Analytical Chemistry* **62**, 1762 (1990)
78. Dingyuan, H. and Jainfei, T., *Analytical Chemistry* **63**, 2078 (1991)
79. Lepine, L. and Archaumbault, J.F., *Analytical Chemistry* **64**, 810 (1992)
80. Bognar, J.A., Knighton, W.B. and Grumsrud, E.P., *Analytical Chemistry* **64**, 2451 (1992)
81. Mock, R.S. and Grimsrud, E.P., *Analytical Chemistry* **60**, 1684 (1988)
82. Louch, D., Motlagh, S. and Paliszgn, J., *Analytical Chemistry* **64**, 1187 (1992)
83. Matz, G. and Kesners, P., *Analytical Chemistry* **65**, 2366 (1993)
84. Blyshak, L.A. Rossi, T.M., Patonen, G. and Warner, I.M., *Analytical Chemistry* **60**, 2127 (1988)
85. Langernfield, J.J., Hawthorne, S.B., Miller, D.J. and Pawliszyn, J., *Analytical Chemistry* **69**, 338 (1993)
86. Ventura, I., Fraisse, D., Cigixach, J. and Rivera, J., *Analytical Chemistry* **63**, 2095 (1991)
87. Evans, K.A., Pubey, S.T., Kravetz, L., Dzidic, I., Gumulka, J., Mueller, R. and Stork, J.R., *Analytical Chemistry* **66**, 699 (1994)
88. Okada, T., *Analytical Chemistry* **62**, 734 (1990)
89. Trehy, H.L., Gledhill, W.E. and Orth, R.G., *Analytical Chemistry* **62**, 2581 (1990)
90. Popenoe, D.D., Morris, S.J., Horn, P.S. and Norwood, K.T., *Analytical Chemistry* **66**, 1620 (1994)
92. Borgerding, A.J. and Hiles, R.A., *Analytical Chemistry* **64**, 1449 (1992)
93. Simms, J.P., Kevugh, T., Ward, S.R., Moore, B.L. and Bondurraga, M.M., *Analytical Chemistry* **60**, 2613 (1988)
94. Shaksher, Z.M. and Seitz, W.R., *Analytical Chemistry* **62**, 1758 (1990)
95. De Corcia, A., Marchett, M., Samperi, R. and Mariomini, A., *Analytical Chemistry* **63**, 1179 (1991)
96. Roblat, A., Xyrafas, G. and Marshall, D., *Analytical Chemistry* **60**, 982 (1988)

97. Hussun, H.N. and Jurs, P.C., *Analytical Chemistry* **60**, 978 (1988)
98. Aga, D.S., Thurman, E.M. and Pomes, H.L., *Analytical Chemistry* **66**, 1495 (1994)
99. Kimata, K., Hosoya, K., Araki, T., Tanaka, N., Barnhard, E.R., Alexander, L.R., Sirimanne, N., McClure, P.C., Grainger, J. and Patterson, D.G., *Analytical Chemistry* **65**, 2502 (1993)
100. Laramee, J.A., Arbogast, B.X. and Dienzer, H.L., *Analytical Chemistry* **60**, 1937 (1988)
101. Taguchi, V.V., Reiner, E.J., Wang, D.T., Meresz, O. and Hallas, B., *Analytical Chemistry* **60**, 1429 (1988)
102. Thieley, D.R. and Olsen, G., *Analytical Chemistry* **60**, 1332 (1988)
103. Bellar, T.A. and Budde, W.L., *Analytical Chemistry* **60**, 2076 (1988)
104. Fernandez Band, P., Linares, M.P., Luque de Castro, M.D. and Valcarcel, M., *Analytical Chemistry* **63**, 1672 (1991)
105. Cappiello, A., Famiglini, G. and Bruner, F., *Analytical Chemistry* **66**, 1416 (1994)
106. Di Corcia, A. and Marchetti, M., *Analytical Chemistry* **63**, 819 (1991)
107. Jones, A.O., *Analytical Chemistry* **63**, 580 (1991)
108. Marvin, C.H., Brindle, I.D., Hall, C.D. and Chiba, M., *Analytical Chemistry* **62**, 1495 (1990)
109. Rule, G.S., Mordelei, A.V. and Henion, J., *Analytical Chemistry* **66**, 230 (1994)
110. Lin Hung, Y. and Vyksner, R.D., *Analytical Chemistry* **65**, 451 (1993)
111. Caldwell, K.A., Ramannjam, V.M., Cai, Z. and Gross, M.L., *Analytical Chemistry* **65**, 2372 (1993)
112. Zangwei, Cai, Sodagopa, O., Ramanujan, N.M., Giblin, D. and Gross, M.L., *Analytical Chemistry* **65**, 21 (1993)
113. Jones, O., *Analytical Chemistry* **62**, 1667 (1990)
114. Thirman, E.M., Meyer, M., Pomes, M., Perry, C.A. and Schwab, A.B., *Analytical Chemistry* **62**, 2043 (1990)
115. Song, H. and Abramson, F.P., *Analytical Chemistry* **65**, 447 (1993)
116. Di Corcia, A., Samperi, R., Mariomini, A. and Stilluto, S., *Analytical Chemistry* **65**, 907 (1993)
117. Johnson, W.E., Fendinger, N.J. and Plimmer, J.R., *Analytical Chemistry* **63**, 1510 (1991)
118. Grate, J.W., Rose Pehrsson, S.L., Venezky, D.L., Klusty, M. and Wohitjen, H., *Analytical Chemistry* **65**, 1868 (1993)
119. Leek, C. and Bagander, L.E., *Analytical Chemistry* **60**, 1680 (1988)
120. Capitan, F., Alonso, E., Avidad, R., Capitan-Valley, L.F. and Vilchez, J.L., *Analytical Chemistry* **65**, 1336 (1993)
121. Grate, J.W., Klusty, M., Barger, W.R. and Snow, A.W., *Analytical Chemistry* **62**, 1927 (1990)
122. Kolesar, E.S. and Walser, R.M., *Analytical Chemistry* **60**, 1731 (1988)
123. Kolesar, E.S. and Walser, R.M., *Analytical Chemistry* **60**, 1737 (1988)
124. Nimura, N., Kinoshita, T., Yoshida, T., Uetake, A. and Nakai, C., *Analytical Chemistry* **60**, 2067 (1988)
125. Kishida, E., Nishimoto, Y. and Kojo, S., *Analytical Chemistry* **64**, 1505 (1992)
126. Kieher, D.J., Vaughan, G.M. and Mopper, K., *Analytical Chemistry* **60**, 1654 (1988)
127. Rijas, P.G. and Pietrzyk, D.J., *Analytical Chemistry* **60**, 1650 (1988)
128. Leek, C. and Bagander, L.E., *Analytical Chemistry* **61**, 1701 (1989)
129. Berglund, I., Dasgupta, P.K., Lopex, J.L. and Nara, O., *Analytical Chemistry* **65**, 1192 (1993)
130. Nordhaus, R. and Anderson, J.M., *Analytical Chemistry* **64**, 2283 (1992)
131. Hirajama, N. and Kuwamoto, T., *Analytical Chemistry* **65**, 141 (1993)
132. Olsen, E.S., Diehl, J.W. and Froelich, M.L., *Analytical Chemistry* **60**, 1920 (1988)

133. Van Loon, W.M.G., Bron, J.S. and de Groot, B., *Analytical Chemistry* **65**, 1726 (1993)
134. Peters, O., *Analytical Chemistry* **66**, 492 (1994)
135. Baykut, G. and Voigt, A., *Analytical Chemistry* **64**, 677 (1992)
136. Hara, H., Okabe, Y. and Kitagawa, T., *Analytical Chemistry* **64**, 2393 (1992)
137. Di Corcia, A. and Samperi, R., *Analytical Chemistry* **62**, 1490 (1990)
138. Dai, J. and Helz, G.R., *Analytical Chemistry* **60**, 301 (1988)
139. Fulton, D.B., Sayer, B.G., Bain, A.D. and Malle, H.V., *Analytical Chemistry* **64**, 349 (1992)
140. Lin, D.P., Falkenberg, C., Payne, D.A., Thakker, J., Tang, C. and Elly, C., *Analytical Chemistry* **65**, 999 (1993)
141. Melcher, R.G., Bakki, D.W. and Hughes, G.H., *Analytical Chemistry* **64**, 2258 (1992)
142. Takeda, S., Wakida, S., Yamone, M., Kawahara, A. and Higashi, K., *Analytical Chemistry* **65**, 2489 (1993)
143. Melcher, R.G. and Morabito, P.L., *Analytical Chemistry* **62**, 2183 (1990)
144. Slivan, L.E., Banor, M.R., Ho, J.S. and Budde, W.L., *Analytical Chemistry* **63**, 1335 (1991)
145. Burkhard, L.P., Durhan, E.J. and Lucasewycz, M.T., *Analytical Chemistry* **63**, 277 (1991)
146. Pratt, K.P. and Pawliszyn, J., *Analytical Chemistry* **64**, 2107 (1992)
147. Ram, R.N. and Sothyanesan, A.G., *Environmental Pollution* **47**, 135 (1987)
148. Kiruagarun, R. and Joy, P., *Ecotoxicology and Environmental Safety* **15**, 171 (1988)
149. McGuire, R.J. and Tkacz, R.J., *Water Pollution Research Journal of Canada* **22**, 227 (1987)
150. Czuba, M., Seagull, R.W., Tran, H. and Cloutier, L., *Ecotoxicology and Environmental Safety* **14**, 64 (1987)
151. Weis, J.S., Gottlieb, H. and Kwiatkowski, J., *Archives of Environmental Contamination and Toxicology* **16**, 321 (1987)
152. Roberts, M.H., *Bulletin of Environmental Contamination and Toxicology* **39**, 1012 (1987)
153. Hanna, C.P. and Tyson, J.F., *Analytical Chemistry* **65**, 653 (1993)
154. Robert, J.M., and Robenstein, P.L., *Analytical Chemistry* **63**, 2674 (1991)
155. Lobinski, R., Bontron, C.F., Candelone, J.P., Hong, S., Lobinska, J.S. and Adams, F.C., *Analytical Chemistry* **65**, 2510 (1993)
156. Shen, W.L., Vela, N.P., Sheppard, B.S. and Carns, J.A., *Analytical Chemistry* **63**, 1491 (1991)
157. Nygren, O., Nilsson, C.A. and Frech, W., *Analytical Chemistry* **60**, 2204 (1988)
158. Lobinski, R., Direx, W.M.R. and Adams, F.C., *Analytical Chemistry* **64**, 159 (1992)
159. Blais, J.S., Momplasir, G.M. and Marshall, W.P., *Analytical Chemistry* **62**, 1161 (1990)
160. Killa, H.M.A. and Robenstein, D.L., *Analytical Chemistry* **60**, 2283 (1988)
161. Tanzer, D. and Henmann, K.G., *Analytical Chemistry* **63**, 1984 (1991)
162. Sasaki, E.T., Sohrin, Y., Hageqawa, H., Kokusen, H., Kihara, C. and Matsui, M., *Analytical Chemistry* **66**, 271 (1994)
163. Aue, W.A., Miller, B. and Xun-Gun, Sen, *Analytical Chemistry* **62**, 2453 (1990)
164. Jong-Hwa, L., *Bulletin of Fisheries Science Institute* **3**, 11 (1987)
165. Moles, A., Babcock, M.M. and Rice, S.D., *Marine Environmental Research* **21**, 49 (1987)
166. Brule, T., *Journal of Marine Biological Association* **67**, 237 (1987)

167. Kocan, R.M., Van Westernhagen, H., Landolt, M.L. and Furstenberg, G., *Marine Environmental Research* **23**, 291 (1987)
168. Wan, M.T., Watts, R.G. and Moul, D.J., *Bulletin of Environmental Contamination and Toxicology* **41**, 609 (1988)
169. Seelye, J.G., Marking, L.L., King, E.L., Hanson, L.H. and Bills, T.D., *North American Journal of Fisheries Management* **7**, 598 (1987)
170. Mitchell, D.G., Chapman, P.M. and Lang, T.J., *Environmental Toxicology and Chemistry* **6**, 875 (1987)
171. Wan, M.T., Moul, D.J. and Watts, R.C., *Bulletin of Environmental Contamination and Toxicology* **39**, 721 (1987)
172. Livingstone, D.R., *Science of the Total Environment* **65**, 3 (1987)
173. Do Lawe, D.M. and Pipe, P.K., *Marine Environmental Research* **22**, 243 (1987)
174. Stromgen, T., *Marine Environmental Pollution* **21**, 239 (1987)
175. Axiak, V. and George, J.J., *Marine Biology* **94**, 241 (1987)
176. Kasymov, A.G. and Gasanov, V.M., *Water, Air and Soil Pollution* **36**, 9 (1987)
177. Reijnders, P.J.H., *Nature* **324**, 456 (1986)
178. Weis, J.S. and Ma, A., *Bulletin of Environmental Contamination and Toxicology* **39**, 224 (1987)
179. Rajeswari, K., Kalarani, V., Reddy, D.C. and Ramamurthi, R., *Bulletin of Environmental Contamination and Toxicology* **40**, 212 (1988)
180. Vijayakumari, P., Reddy, D.C. and Ramamurthi, R., *Bulletin of Environmental Contamination and Toxicology* **38**, 742 (1987)
181. Mallatt, J. and Barron, M.G., *Archives of Environmental Contamination and Toxicology* **17**, 73 (1988)
182. Bengtsson, B.E., *Water, Science and Technology* **20**, 87 (1988)
183. Zischke, J.A. and Arthur, J.W., *Archives of Environmental Contamination and Toxicology* **16**, 225 (1987)
184. Giudici, M., De, N., Migliore, S.M., Guarino, S.M. and Gambardella, C., *Marine Pollution Bulletin* **18**, 454 (1987)
185. Noth, K. and Kumar, N., *Chemosphere* **17**, 465 (1988)
186. Eklund, G., Josefsson, B. and Roos, C., *Journal of High Resolution Chromatography* **1**, 34 (1978)
187. Eklund, G., Josefsson, B. and Roos, C., *Journal of Chromatography* **142**, 575 (1977)
188. Hashimoto, A., Sakinot, H., Yamagani, E. and Tateishi, S., *Analyst (London)* **105**, 787 (1980)
189. Hashimoto, A., Kozima, T., Shakina, H. and Arikey, T., *Water Research* **13**, 509 (1979)
190. Shinohara, R., Kido, A., Okomoto, Y. and Takeshita, R., *Journal of Chromatography* **256**, 81 (1983)
191. Favretto, L., Stancher, B. and Tunis, R., *Analyst (London)* **103**, 955 (1978)
192. Park, K., Williams, T.D., Prescott, J.M. and Hood, D.W., *Science* **138**, 531 (1962)
193. Riley, J.P. and Segar, D.A., *Journal of Marine Biology Association* **50**, 713 (1970)
194. Hosaku, K. and Maita, Y., *Journal of Oceanographic Society, Japan* **27**, 27 (1971)
195. Starikova, N.D. and Korzhikova, R.I., *Okeanogiya* **9**, 509 (1969)
196. Lee, C. and Bada, J.L., *Earth Planetary Science Letters* **26**, 61 (1975)
197. Ziobin, V.S., Perlyuk, M.F. and Orlova, T.A., *Okeanogiya* **15**, 643 (1975)
198. Garrasi, C. and Regens, E.T., Analytische Metoden zur Sablenchromatographischem Bestimmung Von Arminosaurin und Zuckenn Merrwasser und Sediment. Berichte aus dem Projeckt DFG-DE 7413. Litoralforschung—Abwasser in Kustennake, DFG Abschlusskolliguim, Bremerhaven (1976)
199. Dauson, R. and Mopper, K., *Analytical Biochemistry* **83**, 100 (1977)

200. Daumas, R.A., *Marine Chemistry* **4**, 225 (1976)
201. Tatsumato, M., Williams, W.T., Prescott, J.M. and Hood, D.W., *Journal of Marine Research* **19**, 89 (1961)
202. Rittenberg, S.C., Emery, K.O., Hulsemann, J., Regens, E.T., Fay, R.S., Reuter, J.H., Grady, J.R., Richardson, S.H. and Bray, E.E., *J. Sediment Petrol* **33**, 140 (1963)
203. Williams, P.M., *Limnology and Oceanography* **14**, 297 (1969)
204. Sharp, J.H., *Marine Oceanography* **1**, 211 (1973)
205. Mackinnon, M.D., *Marine Chemistry* **7**, 17 (1978)
206. Gerkey, R.M., MacKinnon, M.D., Williams, D.J. and Moore, R.H., *Marine Chemistry* **7**, 289 (1979)
207. Schimmel, H., Schmid, B., Backer, R. and Ballschmitter, K., *Analytical Chemistry* **65**, 640 (1993)
208. Lech, C. and Bagander, L.E., *Analytical Chemistry* **60**, 1680 (1988)
209. Xiao-Hua Yang, Lee, C. and Scranton, M.I., *Analytical Chemistry* **65**, 857 (1993)
210. Vaughan, G.M. and Mopper, K., *Analytical Chemistry* **60**, 1654 (1988)
211. Laughlin, R.B. and Linden, O., *Ambio* **16**, 252 (1987)
212. Davies, I.M., Graham, W.C. and Pirie, S.M., *Marine Chemistry* **7**, 11 (1979)
213. Hayward, M.G. and Riley, J.P., *Analytical Chemistry* **85**, 219 (1976)
214. Jackson, J.A., Blair, W.R., Brinkman, F.E. and Iveson, W.P., *Environmental Science and Technology* **16**, 111 (1982)
215. Valkins, A.O., Seligman, P.F. and Stang, P.M., *Marine Pollution Bulletin* **17**, 319 (1986)
216. Ebdon, L. and Alonso, G., *Analyst (London)* **112**, 1951 (1987)
217. Agmenian, H. and Chau, A.S.Y., *Analytical Chemistry* **50**, 13 (1978)
218. Fitzgerald, W.F. and Lyons, W.B., *Nature (London)* **242**, 452 (1973)
219. Davies, I.M., Graham, W.C. and Pirie, S.M., *Marine Chemistry* **7**, 11 (1979)
220. Yamamoto, J., Kanada, Y. and Hisaka, Y., *International Journal of Environmental Chemistry* **16**, 1 (1983)
221. Department of Environment and National Water Council, UK, *Mercury in Water, Effluents, Soils and Sediments. Additional Methods* (PE-22-AGENV) HMSO, London (1985)
222. Bond, A.M., Bradbury, J.R. and Hanna, P.J., *Analytical Chemistry* **56**, 2392 (1984)
223. Braman, R.S. and Tompkins, M.A., *Analytical Chemistry* **51**, 12 (1979)
224. Hodge, V.F., Seidel, S.L. and Goldberg, E.D., *Analytical Chemistry* **51**, 1256 (1979)
225. Valkirs, A.O., Seligman, P.F. and Stang, P.M., *Marine Pollution Bulletin* **17**, 319 (1986)
226. Valkirs, A.O., Seligman, P.F. and Olsen, G.J., *Analyst (London)* **112**, 17 (1987)
227. Meinema, H.A., Burger, N. and Wiersina, T., *Environmental Science and Technology* **12**, 288 (1978)
228. Jackson, J.A., Blair, W.R., Brinkmann, F.E. and Iveson, W.P., *Environmental Science and Technology* **16**, 111 (1982)
229. Laughlin, R.B., Guard, H.E. and Coleman, W.M., *Environmental Science and Technology* **20**, 201 (1986)
230. Bluden, S.J. and Chapman, A.H., *Analyst (London)* **103**, 1266 (1978)
231. Edwards, J.S. and Francesconi, K.A., *Analyst (London)* **48**, 2019 (1976)
232. Talmi, Y. and Bostick, D.T., *Analytical Chemistry* **47**, 2145 (1975)
233. Penrose, W.R., *Critical Reviews of Environmental Control* **4**, 465 (1974)
234. Andreae, M.O., *Analytical Chemistry* **49**, 820 (1977)
235. Braman, R.S., Johnson, D.L. and Foreback, C.O., *Analytical Chemistry* **49**, 621 (1977)
236. Howard, A.G. and Arbab-Zavar, M.H., *Analyst (London)* **106**, 213 (1981)

237. Hinners, T.A., *Analyst (London)* **105**, 751 (1980)
238. Pierce, F.D. and Brown, H.R., *Analytical Chemistry* **49**, 1417 (1977)
239. Persson, J. and Irgum, K., *Analytica Chimica Acta* **138**, 111 (1982)
240. Haywood, M.G. and Riley, J.P., *Analytica Chimica Acta* **85**, 219 (1976)
241. Yamamoto, M., *Soil Science Society American Proceedings* **39**, 859 (1975)
242. Trim, A.H., *Bulletin of Environmental Contamination and Toxicology* **38**, 681 (1987)
243. Means, J.C., Daniels, C.B. and Baksi, S.M., *Marine Environmental Research* **24**, 327 (1988)
244. Ram, R.N. and Sathyaneson, A.G., *Ecotoxicology and Environmental Safety* **13**, 185 (1987)
245. Braman, R.S. and Tompkins, M.A., *Analytical Chemistry* **51**, 12 (1979)
246. Landy, M.P., *Analytica Chimica Acta* **121**, 39 (1980)
247. Muller, M.D., *Analytical Chemistry* **59**, 617 (1987)
248. Kiemencij, A.M. and Kloosterboer, J.G., *Analytical Chemistry* **48**, 575 (1976)
249. Minagawa, K., Takizawa, Y. and Kifune, I., *Analytica Chimica Acta* **115**, 103 (1980)
250. Potter, H.R., Jarview, A.W. and Markell, R.N., *Water Pollution Control* **76**, 123 (1977)

CHAPTER 7
Pollution of Sedimentary Matter

Pollutants entering freshwaters and the oceans remain partly in solution and partly are adsorbed onto the surface of sedimentary matter. Both sources of pollution, i.e. dissolved and sedimentary, are capable of entering living creatures with possible adverse effects. The concentration of toxicants present in sediments is a measure of its concentration in the water over a period of time and is, therefore, a measure of the risk to creatures. In the case of bottom-feeding creatures there is the additional risk of direct ingestion of sediments via the gills and mouth with consequent adverse effects. Concentrations of dissolved pollutants has been discussed in Chapters 5 and 6 and the concentrations of particulate pollutants are considered below.

Because of the tendency of pollutants to concentrate in sediments their concentrations in the latter can be appreciably higher than in an equivalent volume of water, as discussed in Chapter 2. For these reasons much work has been carried out on the determination of toxicants in sediments.

7.1 METALS IN SEDIMENTS

7.1.1 TOXIC EFFECTS

Cherry et al.[1] carried out a study to identify the toxic sediments of fly ash and bottom ash obtained from the Glen Lyn Power Station Plant, Virginia, in acute laboratory bioassays using a warm-water rainbow trout (*Lepomis macrochirus*) and a cold-water bluegill sunfish species to evaluate the surface availability of trace elements at various pH values. Rainbow trout (*Salmo gairdneri*) were highly sensitive to fly ash when dissolved metal availability was high, but not to high particulate concentrations (up to 2350 mg l^{-1} dissolved solids) when metals were removed. Bluegill were much less sensitive to cadmium, chromium, copper, nickel, lead and zinc. Both species were acutely sensitive below pH 4.0 and above pH 9.1.

Hammer et al.[2] have studied the effect of low dissolved oxygen concentrations in eutropic lakes on the release of mercury from sediments and subsequent bioaccumulation by aquatic plants (*Ceratophyllum demersum*) and clams (*Anodonto grandis*). The mercury concentration in plant and clams in water with reduced dissolved oxygen content $(1.8 \, \text{mg} \, l^{-1})$ was considerably higher than in water with higher dissolved oxygen content $(6.7 - 7.2 \, \text{mg} \, l^{-1})$.

7.1.2   ANALYTICAL COMPOSITION

In Tables 7.1 and 7.2 are summarised concentrations of metallic elements that have been found in freshwaters and seawater. A more detailed breakdown of results obtained for freshwaters is found in Appendix 3, Table A3.1.

List 1 in Table 7.1 shows the results obtained for the toxic elements that have been discussed in various EU directives. List 2 covers the major naturally occurring elements, and List 3 the minor elements, many of which are naturally occurring, most of which are of little toxicological concern.

Table 7.1. Elements in freshwater sediments

	Concentration $(\text{mg} \, \text{kg}^{-1})$	
Element	Rivers	Lakes
(1) *Elements covered in EU directives*		
Al	9890–46 200	26 200–63 800
As	0.22–7.1	1.9–26
Sb		0.01–2.9
Ba		163–2700
Cd	0.06–27.5	3.5–40
Cr	0.48–1143	16–110
Co	1.8–53	3.9–200
Cu	0.07–244	50
Pb	0.11–5060	20–180
Hg	0.91–46.8	1.95–6.8
Ni	1.4–238	1–218
Se	0.09–0.93	0.03–1.0
Ag	1–5.53	0.1–8.05
Ti		800–3800
U		0.78–4.3
V		28–68
Zn	0.31–9040	10–450
(2) *Naturally occurring elements*		
Br		23–96
Ca		12 300–40 000
Cl		20–609
Fe	16.9–31 000	14 700–30 600
Li		50
Mg		5 900–16 800
Mn	0.34–9640	214–4500
P	675–1870	
Na		3000–9200
Sr		10–242

Table 7.1. (*continued*)

Element	Concentration (mg kg^{-1})	
	Rivers	Lakes
(3) *Minor elements (few or no toxicity data)*		
Ru		19–49
Cs		0.5–14
Au		0.25–19
Th		4.0–9.4
Hf		1.7–12
Zr		55–488
In		5.3–19
Ru		45–500
Sc		3.3–9.2
Ta		0.4–1.4
Tm		0.19–7.4
Ce		53–160
Yb		2.3–9.3
Dy		5.3–15
Gd		6.4–22
La		28–73
Tb		0.95–2.4
Nd		15–137
Sm		7.9–28
Ir		0.5–48
Os		1–4.5
Pt		0.3–8.1

To take the case of cadmium, it is seen in Chapter 2 (Table 2.2) that in this element a range of values of 800–4000 has been obtained for the concentration ratio

$$\frac{\text{mg kg}^{-1} \text{ (in sediment)} \times 1000}{\mu\text{g l}^{-1} \text{ (in water)}} = \frac{\mu\text{g kg}^{-1}}{\mu\text{g l}^{-1}}$$

The range of cadmium content found in river sediments (Table 7.1) is 0.06–27.5 mg kg^{-1}. When the river water is relatively unpolluted, i.e. cadmium content of sediment 0.06 mg kg^{-1},

$$\frac{\text{mg kg}^{-1} \text{ (in sediment)} \times 1000}{\mu\text{g l}^{-1} \text{ (in water)}} = 800\text{–}4000$$

Concentration of cadmium in water (0.06 mg kg^{-1} Cd in sediment)

$$= \frac{0.06 \times 1000}{800} - \frac{0.06 \times 1000}{4000} = 0.075 - 0.015 \,\mu\text{g l}^{-1}$$

Table 7.2. Metals in marine sediments

Element	Location	Concentration (mg kg^{-1})	Ref.
Bismuth	Narragonsett Bay, USA	Surface 0.40 49–54 mm core 0.27	3
	Pacific	0.1	3
Mercury		Sand < 0.1–1.4 Clay < 0.1–0.8	4
	River Loire estuary, salinity 20–35%	13.2	5
	River Loire 0–10 km, upstream of estuary	28.0	5
	River Loire 10–15 km, upstream of estuary	22.9	5
	River Loire 15–30 km, upstream of estuary	46.8	5
Tin	Narragonsett Bay, USA	1 cm core 20 80 cm core 1	6 6
Lanthanum	Deep-sea sediments	65.1	7
Cerium		91.0	
Neodynium		92.5	
Samerium		22.9	
Europium		5.7	
Gadolinium		25.2	
Dysprosium		23.0	
Erbium		13.4	
Ytterbium		13.1	

Table 7.3. Toxic effects associated with high metal levels in sediments

Element	As[3]				Cu			
Type of fish	n/s				s			
Concentration of metal in sediments (mg kg^{-1})	0.22		7.1		0.07		244	
Concentration factor $\dfrac{\text{mg kg}^{-1}(\text{sed}) \times 1000}{\mu\text{g l (water)}}$ (from Chapter 2, Table 2.2)	3700	37 000	3700	37 000	15 714	430 000	15 714	430 000
Concentration of metal in water (μg l^{-1})	0.06	0.006	1.9	0.19	0.0044	0.0001	16.2	0.57
S_x	80	80	80	80	4	4	4	4
Concentration of metal in water > S_x, i.e. toxic effects expected							Yes	

When the river water is more highly polluted, i.e. cadmium content of sediment 27.5 mg kg^{-1}, concentration of cadmium in water (25.5 mg kg^{-1} Cd in sediment)

$$= \frac{27.5 \times 1000}{800} - \frac{27.5 \times 1000}{4000} = 34.3 - 6.87 \ \mu g\,l^{-1}$$

The S_x value (i.e. the maximum safe concentration of cadmium in water for survival of non-salmonids for periods exceeding one year (see Table 1.5, Chapter 1) is 4 $\mu g\,l^{-1}$.

Thus it is seen that waters over sediments containing 0.06 mg kg^{-1} cadmium would enable fish to survive long term as the cadmium content of the water would be 0.075–0.015 $\mu g\,l^{-1}$, i.e. below the safe limit of 4 $\mu g\,l^{-1}$ while waters over sediments containing 27.5 mg kg^{-1} would not enable fish to survive long term as the cadmium content of the water would be 34.3–6.87 $\mu g\,l^{-1}$, i.e. above the safe limit of 4 $\mu g\,l^{-1}$.

Similar calculations show that at the higher concentrations of metals in sediments, reflecting as they do higher averages of metal concentrations in water over a period of time, toxic effects towards fish would also be expected for copper, lead and zinc at the higher end of the range, while toxic effects might not be expected for these elements at lower levels or for arsenic and nickel at any of the concentrations studied (Table 7.3).

7.1.3 ANALYTICAL METHODS

Analytical methods that have been employed for the determination of metals in the sediments originating in non-saline waters or estuary waters are reviewed respectively in Tables 7.4 and 7.5. As summarised below, the principal

Table 7.3. (*continued*)

Pb				Ni				Zn			
	n/s				n/s				n/s		
0.11			5060	1.4			238	0.31			9040
26 829	136 000	26 829	136 000	2133	32 000	2133	32 000	4060	102 500	4060	102 500
0.0041	0.000 81	189	37.2	0.656	0.044	111.6	7.43	0.076	0.0030	2226	88.2
20	20	20	20	220	220	220	220	23	23	23	23
		Yes	Yes							Yes	Yes

n/s—non-salmonid
s—salmonid

Table 7.4. Methods of analysis of metals and non-metals in freshwater (non-saline) sediments

Determined	Techniques	Dectection limit ($mg\,kg^{-1}$ unless otherwise stated)	Ref.
Arsenic	Conversion to arsenic hydride, conversion to diethyldithio carbamate and spectrophotometric estimation	–	8, 9
	Continuous flow semi-automated inductively coupled emission spectrometry	$0.02\ \mu g\,l^{-1}$	12
Barium	Direct injection graphite furnace atomic absorption spectrometry with vanadium/silicon modifier	$100\ pg\,l^{-1}$	10
Bismuth	Electrothermal atomic absorption spectrometry	1	11
Cadmium	Zeeman atomic absorption spectrometry	0.1	13
	Graphite furnace atomic absorption spectrometry	–	14
Copper	Potentiometric stripping analysis	3	15
Gallium	Graphite furnace absorption spectrometry	$\mu g\,kg^{-1}$	16
Lead	Graphite furnace atomic spectroscopy	3	14
	Potentiometric stripping analysis	3	15
Mercury	Isotope dilution inductively coupled plasma atomic emission spectroscopy with 201 Hg spike	$2\ ng\,l^{-1}$	17
	Differential pulse anodic scanning voltammetry	$0.02\ \mu g\,l^{-1}$	18
	Gold film mercury analyser	–	19
Selenium	Continuous flow semi-automated inductively coupled plasma atomic emission spectrometric method	$0.03\ \mu g\,l^{-1}$	12
Silver	Zeeman atomic absorption spectrometry	0.1	13
Tin	Flame atomic absorption spectrometry	0.0005	20
	Graphite furnace atomic absorption spectrometry	2.5	21
	Inductively coupled plasma atomic emission spectrometry	1 pg as Sn absolute	22
Vanadium	Differential pulse voltammetry	–	23

Table 7.4. (*continued*)

Determined	Techniques	Dectection limit (mg kg^{-1} unless otherwise stated)	Ref.
Cd, Cu, Fe, Mn, Pb, Zn	Flame or flameless atomic absorption spectrometry	–	24
Cd, Co, As	Atomic absorption spectrometry	–	25
Cu, Fe, Cr, Mn, Pb, Zn	Atomic absorption spectrometry	–	26, 27
Pb, Cd, Zn	Electrothermal atomic absorption spectrometry	–	28
Various metals	Graphite furnace atomic absorption spectrometry	–	29, 30
Various metals	Zeeman atomic absorption spectrometry	–	30
Al, Ba, Ca, Cd, Cr, Fe, K, Mg, Mn, Na, P, Si, Sr, Ti, U, Zn	Parr bomb digestion inductively coupled plasma atomic emission spectrometry	–	31
Various metals	Inductively coupled plasma atomic emission spectrometry	–	32, 33
Fe, Mn, Zn, Cu, Cr, Ni, Pb, Al	Plasma emission spectrometry	–	25
Re, Ir, Pt	Flow injection isotope dilution inductively coupled plasma mass spectrometry	–	34
Sb, As	Selective hydride generation with gas chromatography photoionization detector	10 pmole l^{-1} (As) 3.3 pmole l^{-1} (Sb)	35
Al, As, Ba, Br, Ca, Ce, Co, Cr, Cs, Dy, Eu, Fe, Gd, Hf, K, La, Mn, Na, Nd, Ni, Rb, Sb, Sc, Sm, Ta, Tb, Tl, U, V, W, Zn	Neutron activation analysis	–	36
As, Ba, Co, Cr, Eu, Fe, K, La, Mn, Na, Sb, Sc, Se, U, W	Neutron activation analysis	–	37

continued overleaf

Table 7.4. (*continued*)

Determined	Techniques	Dectection limit (mg kg⁻¹ unless otherwise stated)	Ref.
Various metals	Neutron activation analysis	–	38
25 metals	X-ray fluorescence spectroscopy	–	39
Th, U, rare earths (11)	γ-ray spectrometry	–	40
Ca, Sr, Fe, Al, Cu, Mn, Ni, Zn, Co, Pb, Cr, V	Sequential extraction method	–	41
Total nitrogen	Hydrogen peroxide digestion, kjeldahl digestion	–	42
Organic carbon	Microcombustion techniques	–	43–45
Sediment oxygen demand	–	–	46
Sulphide	Spectrophotometric	–	47

Table 7.5. Methods for the determination of metals and non-metals in marine and estuary sediments

Determined	Type of sediment	Technique	Detection limit (mg kg⁻¹ unless otherwise stated)	Ref.
Arsenic	Marine	Reduction to arsine, spectrophotometric	0.05	48
		Acid reduction, derivatisation with 2, 3-mercaptopropanol-gas chromatography with electron capture detector	–	49
		Hydride generation–graphite furnace atomic absorption spectrometry	5	50
Bismuth		Wavelength modulated inductively coupled plasma echelle spectrometry	–	51
		Hydride generation (LiBH)–flameless atomic absorption spectrometry	–	52
Boron		Spectrophotometric as carminic acid derivative	–	53

Table 7.5. (*continued*)

Determined	Type of sediment	Technique	Detection limit (mg kg^{-1} unless otherwise stated)	Ref.
Cadmium		Isotope dilution inductively coupled plasma mass spectrometry	–	54
Carbon (organic)	Estuary	Potassium persulphate oxidation, ignition at 560 °C	–	55
	Marine estuary	Persulphate oxidation–non-dispersive infrared spectroscopy	–	56
	Marine	Carbon dioxide evolution and measurement	–	57
Iridium	Marine	Graphite furnace atomic absorption spectrometry	pg level	184
Lead		Isotope dilution inductively coupled plasma mass spectrometry	–	54
		Wavelength modulated inductively coupled plasma echelle spectrometry	–	51
Mercury		Non-dispersive atomic fluorescence spectroscopy	0.04	50
		Reduction to mercury with tin-atomic absorption spectrometry	ng level	58
Molybdenum		Dithiooxamide spectrophotometric method	2 ng level	59
Oxygen uptake	Estuary	Measurement	–	60
Platinum	Marine	Graphite furnace atomic absorption spectrometry	pg level	58
Selenium	Marine	Conversion to hyride (by NaBH$_4$)—graphite furnace atomic absorption spectrometry	0.001	61
		Conversion to 5-nitroplaz-selenol-gas chromatography with electron capture detector	2 pg level	62
	Estuary	Conversion to hydride (NaBH$_4$), volatilisation, with atomic absorption detection	–	63
Silver	Marine	Zeeman atomic absorption spectrometry	–	64

continued overleaf

Table 7.5. (*continued*)

Determined	Type of sediment	Technique	Detection limit (mg kg^{-1} unless otherwise stated)	Ref.
Thallium		Conversion to tetrachloro-thallate, graphite furnace atomic absorption spectrometry	–	65
Pb, Cd, Zn,	Marine	Flame atomic absorption spectro-metry graphite furnace atomic absorption spectrometry	–	66
Al, Fe, Ca, Mg, Na, P, Be, Co, Cu, Mn, Ni, Pb, V, Zn, As		Inductively coupled plasma atomic emission spectrometry	0.5 (Be) to 5 mg kg^{-1} (As)	67
As, Pb, Bi		Continuous hydride generation (AlBH$_4$) inductively coupled plasma atomic emission spectrometry	0.1	68, 69
Al, Ti, Fe, Mn, Zn		Acid digestion–flame atomic absorption spectrometry	–	70
Cu, Cd, Cr, As		Graphite furnace atomic absorption spectrometry	–	70
12 Trace elements		Analysis of γ-ray spectrum after photoactivation	–	71
Co, Cr, Cs, Fe, Rb, Sb, Se, Sr		Neutron activation analysis	–	72
Cd, Pb, Cu, Ni, Co, Be, Cr	Estuarine	Digestion, graphite furnace atomic absorption spectrometry	–	73
Al, P, Si, K, Ca, Ti, Fe, Cr, Mn, Ni, Cu, Zn	Estuarine	X-ray secondary emission spectrometry	–	74
Cr, Mn, Fe, Co, Ni, Cu, Zn, Cd, Pb		Acid digestion, atomic absorption spectrometry	–	75
Various metals		Inductively coupled plasma atomic emission spectrometry	–	76–78
Cr, Mn, Fe, Co, Ni, Cu, Zn, Cd, Pb		Metaborate fusion–plasma emission spectrometry	–	79
28 elements		Neutron activation analysis	0.37 ng to 1 ng absolute	80

techniques for the determination of metals in sediments include atomic absorption spectrometry (including electrothermal, graphite furnace and Zeeman spectrometry) (detection limits ranging from 0.0005 to 2.5 mg kg^{-1}), inductively coupled plasma atomic emission spectrometry (detection limit 0.5 to 5 mg kg^{-1}), conversion of metal to a volatile derivative followed by gas chromatography (applicable at the 2 pg metal level), neutron activation analysis, X-ray spectrometry and X-ray secondary emission spectrometry. Metals which are capable of being converted to volatile hydride (e.g. arsenic, antimony, bismuth and selenium) can be analysed in their volatile form by gas chromatography, atomic absorption spectrometry, graphite furnace atomic absorption spectrometry or inductively coupled plasma atomic emission spectrometry (detection limits 0.01 − 5 mg kg^{-1} depending on the element or the technique concerned).

Review of methods and detection limits applicable to determination of metals in freshwater sediments (non-saline)

Method		Detection limit (mg kg^{-1} unless otherwise stated)
Electrothermal atomic absorption spectrometry	Bismuth	1
	Mercury	
	Tin	0.0005
	Cadmium	
	Copper	
	Iron	
	Manganese	
	Lead	
	Zinc	
	Cobalt	
	Arsenic	
	Chromium	
Graphite furnace atomic absorption spectrometry	Barium	100 pg l^{-1}
	Cadmium	0.1
	Gallium	μg kg^{-1} level
	Lead	
	Tin	2.5
Zeeman atomic absorption spectrometry	Cadmium	0.1
	Mercury	0.1

Method		Detection limit ($mg\,kg^{-1}$ unless otherwise stated)
Inductively coupled plasma atomic emission spectroscopy	Arsenic	0.02 $\mu g\,l^{-1}$
	Selenium	0.03 $\mu g\,l^{-1}$
	Tin	1 pg as Sn
	Aluminium	
	Barium	
	Calcium	
	Cadmium	
	Chromium	
	Iron	
	Potassium	
	Magnesium	
	Manganese	
	Sodium	
	Phosphorus	
	Silicon	
	Strontium	
	Titanium	
	Vanadium	
	Zinc	
Isotope dilution inductively coupled plasma atomic emission spectrometry	Mercury	2 $ng\,l^{-1}$
	Rhenium	
	Iridium	
	Platinum	
Plasma emission spectrometry	Iron	
	Manganese	
	Zinc	
	Copper	
	Chromium	
	Nickel	
	Lead	
	Aluminium	
X-ray spectrometry	Thorium	
	Uranium	
	Rare earths	
Hydride generation gas chromatography photoionisation detector	Antimony	10 $p\,mole\,l^{-1}$
	Arsenic	3.3 $p\,mole\,l^{-1}$

Method		Detection limit (mg kg^{-1} *unless otherwise stated*)
Neutron activation analysis	Aluminium	
	Arsenic	
	Barium	
	Bromine	
	Calcium	
	Cerium	
	Cobalt	
	Chromium	
	Caesium	
	Europium	
	Iron	
	Cadmium	
	Hafnium	
	Potassium	
	Lanthanum	
	Manganese	
	Sodium	
	Neodynium	
	Nickel	
	Rubidium	
	Antimony	
	Scandium	
	Samerium	
	Tantalum	
	Terbium	
	Thorium	
	Uranium	
	Vanadium	
	Tungsten	
	Zinc	
Anodic scanning voltammetry	Mercury	0.02 μg l^{-1}
Differential pulse voltammetry	Vanadium	
Potentiometric stripping analysis	Copper	3
	Lead	3
Hydride generation spectrometry	Arsenic	

Review of methods of detection limits applicable to the determination of metals in marine and estuarial sediments

Method		Detection limit ($mg\,kg^{-1}$ unless otherwise stated)
Flame atomic absorption spectrometry	Lead	
	Cadmium	
	Zinc	
	Aluminium	
	Titanium	
	Iron	
	Manganese	
	Chromium	
	Nickel	
	Copper	
Conversion to hydride—atomic absorption spectrometry	Selenium	
	Bismuth	
Reduction to mercury—atomic absorption spectrometry	Mercury	ng level
Graphite furnace atomic absorption spectrometry	Lead	
	Cadmium	
	Zinc	
	Platinum	pg level
	Iridium	pg level
	Copper	
	Nickel	
	Cobalt	
	Beryllium	
	Chromium	
	Arsenic	
Conversion to hydride—graphite furnace atomic absorption spectrometry	Selenium (as SeH_2)	0.0014
	Arsenic (as AsH_3)	5
Conversion to tetrachlorothallate—graphite furnace atomic absorption spectrometry	Thallium	
Zeeman atomic absorption spectrometry	Silver	

Method		Detection limit (mg kg⁻¹ *unless otherwise stated*)

Method		Detection limit (mg kg^{-1} *unless otherwise stated*)
Inductively coupled plasma atomic emission spectometry	Aluminium	
	Iron	
	Calcium	
	Magnesium	
	Sodium	
	Phosphorus	all between
	Beryllium	0.5 and 5
	Cobalt	
	Copper	
	Manganese	
	Nickel	
	Lead	
	Vanadium	
	Zinc	
	Arsenic	
Hydride generation inductively coupled plasma atomic emission spectrometry	Arsenic	0.1
	Antimony	0.1
	Bismuth	0.1
Wavelength modulated inductively coupled plasmas echelle spectrometry	Bismuth	
	Lead	
Isotope dilution inductively coupled plasma mass spectrometry	Cadmium	
Gas chromatography derivitivisation with 2-,3-mercapto propanol-gas chromatography with electron capture detection	Arsenic	
Conversion to nitropiazselenol—gas chromatography with electron capture detection	Selenium	2 pg level
dc Plasma emission spectrometry	Chromium	
	Manganese	
	Iron	
	Cobalt	
	Nickel	
	Copper	
	Zinc	
	Cadmium	
	Lead	

Method		*Detection limit (mg kg^{-1} unless otherwise stated)*
X-ray secondary emission spectrometry	Aluminium	
	Phosphorus	
	Silicon	
	Potassium	
	Calcium	
	Titanium	
	Iron	
	Chromium	
	Manganese	
	Nickel	
	Copper	
	Zinc	
X-ray spectrometry neutron activation analysis	12 metals	0.4–1 ng absolute
	Cobalt	
	Chromium	
	Caesium	
	Rubidium	
	Antimony	
	Scandium	
	Strontium	
Spectrophotometry		
As carminic acid	Boron	
As dithiooxamide	Molybdenum	2 ng level
Conversion to hydride spectrophotometry	Arsenic	0.05
Atomic fluorescence spectroscopy	Mercury	0.04

7.2   ORGANOMETALLIC COMPOUNDS IN SEDIMENTS

7.2.1   ANALYTICAL COMPOSITION

Butyl and cyclohexyl tin compounds have been found in river and lake sediments (Table 7.6). These probably originate from the use of organotin antifoulants on boats and pier works[81]. Further information on the occurrence of organometallic compounds is given in Appendix 3 (Table A3.2).

Table 7.6. Organotin compounds in river and lake
sediments

Compound	Concentration (mg kg^{-1})
$BuSn^{3+}$	0.055
$BuSn^{2+}$	0.14
$BuSn^+$	0.28
$Cyclohexyl_2Sn^{2+}$	0.01
$Cyclohexyl_3Sn^+$	0.075

7.2.2   ANALYTICAL METHODS

Methods that have been used for the determination of organometallic compounds
in the sediments originating in non-saline freshwaters and in marine and estuary
waters are reviewed, respectively, in Tables 7.7 and 7.8. Major techniques as
reviewed below include cold vapour atomic absorption spectrometry (detection
limit organomercury 0.0025 mg kg^{-1}), gas chromatography with various types of
detector including atomic absorption spectrometry (detection limits organo-lead
and organotin 0.01 mg kg^{-1}).

The technique of converting organotin and organoarsenic compounds to
hydride derivatives prior to detection with an atomic absorption detector has

Table 7.7. Methods for the determination of organometallic compounds in freshwater
(non-saline) sediments

Compound	Technique	Detection limit (mg kg^{-1} unless otherwise stated)	Ref.
Organotin compounds			
Mono-, di-, tri- and tetra-substituted organotin compounds	Ethylation with EtMgBr, then high-resolution gas chromatography with flame photometric and mass spectrometric detection	–	82
Alkyl and aryltin compounds	Benzene extraction—conversion to hydrides with NaBH$_4$, then gas chromatography with electron capture detection	0.02	83
Butyltin compounds	Conversion to Bu$_3$MeSn, capillary gas chromatography with flame photometric detectors, also mass spectrometric detector	–	84

Table 7.7. (*continued*)

Compound	Technique	Detection limit ($mg\,kg^{-1}$ unless otherwise stated)	Ref.
Tributyltin, dibutyltin monobutyltin compounds	Capillary gas chromatography with helium microwave-induced plasma emission spectrometric detector	0.05 pg (as Sn) $13-32\,ng\,l^{-1}$	85
Butyltin and phenyltin compounds	Derivativisation, supercritical fluid chromatography	–	86
Methyltin compounds	Conversion to tin-hydrides using $NaBH_4$, ion monitoring with mass spectrometry	3–5 pg level	87
Organolead compounds Organolead	Gas chromatography with atomic absorption spectrometric detector	–	89
Methyl and ethyl lead compounds	Conversion to tetra-alkyl lead compounds using nBu MgCl then gas chromatography with atomic absorption spectrometric detector	0.015	89
Organomercury compounds Methylmercury compounds	Cold vapour atomic absorption spectrometry	$0.1\,\mu g\,l^{-1}$	90–92, 93
	Steam distillation—cold vapour atomic absorption spectrometry	0.0025	94
Methylmercury compounds	Extraction–gas chromatography	–	95, 96
Alkyl and arylmercury compounds	Conversion to chlor-derivatives–gas chromatography	1.001	97
Methylmercury, ethylmercury, methoxyethylmercury	Conversion to iodide, benzene extraction, gas chromatography with electron capture detector	–	98

been applied to achieving the remarkable detection limit of 0.6×10^{-6} $mg\,kg^{-1}$ in the case of organotin compounds. Conversion to hydride followed by purge and trap analysis and gas chromatography with a mass spectrometric detector has been applied to the determination of $0.002-0.02$ $mg\,kg^{-1}$ organotin compounds.

Table 7.8. Methods for the determination of organometallic compounds in marine and estuary sediments

Determined	Type of sediment	Technique	Detection limit (mg kg^{-1} unless otherwise stated)	Ref.
Organotin compounds				
Mono-, di- and tri-methyltin, mono-, di- and tri-*n*-butyltin, mono- and di-ethyltin, phenyltin	Marine	Reaction with sodium borohydride to form tin hydrides, controlled evaporation and detection by atomic absorption spectrometry	–	99
Methyltin, butyltin	Estuary	Reaction with sodium borohydride to form tin hydrides, controlled evaporation and detection by atomic absorption	0.6×10^{-6}	100
Methyltin	Marine	Reaction with sodium borohydride to form tin hydrides, purge and trap analysis followed by gas chromatography with mass spectrometric detection	3–5 pg (as Sn) sub μg kg^{-1}	101
Organolead compounds				
Tetramethylead, trimethyl ethyl lead, dimethydiethyl lead, methyltriethyl lead, tetraethyl lead	Marine	Solvent extraction, gas chromatography with atomic absorption spectrometry detection	0.01	102
Organoarsenic compounds				
	Marine	Ion exchange chromatography, reduction to arsines with sodium borohydride, controlled evaporation and detection by atomic absorption spectrometry	–	103

Conversion of organometallic compounds to volatile derivatives prior to gas chromatography has been applied in the case of organomercury, lead and tin, achieving detection limits of $0.001-0.015$ mg kg^{-1}. Derivativisation followed by supercritical fluid chromatography has been applied to organotin compounds.

Review of methods and detection limits applicable to determination of organo-metallic compounds in freshwater (non-saline) sediments

		Detection limit (mg kg^{-1} *unless otherwise stated*)
Gas chromatography	Methylmercury compounds	
Conversion to chloro derivative—glc	Alkyl and aryl mercury compounds	
With atomic absorption spectrometric detector	Organolead compounds Tetralkylead compounds	0.015
Conversion to hydride, with electron capture detector	Alkyl and aryl tin compounds	0.02
Conversion to iodide derivative, with electron capture detector	Methylmercury, ethyl, mercury and methoxyl and ethyl mercury	
Conversion to butyl methyl tin, capillary glc with flame photometric or mass spectrometric detector	Butyltin compounds	
Conversion to ethyl derivatives, gas chromatography with flame photometric or mass spectrometric detectors	Mono-, di-, tri- and tetra-substituted organo tin compounds	
Capillary gas chromatography with helium microwave-induced plasma emission spectrometric detector	Mono-, di- and tributyl tin compounds	$13-32$ ng l^{-1}
Supercritical fluid chromatography	Butyltin and phenyltin compounds	
Conversion to hydride mass spectrometric detector	Methyltin compounds	$3-5$ pg Sn level

		Detection limit $(\text{mg kg}^{-1}$ *unless otherwise stated*)
Cold vapour atomic absorption spectrometry	Methylmercury compounds	0.0025

Review of methods and detection limits applicable to determination of organo-metallic compounds in marine and esturial sediments

Atomic absorption spectrometry

Reduction to tin hydrides, controlled evaporation—atomic absorption spectrometry	Mono-, di-, trimethyl tin Mono-, di-, tri-*n*-butyltin Diethyltin, phenyl tin	0.6×10^{-6}
Reduction to arsenic hydrides, controlled evaporation—atomic absorption spectrometry	Organoarsenic compounds	
Gas chromatography reduction to tin hydrides, purge and trap combined with gas chromatrography and mass spectrometric detection	Methyltin compounds	sub $\mu\text{g kg}^{-1}$
Gas chromatography—with atomic absorption detector	Tetramethyl lead Trimethyl ethyl lead Dimethyl diethyl lead Methyl triethyl lead Tetra ethyl lead	0.01

7.3   ORGANIC COMPOUNDS IN SEDIMENTS

7.3.1   TOXIC EFFECTS

Following an aviation kerosin spill, hydrocarbons were detected in trout stream sediments and fish up to 14 months after the spill[104]. After a fire at a weed treatment plant in 1970 a large area of mixed forested ecosystem became contaminated with polycyclic aromatic hydrocarbons and creosote[105]. High polyaromatic concentrations in stream sediments adversely affected micro and meio benthic communities at all trophic levels. Stein *et al.*[106] have studied the uptake by bethnic fish (English sole, *Parophrys vetulus*) of benzopyrene and polychlorinated biphenyls from sediments. Accumulation of contaminants from sediments was a significant route of uptake by English sole. It has been shown that microorganisms in river sediments rapidly dechlorinated polychlorobiphenyls[107].

Sediments containing $50-1600$ mg kg^{-1} of triphenyl phosphate altered the drift dynamics of benthic invertebrates. Invertebrates exposed to contaminated sediments drifted almost immediately when threshold toxicity was reached[108].

7.3.2 ANALYTICAL COMPOSITION

Some data on the occurrence of organic compounds in river, lake and marine sediments are given in Table 7.9. References to further information are available in Appendix 3 (Table A3.3). Chlorinated aliphatic and aromatic compounds, detergents and organic phthalate esters are among the compounds that have been detected in sedimentary matter.

Table 7.9. Organic compounds in river, lake and marine sediments

Compound	Type of sediment	Concentration (mg kg^{-1})	Ref.
Aromatic hydrocarbons	Freshwaters	0.001–3	109
1,3-dihexachloro butadiene	River and lake	0.05	110, 111
1,3,5-trihexachloro butadiene		0.25	110, 111
1,2,4-trihexachloro butadiene		0.07	110, 111
1,2,3-trihexachloro butadiene		0.10	110, 111
1,2,3,5-tetrahexachloro butadiene		0.01	110, 111
1,2,3,4-tetrahexachloro butadiene		0.27	110, 111
Pentachlorobutadiene		0.15	110, 111
Hexachlorobutadiene		1.2	110, 110
1,3-dichlorobenzene	Marine	0.031	110
1,4-dichlorobenzene		0.081	110
1,3,5-trichlorobenzene		0.004	110
1,2,4-trichlorobenzene		0.020	110
1,2,3,5-tetrachlorobenzene		0.004	110
1,2,4,5-tetrachlorobenzene		< 0.001	110
Chlorobenzenes	Estuary	0.003–0.07	112
Perchlorobenzene		0.004	110
Hexachlorobenzene		0.007	110
Diethylhexylphthalate		0.1–70.5	113
Dibutyl phthalate		< 0.1–15.5	113
Alkyl benzene sulphonates		16.9–96.3	114
Methylene blue active substances		107–288	114
Nitrogen containing aromatics	Marine	200–1200	115
Fluorescent whitening agents		0.25–1.35	114
Total organic carbon		2.4–65.6	111, 117
		1.4–6.2	118
Total phosphorus		610–1870	119
Total sulphur		229	120

7.3.3 ANALYTICAL METHODS

Methods that have been used for the determination of organic compounds in the sediments originating in non-saline freshwaters and in marine or estuary waters are reviewed, respectively, in Tables 7.10 and 7.11. The major techniques reviewed below include gas chromatography in some cases using a mass spectrometer to identify and determine the organics, variants on the gas chromatographic technique including pyrolysis gas chromatography and purge and trap gas chromatography, high-performance liquid chromatography, ion-pair chromatography and supercritical fluid chromatrography for less volatile sample constituents (detection limits in the 0.0001–0.001 mg kg^{-1} range were achieved for the above-mentioned methods) also ultraviolet and fluorescence spectroscopy and proton nuclear magnetic resonance spectroscopy.

Review of methods and detection limits applicable to determination of organic compounds in freshwater (non-saline) and sediments

		Detection limit (mg kg^{-1} *unless otherwise stated*)
Gas chromatography	Trichloroethylene	
	Tetrachloroethylene	
	Chloroform	
	Carbon tetrachloride	
	Chlorobiphenyls	
	Higher chloro aromatics	0.0005
	Chlorobenzenes	0.004–0.01
With capillary column glc	Polyaromatic hydrocarbons	
	Chlorobenzene	
	Hexachlorobenzene	
	Chloroanisoles	0.002
	2-chloromethyl anisoles	0.002
Pyrolysis–glc	Polyaromatic hydrocarbons	
With ether capture detector	Chlorophenol	0.001
With nitrogen–phosphorus specific detector	Organophosphorus Insecticides	0.0001–0.0002
With photometric detector	Trialkyl and triaryl phosphates	
With mass spectrometric detector	Linear alkyl benzene sulphonates	
	Polyaromatic hydrocarbons	10
	Polychlorobiphenyls	
	Polychloro-dibenzo-*p*-dioxins	

		Detection limit (mg kg^{-1} *unless otherwise stated*)
Purge and trap gas chromatography	Chloroform Chlorobenzenes	
High-performance liquid chromatography	Polyaromatic hydrocarbons	
Ion-pair chromatography	Organosulphur compounds including:	2 p mole absolute
With electrochemical detector	Cysteine Monothioglycerol Glutathion Mercaptopyruvic and 3-mercaptopropanoic acid 2-mercaptopropanoic acid	
Supercriticial fluid chromatography	Polyaromatic hydrocarbons Polychlorobiphenyls Atrazine Cyanazine Desethylatrazine Deisopropylatrazine Metolchlor Dichlorvos Diazinon Ronnel Parathion ethyl Methidathion Tetrachlorovinphos Endrin Endrin aldehyde p,p'-DDT Mirex Dechlorobiphenyl	
Mass spectrometry electron capture negative ion chemical *ionisation mass spectrometry*	Polychlorobiphenyls	
Liquid chromatography—particle *beam mass spectrometry*	Phenoxy acetic acid and ester herbicides	0.1

		Detection limit ($mg\,kg^{-1}$ *unless* *otherwise stated*)
Flame atomic absorption spectrometry	Fulvic acid Fulvic acids (metal complexes)	
Spectrofluorimetry	Polyaromatic hydrocarbons	
Ultraviolet spectroscopy	Polyaromatic hydrocarbons	
Spectrophotometric methods	Herbicides	
Photoacoustic luminescence analysis	Polychlorobiphenyls	

Review of methods and detection limits applicable to determination of organic compounds in marine and estuarial sediments

Gas chromatography	Chlorobenzenes
Acetylation—gas chromato- graphy	Chlorophenols
Capillary gas chromatography with flame ionisation detection	Polyaromatic hydrocarbons
With flame photometric detection	Organosulphur compounds
With non-spectrometric detection	Polychlorobiphenyls Basic nitrogen compounds *n*-alkanes, pristane, phytane
With nitrogen–phosphorus specific detection	Nitrogen-containing aromatics
High-performance liquid *chromatography*	Aromatic hydrocarbons Chlorohydrocarbons Humic acids Fulvic acids
Ultraviolet spectroscopy	Aromatic hydrocarbons
Fluorescence specroscopy	Oils Polyaromatic hydro- carbons Humic acids Fulvic acids
Proton nuclear magnetic resonance spectroscopy	Humic acids Fulvic acids

Table 7.10. Methods for the determination of organic compounds in non-saline sediments

Compound	Technique	Detection limit (mg kg^{-1} unless otherwise stated)	Ref.
Polyaromatic hydrocarbons	Thermal desorption–gas chromatography–mass spectrometry	–	121
	Capillary gas chromatography		122
	High-performance liquid chromatography	0.0001–0.001	123
	Flash evaporation, pyrolysis gas chromatography with mass spectrometric detector		124
	High-resolution spectrofluorimetry	–	125
	Ultraviolet spectroscopy	–	126
	Supercritical fluid chromatography	–	127–130
Linear alkyl benzene sulphonates	Gas chromatography–mass spectrometry	0.5	131
Alkyl phthalates	Gas chromatography	0.1 ng absolute	132
Humic and fulvic acids (metal derivatives)	Flame atomic absorption spectrometry	–	133
Trichloroethylene, tetrachloroethylene chloroform, carbon tetrachloride	Gas chromatography	–	134
Chlorobenzenes	Gas chromatography	0.0004–0.01	135
Higher chloro aromatics, e.g. 29 α- and γ-hexachloro-cyclohexanes, dichlorobenzenes		0.0005	136–138
Chloroform, chlorobenzene	Purge and trap gas chromatography	–	139
Chlorobenzene, hexachlorobenzene	Capillary gas chromatography	–	140
Polychlorobiphenyls	Gas chromatography	–	141–5
Polychlorinated biphenyls	Thermal desorption–gas chromatography–mass spectrometry	10	146, 147
	Electron capture negative ion chemical ionisation mass spectrometry		148
	Supercritical fluid chromatography	–	127, 128
	Photoactivated luminescence analysis	–	149

Analyte	Method	Detection limit	Ref.
Chlorophenols	Gas chromatography with electron capture detector	0.001	150
Chloroanisoles, 2-chloromethyl anisoles	Capillary gas chromatography	0.002	151
Polychlorodibenzo-p-dioxins	High-resolution gas chromatography–low resolution mass spectrometry	–	152
	Gas chromatography–mass spectrometry	–	153
Pesticides and herbicides; atrazine, cyanazine, desethylatrazine, deisopropyl atrazine, metolchlor	Supercritical fluid extraction	–	154
Dichlorvos, diazinon, Ronnel, parathion ethyl, methidathion, tetrachlorovinphos endrin, endrin aldehyde, p,p'DDT, Mirex, decachlorobiphenyl	Supercritical fluid extraction, comparison with classical extraction methods	–	155
Triazine metabolites	Automated solid phase extraction with methanol:water 4:1 v/v, collection on octadecyl resin	0.0001	156
Herbicides	Spectrophotometric and thin-layer chromatography	–	157
Chlorinated phenoxy and ester herbicides	Liquid chromatography, particle beam mass spectrometry and liquid chromatography UV absorption spectrometry	0.1	158
Organosulphur compounds. Cysteine, monothioglycerol, glutathione, mercapto-pyruvic acid, 3-mercaptopropanoic acid, 2-mercaptopropanoic acid	Ion-pair chromatography with electrochemical chemical detector	2 p mole absolute	159
Organophosphorus insecticides	Gas chromatography with N, P detector	0.0001–0.0002	160
Trialkyl and triaryl phosphates	Gas chromatography with photometric detector	–	161
Miscellaneous polyaromatic hydrocarbons phenols, organochlorine pesticides	Microwave-assisted solvent extraction	–	162
Multi-component organic mixtures	Supercritical fluid extraction study of extraction conditions	–	163

Table 7.11. Methods for the determination of organic substances in marine and estuarine sediments

Determined	Type of sediment	Technique	Ref.
n-alkanes, pristane, phytane, pentacyclic triterpanes	Marine	Gas chromatography–mass spectrometry	164
Oils		Fluorescence spectroscopy	165, 166
Aromatic hydrocarbons		Ultraviolet spectroscopy Silica gel chromatography High-performance liquid chromatography	167 168 169
Polyaromatic hydrocarbons	Estuarine Marine	Flame ionisation capillary gas chromatography Thin-layer chromatography spectrofluorimetry, gas chromatography	170 171, 172
	Marine	High-resolution spectro-fluorimetry	188
Chlorohydrocarbons	Estuarine	High-performance liquid chromatography	169
Chlorobenzenes		Solvent extraction gas chromatography	170 185
Chlorophenols	Marine	Acetylation–capillary gas chromatography with electron capture detection	171 186
Humic and fulvic acids	Marine	Proton NMR spectroscopy Fluorescence and absorption spectroscopy Infrared spectroscopy Reversed phase – liquid chromatography	187 173 174 175
Polychlorinated biphenyls		Solvent extraction – gas chromatography–mass spectrometry Solvent extraction, clean-up gas chromatography	176, 177 178
Chlorinated insecticides		Measurement of radioactivity of labelled compounds	179
Nitrogen-containing aromatic compounds		Capillary column gas chromatography with *N* specific detection	180
Basic nitrogen compounds		Gas chromatography–mass spectrometry	181
Organosulphur compounds		Solvent extraction, gas chromatography with flame photometric detector	182
FDA priority pollutants		Clean-up gas chromatography	183

REFERENCES

1. Cherry, D.S., Van Hassel, J.H., Ribbe, P.H. and Cairns, J., *Water Resources Bulletin* **23**, 293 (1987)
2. Hammer, M.T., Merkowsky, A.T. and Huang, P.M., *Archives of Environmental Contamination and Toxicology* **17**, 257 (1988)
3. Pillay, K.K.S., Thomas, C.C., Sondel, J.A. and Hyche, C.M., *Analytical Chemistry* **43**, 1419 (1971)
4. Jirka, A.M. and Carter, M.J., *Analytical Chemistry* **50**, 91 (1978)
5. Fresnet-Rabin, M. and Ottman, F., *Estuarine and Marine Coastal Science* **7**, 425 (1978)
6. Hodge, W.F., Seidel, S.L. and Goldberg, D., *Anal. Chem.* **51**, 1256 (1979)
7. Edenfield, H. and Greaves, M.J., in Wong, C.S. (ed.), *Trace Metals in Seawater*, Proceedings of a NATO Advanced Research Institute on Trace Metals in Seawater, 30 March to 3 April 1981, Sicily, Italy, Wiley, New York (1981)
8. Cutter, G.A., Electric Power Research Institute, Palo Alto, California. Report EPRI EA-5641, *Specification of Selenium and Arsenic in Natural Waters and Sediments: Selenium Speciation.*
9. Sandhu, A.S., *Analyst (London)* **106**, 311 (1981)
10. Bishop, J.K.B., *Analytical Chemistry* **62**, 553 (1990)
11. Zhe-Ming, N., Xiao-Chun, L. and Heng-Bin, H., *Analytica Chimica Acta* **186**, 147 (1986)
12. Goulden, P.D., Anthony, D.A.J. and Austen, K.D., *Analytical Chemistry* **53**, 2027 (1981)
13. Lum, K.R. and Edgar, D.G., *Analyst (London)* **108**, 918 (1983)
14. Sakata, M. and Shimada, O., *Water Research* **16**, 231 (1982)
15. Madsen, S.P., Drabach, I. and Sorensen, J., *Analytica Chimica Acta* **151**, 479 (1983)
16. Xiao-Quan, S., Zhi-Neng, Y. and Zhe-Ming, N., *Analytical Chemistry* **57**, 857 (1985)
17. Smith, R.G., *Analytical Chemistry* **65**, 2485 (1993)
18. Hatle, M., *Talanta* **34**, 1001 (1987)
19. Mudrock, A. and Kokitich, E., *Analyst (London)* **112**, 709 (1987)
20. Dogan, S. and Haerdi, W., *International Journal of Environmental Analytical Chemistry* **8**, 249 (1980)
21. Lang-Zhu, J., *Atomic Spectroscopy* **5**, 91 (1984)
22. Brzenzinska Pandyn, A. and Van Loon, J.C., *Fresenius Zeitschrift für Analytische Chemie* **33**, 707 (1988)
23. Hasebe, K., Kakizaki, T. and Yoshida H., *Fresenius Zeitschrift für Analytische Chemie* **322**, 486 (1985)
24. Van Valin, R. and Morse, J., *Marine Chemistry* **11**, 535 (1982)
25. Welte, B., Bles, N. and Monteil, A., *Environmental Technology Letters* **4**, 223 (1983)
26. Mahan, K.I., Fuderaro, T.A., Garza, T.I., Martinez, R.M., Maroncy, M.R., Trivisanne, M.R. and Willging, E.M., *Analytical Chemistry* **59**, 936 (1987)
27. Revers, F.R. and Hasty, E., Recovery study using an elevated pressure temperature microwave dissolution technique. Presented at 1987 Pittsburgh Conference and Exposition on Analytical Chemistry and Applied Spectroscopy, March (1987)
28. Legret, M., Divet, L. and Demare, D., *Analytica Chimica Acta* **175**, 203 (1985)
29. Bettinelli, M., Pastorelli, N. and Borani, U., *Analytica Chimica Acta* **185**, 109 (1986)
30. Van Son, M. and Muntan, H. *Fresenius Zeitschrift für Analytische Chimie* **328**, 390 (1987)
31. Que-Hee, S.G. and Boyle, J.R., *Analytical Chemistry* **60**, 1033 (1988)
32. Kanda, Y. and Taira, M., *Analytica Chimica Acta* **207**, 269 (1988)

33. Kheboian, C. and Bauer, C.F., *Analytical Chemistry* **59**, 1417 (1987)
34. Colodner, D.C., Boyle, E.A. and Edmund, J.M. *Analytical Chemistry* **65**, 1419 (1993)
35. Cutters, L.S., Cutter, G.A. and San-Diego McClure, M.L.C., *Analytical Chemistry* **63**, 1138 (1991)
36. Madoro, M. and Moauro, A., *Journal of Radioanalytical and Nuclear Chemistry* **90**, 129 (1985)
37. Boniforti, R. Madaro, M. and Moauro A. *Journal of Radioanalytical and Nuclear Chemistry* **84**, 441 (1984)
38. Cheam, V. and Chau, A.S.Y., *Analyst (London)* **109**, 775 (1984)
39. Prange, A., Knoth, J., Stossel, R.P., Baddeber, B. and Kramer, K., *Analytica Chimica Acta* **195**, 275 (1987)
40. Labresque, J.J., Rosales, P.A. and Mejias, G., *Applied Spectroscopy* **40**, 1232 (1986)
41. Xiao-Quan, S. and Bin Chen, *Analytical Chemistry* **65**, 802 (1993)
42. Muhlhauser, H.A., Sota, L. and Zabradrik, P., *International Journal of Environmental Analytical Chemistry* **28**, 2115 (1987)
43. Suzuki, J., Yokoyama, Y., Unno Y. and Susuki, S., *Water Research* **17**, 431 (1983)
44. Whitfield, P.H. and McKinley, J.W., *Water Research Bulletin* **17**, 381 (1981)
45. Charles, M.J. and Simmons, M.S., *Analyst (London)* **111**, 385 (1986)
46. Markert, B.E., Tesmer, M.G. and Parker, P.E., *Water Research* **17**, 603 (1983)
47. Davison, W. and Lishman, J.P., *Analyst (London)* **108**, 1235 (1983)
48. Maher, W.A., *Analyst (London)* **108**, 939 (1983)
49. Brzezinska-Pandyn, A., Van Hoon, J. and Hancock, R., *Atomic Spectroscopy* **7**, 72 (1986)
50. Hutton, R.C. and Preston, B., *Analyst (London)* **105**, 981 (1980)
51. McLaren, J.W. and Berman, S.S., *Applied Spectroscopy* **35**, 403 (1981)
52. Lee, D.S., *Analytical Chemistry* **54**, 1682 (1982)
53. Kiss, E., *Analytica Chimica Acta* **211**, 342 (1988)
54. McLaren, J.W., Beauchemin, D. and Berman, S.S., *Analytical Chemistry* **59**, 610 (1987)
55. Dankers, N. and Laane, R., *Environment Technology Letters* **4**, 283 (1983)
56. Mills, G.L. and Quinn, L.G., *Chemical Geology* **25**, 165 (1979)
57. Weliky, K., Suess, E., Muller, P.J. and Fischer, K., *Limnology and Oceangraphy* **28**, 1252 (1983)
58. Abo-Rady, M.D.K., *Fresenius Zeitschrift für Analytische Chemie* **299**, 187 (1979)
59. Pavlova, U.K. and Yatsimirskii, K.B., *Zhur. Analit. Khim.*, **24**, 1347 (1969)
60. Nothlick, L. and Renter, W., *Deutsche Gewasserkundliche Mitteilungen* **26**, 162 (1982)
61. Willie, S.N., Sturgeon, R.E. and Berman, S.S., *Analytical Chemistry* **58**, 1140 (1983)
62. Siu, K.W. and Berman, S.S., *Analytical Chemistry* **55**, 1603 (1983)
63. Itoh, K., Chikuma, M. and Tanaka, H., *Fresenius Zeitschrift für Analytische Chemie* **330**, 600 (1988)
64. Bloom, N., *Atomic Spectroscopy* **4**, 204 (1983)
65. Riley, J.P. and Siddique, S.A., *Analytica Chimica Acta* **181**, 117 (1986)
66. Pilkington, E.S. and Warren, L.T., *Environmental Science and Technology* **13**, 295 (1979)
67. McLaren, J.W., Berman, S.S., Boyko, V.J. and Russell, D.S., *Analytical Chemistry* **58**, 1802 (1986)
68. de Oliveira, E., McLaren, J.W. and Berman, S.S., *Analytical Chemistry* **55**, 2047 (1983)
69. Goulden, P.D., Anthony, F.H.J. and Austen, K.D., *Analytical Chemistry* **53**, 2027 (1981)

70. Belhomme, J.M., Erb, R., Dequidt, J. and Philippo, A., *Revue Français des Sciences de l'Eau* **1**, 205 (1982)
71. Lansberger, S. and Davidson, W.F., *Analytical Chemistry* **57**, 197 (1985)
72. Piper, D.T. and Goles, G.G., *Analytica Chimica Acta* **47**, 560 (1969)
73. Sturgeon, R.E., Desanliniers, J.A.H., Berman, S.S. and Russell, D.S., *Analytica Chimica Acta* **134**, 283 (1982)
74. Baker, E.T. and Piper, D.Z., *Deep Sea Research* **23**, 1181 (1976)
75. Sinex, S.A., Cantillo, A.Y. and Helz, G.R., *Analytical Chemistry* **52**, 2342 (1980)
76. Cantillo, A.Y., Sinex, S.A. and Helz, G.R., *Analytical Chemistry* **56**, 33 (1984)
77. McQuaker, N.R., Kluckner, P.D. and Chang, G.N., *Analytical Chemistry* **51**, 888 (1979)
78. McLaren, J.W., Berman, S.S., Boyko, V.J. and Russell, D.S., *Analytical Chemistry* **53**, 1802 (1981)
79. Suhr, N.H. and Ingasmello, C.O., *Analytical Chemistry* **38**, 730 (1968)
80. Ellis, K.M. and Chattopadhya, V.A., *Analytical Chemistry* **51**, 942 (1979)
81. Muller, M.D., *Analytical Chemistry* **59**, 617 (1987)
82. Muller, M.D., *Analytical Chemistry* **59**, 619 (1987)
83. Hattori, Y., Kabayashi, A., Takemoto, S., Takami, K., Kuge, Y., Sigimae, A. and Nakamoto, N., *Journal of Chromatography* **315**, 341 (1984)
84. Muller, M.D., *Fresenius Zeitschrift für Analytische Chemie* **317**, 32 (1984)
85. Lobinski, R., Dirlex, W.M.R. and Adams, F.C., *Analytical Chemistry* **64**, 159 (1992)
86. Cal, Y., Aizaga, R. and Bayona, J.M., *Analytical Chemistry* **66**, 1161 (1994)
87. Gilmour, C.C., Tuttle, J.H. and Means, J.C., *Analytical Chemistry* **58**, 1848 (1986)
88. Reisinger, K., Stoeppler, M. and Nurnberg, H.V., *Nature (London)* **291**, 228 (1981)
89. Chau, Y.K., Wong, P.T.S., Bengert, G.A. and Dunn, J.L., *Analytical Chemistry* **56**, 271 (1984)
90. Jurka, A.M. and Carter, M.J., *Analytical Chemistry* **50**, 91 (1978)
91. Masunaga, K. and Takahashi, T., *Analytica Chimica Acta* **87**, 487 (1976)
92. Longmyhr, F.J. and Aamodt, J., *Analytica Chimica Acta* **87**, 483 (1976)
93. Craig, P.J. and Morton, S.F., *Nature (London)* **261**, 125 (1976)
94. *Official Methods of Analysis of AOAC* (11th edition), 418 (1976)
95. Bartlett, P.D., Craig, P.J. and Morton, S.F., *Nature (London)* **267**, 606 (1977)
96. Longbottom, J.E., Dressman, R.C. and Lichtenberg, J.J., *Journal of Association of Official Analytical Chemists* **56**, 1297 (1973)
97. Cappon, C.R. and Crispin-Smith, V., *Journal of Analytical Chemistry* **49**, 365 (1977)
98. Ealy, J.A., Shultz, W.D. and Dean, J.A., *Analytica Chimica Acta* **64**, 235 (1973)
99. Hodge, V.F., Seider, S.L. and Goldberg, D., *Analytical Chemistry* **91**, 1256 (1979)
100. Randall, L., Han, J.S. and Weber, J.H., *Environmental Technology Letters* **7**, 571 (1986)
101. Sinex, S.A., Cantillo, A.Y. and Helz, G.R., *Analytical Chemistry* **52**, 2342 (1980)
102. Chau, Y.K., Wong, P.T.S., Bengert, G.A. and Kramer, O., *Analytical Chemistry* **51**, 186 (1979)
103. Maher, W.A., *Analytica Chimica Acta* **126**, 157 (1981)
104. Guiney, P.D., Sykora, J.L. and Keleti G., *Environmental Toxicology and Chemistry* **6**, 105 (1987)
105. Catello, W.J. and Gambrell, R.P., *Chemosphere* **16**, 1053 (1987)
106. Stein, J.E., Hom, E., Casillas, E., Friedman, A. and Varanasi, U., *Marine Environmental Research* **22**, 123 (1987)
107. Quensen, J.F., Tiedje, J.M. and Boyd, S.A., *Science* **242**, 752 (1988)
108. Fairchild, J.F., Boyle, T., English, W.R. and Rabeni, C., *Water, Air and Soil Pollution* **36**, 271 (1987)

109. Hargrave, B.T. and Phillips, G.A., *Environmental Pollution* **8**, 193 (1975)
110. Onuska, F.T. and Terry, A., *Analytical Chemistry* **57**, 801 (1985)
111. Lee, H.B., Hong You, R. and Chau, A.S., *Analyst (London)* **111**, 81 (1986)
112. Onusaka, F.T. and Terry, K.A., *Analytical Chemistry* **57**, 801 (1985)
113. Schwartz, H.W., Anzion, G.S.M., Von Vleit, H.P.M., Peereboom, J.W.L. and Brinkman, U.A.T., *International Journal of Environmental Analytical Chemistry* **6**, 133 (1979)
114. Uchiyama, M., *Water Research* **13**, 847 (1979)
115. Krone, C.A., Burrows, D.W., Brown, D.W., Robisch, P.A., Friedman, A.J. and Halins, D.C., *Environmental Science and Technology* **20**, 1144 (1986)
116. Mills, G.L. and Quinn, L.G., *Chemical Geology* **25**, 155 (1979)
117. McQuaker, N.R. and Fung, T., *Analytical Chemistry* **47**, 1435 (1975)
118. Kerr, R.A., *The Isolation and Partial Characterisation of Dissolved Organic Matter in Seawater*, PhD thesis, University of Rhode Island, Kingston (1977)
119. Aspila, K.I., Agemian, H. and Chau, A.S.Y., *Analyst (London)* **101**, 187 (1976)
120. Landers, D.H., David, M.P. and Mitchell, M.J., *International Journal of Environmental Analytical Chemistry* **14**, 245 (1983)
121. Robbat, A., Liu Tyng-Yuni and Abraham, B.M., *Analytical Chemistry* **64**, 1477 (1992)
122. Giger, W. and Schnaffrer, C., *Analytical Chemistry* **50**, 243 (1987)
123. Marcomini, A., Striso, A. and Pavoni, B., *Marine Chemistry* **21**, 15 (1987)
124. DeLecuw, J.W., De Leer, E.W.B., Damste, J.S.S. and Schnyl, P.J.W., *Analytical Chemistry* **58**, 1852 (1986)
125. Garrignes, P. and Ewald, M., *Chemosphere* **16**, 485 (1987)
126. Lee, H.K., Wright, C.J. and Swallow, W.H., *Environmental Pollution* **49**, 167 (1988)
127. Hawthorne, S.B., *Analytical Chemistry* **64**, 1614 (1992)
128. Lagenfield, J.J., Hawthorne, S.B., Miller, D.I. and Pawliszyn, J., *Analytical Chemistry* **65**, 338 (1993)
129. Rendt, S. and Hafler, F., *Analytical Chemistry* **66**, 1808 (1994)
130. Burford, M.D., Hawthorne, S.B. and Miller, D.J., *Analytical Chemistry* **66**, 1750 (1994)
131. Trehy, M.J., Gledhill, W.E. and Orth, R.G., *Analytical Chemistry* **62**, 2581 (1990)
132. Thuren, A., *Bulletin of Environmental Contamination and Toxicology* **36**, 33 (1986)
133. Kleuke, T., Oskierski, K.W., Poll, K.G. and Reichel, B., *Wasser Abiwasser* **127**, 650 (1986)
134. Murray, A.J. and Riley, J.P., *Nature (London)* **242**, 37 (1973)
135. Onuska, F.I. and Terry, K.A., *Analytical Chemistry* **57**, 801 (1985)
136. Bierl, R., *Fresenius Zeitschrift für Analytische Chemie* **330**, No. 4/5 (1988)
137. Wegman, R.C.C. and Hafster, A.W.M., *Water Research*, **16**, 1265 (1982)
138. Hellman, H., *Deutsch Gewasserkundliche Mitteilungen* **29**, 111 (1985)
139. Charles, M.J. and Simmons, M.S., *Analytical Chemistry* **59**, 1217 (1987)
140. Lee, H.B., Hang-yen, R. and Chau, A.S.Y., *Analyst (London)* **111**, 81 (1986)
141. Kerkhoff, M.A.J., de Vries, A., Wegman, R.C.C. and Hotske, A.W.M., *Chemosphere* **11**, 165 (1982)
142. Kominor, R.J., Onuska, F.L. and Terry, K.A., *Journal of High Resolution Chromatography and Chromatography Communication* **8**, 585 (1985)
143. Brown, J.F., Bedard, D.L., Brannan, M.J., Carnahan, F.C., Feng, H. and Wagner, R.E., *Science* **236**, 709 (1987)
144. Maris, F.A., Noroozian, E., Otten, R.R., Van Dijek, R.C.J.M., De Jong, G.J. and Brinkman, T., *Journal of High Resolution Chromatography and Chromatography Communication* **11**, 197 (1988)

145. Alford Stevens, A.L., Budde, W.L. and Bellar, T.A. *Analytical Chemistry* **57**, 2452 (1985)
146. Rabbat, D., Liu Tyng-Yuni and Abraham, B.M., *Analytical Chemistry* **64**, 358 (1992)
147. McMurtrey, K.D., Wildman, N.J. and Tal, H. *Bulletin of Environmental Contamination and Toxicology* **31**, 734 (1983)
148. Yu Ma, C. and Bayne, C.K., *Analytical Chemistry* **65**, 772 (1993)
149. Vo Dink, T., Pal, A. and Pal, T., *Analytical Chemistry* **66**, 1264 (1994)
150. Lee, H.B., Stokker, Y.D. and Chau, A.S.Y., *Journal of Association of Official Analytical Chemists* **70**, 1003 (1987)
151. Lee, H.B., *Journal of Association of Official Analytical Chemists* **71**, 803 (1988)
152. Smith, L.M., Stalling, D.L. and Johnson, J.L., *Analytical Chemistry* **56**, 1830 (1984)
153. Tong, H.Y., Giblin, D.E., Lapp, R.L., Monson, S.J. and Gross, M.L., *Analytical Chemistry* **63**, 1772 (1991)
154. Steinheimer, T.R., Pfeiffer, R.L. and Scoggin, K.D., *Analytical Chemistry* **66**, 645 (1994)
155. Snyder, J.L., Grob, R.L., McNally, M.E. and Oostdyk, T.S., *Analytical Chemistry* **64**, 1940 (1992)
156. Mills, M.S. and Thurman, E.M., *Analytical Chemistry* **64**, 1985 (1992)
157. Spengler, D. and Jumar, A., *Pflanzemschutz* **7**, 151 (1971)
158. Kim Insuk, Sasin, O.S., Stephens, R.V., Wong, J. and Brown, M.A., *Analytical Chemistry* **63**, 819 (1991)
159. Shea, D., *Analytical Chemistry* **60**, 1449 (1988)
160. Kjolholt, J., *Journal of Chromatography* **325**, 231 (1985)
161. Ishikawa, S., Taketami, M. and Shinohara, R., *Water Research* **19**, 119 (1985)
162. Lopez Avila, V., Young, R. and Beckert, W.F., *Analytical Chemistry* **66**, 1097 (1994)
163. Fahing, T.M., Panlatis, M.E., Johnson, D.M. and McNally, M.E.P., *Analytical Chemistry* **65**, 1462 (1993)
164. Brown, D., Colsmio, A., Ganning, B., Naf, C., Zabuhr, Y. and Ostman, C., *Marine Pollution Bulletin* **18**, 380 (1987)
165. Zitko, V. and Carson, V.W., Tech. Report Fisheries Research Board, Ottawa, Canada. No. 217 (1970)
166. Scarrett, D.J. and Zitko, V., *J. Fisheries Research Board, Canada* **29**, 1347 (1972)
167. Hennig, H.F.O., *Marine Pollution Bulletin* **10**, 234 (1979)
168. Takada, H. and Ishimatari, R., *Journal of Chromatography* **346**, 281 (1985)
169. Krahn, M.N., Moore, L.K., Bogar, R.G., Wigren, C.A., Chau, S.L. and Brown, D.W., *Journal of Chromatography* **437**, 161 (1988)
170. Readman, J.W., Preston, M.R. and Mantoura, R.F.C., *Marine Pollution Bulletin* **17**, 298 (1986)
171. Dunn, B.P., *Environmental Science and Technology* **10**, 1018 (1976)
172. Dunn, B.P. and Stich H.F.J., *Fisheries Research Board, Canada* **33**, 2040 (1976)
173. Hayase, K., *Geochimica and Cosmichemica Acta* **49**, 159 (1985)
174. Raspar, B., Nurnberg, H.W., Valentia, M. and Branica, M., *Marine Chemistry* **15**, 217 (1984)
175. Hayase, K., *Journal of Chromatography* **295**, 530 (1984)
176. Teichman, J., Bevenue, A. and Hylin, J.W., *Journal of Chromatography* **151**, 155 (1978)
177. Jensen, S., Renberg, L. and Rentergard, L., *Analytical Chemistry* **49**, 316 (1977)
178. Japenga, J., Wagenaar, W.J., Smedes, F. and Salomons, W., *Environmental Technology* **8**, 9 (1987)
179. Picer, N., Picer, M. and Strohal, P., *Bulletin of Environmental Contamination and Technology* **14**, 565 (1975)

180. Krone, C.A., Burrows, D.W., Brown, D.W., Robisch, P.A., Friedman, A.J. and Halins, D.C., *Environmental Science and Technology* **20**, 1144 (1986)
181. Kido, A., Shinohara, R., Eto, S., Koga, M. and Hori, T., *Japan Journal of Water Pollution Research* **2**, 245 (1979)
182. Bates, T.S. and Carpenter, R., *Analytical Chemistry* **51**, 551 (1979)
183. Ozretich, R.J. and Schroeder, W.P., *Analytical Chemistry* **58**, 2041 (1986)
184. Hodge, V., Stallard, M., Kiode, M. and Goldberg, E.D., *Analytical Chemistry* **58**, 616 (1986)
185. Onuska, F.I. and Terry, K.A., *Analytical Chemistry* **57**, 801 (1985)
186. Xie, T.H., *Chemosphere* **12**, 1183 (1983)
187. Pontanen, E.L. and Morris, R.J., *Marine Chemistry* **17**, 115 (1985)
188. Sabel, A., Jarosz, J., Martin-Bouer, M., Paturel, L. and Vial, M., *International Analytical Chemistry* **28**, 171 (1987)
189. Siu, K.W.M., Roberts, S.Y. and Berman, S.S., *Chromatographia* **19**, 398 (1984)

CHAPTER **8**
Pollution of Sea Creatures, Phytoplankton, Algae and Weeds

The concentration of substances picked up from water by sea creatures such as fish and invertibrates and by algal and plant matter is dependent upon the concentration of the substances in the water and, to some extent, upon its concentration in sedimentary matter. Many creatures bioaccumulate toxicants from the water as discussed in Chapter 2 and as a consequence their concentration in the organism is many times higher than that present in the water. Once the concentration of toxicant in the organism exceeds a certain level then harmful effects or mortalities occur.

Analysis of creatures is therefore a very useful means of ascertaining the cause of adverse effects or death in creatures and analysis of plant and algal material is valuable for obtaining an early warning that excessive levels of toxicants may be present. Much work has been carried out on the determination of toxicants in creatures and plant life and this is discussed below.

8.1 METALS IN FISH

8.1.1 TOXIC EFFECTS

As discussed in Chapter 4, the concentrations of metals in fish organs are a useful indicator of the cause of mortalities, while the metal content of gill, muscle and skin do not provide any such indicator. Thus, the data in Table 8.1 show the maximum concentrations of chromium, zinc, copper and cadmium found in opercle, liver and kidney organs taken from environmental fish samples that would lead to fish mortalities, while fish would survive at the lowest concentrations encountered in environmental fish samples.

Cadmium content of muscle taken from juvenile *Tilapia aurea* fish exposed to water containing 6.8–52 μg l^{-1} cadmium ranged from 0.12 mg kg^{-1} at the 6.8 μg l^{-1} level to 0.92 mg kg^{-1} at the 52 μg l^{-1} level in water. Few national or international authorities have set limits for cadmium in foodstuffs, and Norway and the Netherlands are reported to have set a limit of 0.5 mg kg^{-1} in fish[1].

Starved or fed yearling roach (*Rutilus rutilus*) exposed to sublethal copper contamination (80 μg l^{-1} copper) for 7 days accumulated 19 μg copper g^{-1} (dry

Table 8.1. Effect of metal content of fish organs on mortality

Exposure time	Laboratory tests on rudd fish				Maximum concentrations found in organs of wide variety of fish		Minimum concentrations found in fish organs of wide variety of fish	
	10 weeks	3 weeks	12 h	<12 h	mg kg⁻¹ dry weight	Comments (see Appendix 4, Table A4.1)	mg kg⁻¹ dry weight	Comments
Element: chromium								
Concentration (μg l^{-1}) of metal in water	3	16	20	80–145				
Condition of animal (rudd fish)	Good	Good	Good	100% mortality				
Concentration of metal in organ (mg kg^{-1} dry weight)								
Opercle	< 0.2	< 2	8.3	20–26	26	Mortalities during 12 h exposure	8.3	No mortalities in 12 h exposure
Liver	< 0.2	< 2	5.6	15–18	18.4	"	5.6	"
Kidney	< 0.2	< 2	10.3	24–27	23.7	"	10.3	"
Element: zinc								
Concentration (μg l^{-1}) of metal in water	180	800	1600	7500–18 000				
Condition of animal	Good	Good	Good	100% mortality				
Concentration of metal in organ (mg kg^{-1} dry weight)								
Opercle	48	20	115.3	91–174	120	Mortalities during 12 h exposure	No data	

Liver	120	15	42.5	34–63	150	"
Kidney	29	28	154.6	92–216	57	No data
Element: copper						
Concentration ($\mu g\,l^{-1}$) of metal in water	11	50	—	250–1600	No data	13
Condition of animal	Good	Good	—	100% mortality	Mortalities during 12 h exposure	No mortalities
Concentration of metal in organ ($mg\,kg^{-1}$ dry weight)						
Opercle	12	31	—	52–104	12.4	
Liver	7	20	—	22–40	62	1.7
Kidney	6	28	—	30–100	6	0.7
Element: cadmium						
Concentration ($\mu g\,l^{-1}$) of metal in water	3	250	—	1100–11 000	No data	No data
Condition of animal	Good	Good	—	100% mortality	Mortalities during 12 h exposure	No mortalities
Concentration of metal in organ ($mg\,kg^{-1}$ dry weight)						
Opercle	9.5	9		6–29	9.5	
Liver	5	10		4–12	9	0.2
Kidney	4	14		14–28	7.1	3

weight in gill tissue) but only starved fish accumulated significant quantities of copper in water (95 mg kg^{-1} copper, dry weight). Refeeding after cessation of copper exposure resulted in a significant loss of copper from the liver which fell to 70 mg kg^{-1} copper dry weight[2].

Juvenile rainbow trout (*Salmo gairdneri*) upon exposure to 55 μg l^{-1} copper in water from 28 days led to uptakes of copper in the whole body from 1.2 μg g^{-1} copper on day 1 to 6.6 mg kg^{-1} copper on day 28. Liver copper increased from 25 μg^{-1} copper dry weight on day 0 to 69 mg kg^{-1} on day 2 and 113 mg kg^{-1} copper on day 28, both dry weight[3].

Mosquito fish (*Gambusia affinis*) in a reservoir at San Joaquin Valley, California, were found to contain 30 μg selenium kg^{-1} (as selenate) originating in drainage waters. All other species of fish had died[4].

Concentrations of zinc found in *Tilapia zilli* gills, liver and muscle after 4 days' exposure to zinc were, respectively, 38 000, 23 000 and 2000 mg zinc kg^{-1} dry weight. The corresponding figures for *Clarius lazera* were 49 000, 34 000 and 5000 mg kg^{-1} dry weight[426].

8.1.2 ANALYTICAL COMPOSITION

The concentrations of metals found in whole fish and in fish organs from a variety of sources are summarised in Table 8.2 and reviewed in detail in Appendix 4. It will be observed that a wide range of concentrations occur in fish or in organs and that, in general, the highest concentrations of metals are found in fish organs rather than whole fish tissue, and this is particularly so for cadmium and lead[5].

The US Safe Water Drinking Act points out that concentrations exceeding 0.05 μg l^{-1} of mercury and 0.4 μg l^{-1} of cadmium in water will cause levels of these elements in fish and other creatures that may be harmful to aquatic life and human consumers.

8.1.3 ANALYTICAL METHODS

A wide variety of analytical methods have been employed for the determination of metals and non-metals in fish (Table 8.3). The major techniques employed (reviewed below) include electrothermal atomic absorption spectrometry and its cold vapour and hydride generation variants, the more recently developed inductively coupled plasma atomic emission technique and neutron activation analysis. These methods have adequate sensitivity for environmental testing, i.e. below 1 mg kg^{-1}, and in some cases below 0.1 mg kg^{-1}. Other techniques employed for particular elements or groups of elements include gas chromatography (mercury), high-performance liquid chromatography (arsenic), anodic scanning voltammetry (mercury, selenium, copper, lead and cadmium).

Table 8.2. Metal content of whole fish and fish organs

Element	(A) Whole fish tissue				(B) Fish organs				Maximum value in organs
	Minimum reported value	Fish type	Maximum reported value	Fish type	Minimum reported value (See Appendix 4, Table A4.1)	Organ	Maximum reported value	Organ	Maximum value in whole creature
As Inorganic	0.02	Herring and haddock	0.44	Smelt					
As Total	1.1	Herring	2.9	Tuna					
Cd	0.02	Sardine	0.17	Horse mackerel	0.038	Gill	9.5 / 10.9 / 9.0 / 7.1	Opercle / Skin / Liver / Kidney	64
Cr	0.10	Grey mullet	2.2	Rainbow trout	0.8	Muscle	23.7 / 26.0 / 18.4	Kidney / Opercle / Liver	12.6
Cu	0.39 / 0.53	Flathead / Rainbow trout	3.46 / 2.18	Crayfish / Sardine	0.6	Gill	48 / 62	Perch Liver / White fish Liver	13.9–17.9
Pb	0.12	Striped mullet	1.36	Grey mullet	0.12	Muscle	36	Kidney	26.5
Mn	0.22	Striped mullet	1.63	Sardine					
Hg	0.09	Chub, crappie	2.4 / 7.23	Carp / Unidentified					
Ni	0.15	Rainbow trout	0.2	Rainbow trout	0.34	Trout	1.9	Kidney	9.5
Se	0.19	Shark	0.55	Coho salmon					
Ag	0.02 / 0.04	Whale meat / Trout	0.04	Trout					
Zn	6.3	Sardine / Striped mullet	39	Trout	12.6	Liver	150 / 120 / 57	Liver / Opercle / Kidney	2.5–3.8

Table 8.3. Methods for the determination of metals and non-metals in fish

Determined	Technique	Detection limit ($mg\,kg^{-1}$ unless otherwise stated)	Ref.
Arsenic	Hydride (AsH_3) generation – atomic absorption spectrometry	0.3 0.002	6, 7 8, 9
	Individually coupled plasma emission spectrometry	–	10
	Hydride generation (AsH_3) – spectrophotometric analysis	–	11
	High-performance liquid chromatography with inductively coupled plasma atomic emission spectrometric detection	0.3 ng As absolute	12
Antimony	Inductively coupled plasma atomic emission spectrometry	–	10
Cadmium	Flameless graphite furnace atomic absorption spectrometry	0.2 pg in injected portion (0.6 pg fish per 10 ml extract)	13–16
	Atomic absorption spectrometry	–	17, 18
Copper	Spark source mass spectrometry, atomic absorption spectrometry	–	19 18
Lead	Atomic absorption spectrometry	–	17, 18
	Graphite furnace atomic absorption spectrometry	0.15	20, 21
Mercury	Cold vapour atomic absorption spectrometry	0.005	6
		0.01	22–5
	Conversion to methyl mercury – gas chromatography	–	26
	Derivativisation with 2_12^1dimethyl 2-silapentane-5-sulphonate-gas chromatography	–	27
Mercury	Anodic scanning voltrammetry using gold disk electrode	–	28
	Neutron activation analysis	–	29, 30
Selenium	Hydride generation atomic absorption spectrometry	0.2	6
	Inductively coupled plasma atomic emission spectrometry	–	10, 31

Table 8.3. (*continued*)

Determined	Technique	Detection limit (mg kg^{-1} unless otherwise stated)	Ref.
Tin	Flameless atomic absorption spectrometry	–	32
Vanadium	Cation-exchange chromatography Neutron activation analysis	0.03	33
Zinc	Atomic absorption spectrometry	–	18
Co, V, Pb	Ion exchange spectrophotometric method	0.4 pg metal (injected portion)	34
Cr, Cu, Zn, Cd, Ni, Pb	Atomic absorption spectrometry	0.02 (Cd) 0.05 (Ni) 0.1 (Pb) 0.2 (Cr, Cu, Zn)	35
Cd, Zn, Pb, Cu, Ni, Co, Hg		–	36
Cd, Pb, Cu, Mn, Zn, Cr, Hg		–	37
Cd, As, B Cr, Hg, Mo, Ni, Pb, Se	Inductively coupled plasma atomic emission spectrometry	–	38
Se, Cu, Pb, Cd	Differential pulse anodic stripping voltrammetry	–	39
Co, Cr, Se, Ag, Rb, Ni, Zn	Neutron activation analysis	–	40
Al, Au, Br, Ca, Cl, Co, Cr, Se, V, Cu, Fe, I, K, Mg, Mn, Na, Rb, W		–	41
Total nitrogen	Kjeldahl digestion	–	42–4
Organic halogen	Oxygen flask combustion	–	45
Total phosphorus	Ashing—spectrophotometric	–	43, 44
	Solvent extraction – gas chromatography with flame photometric detector	–	46

Review of methods and detection limits applicable to the determination of metals and non-metals in fish

		Detection limit (mg kg^{-1} unless otherwise stated)
Electrothermal atomic absorption spectrometry	Lead	0.15
	Tin	
	Chromium	0.2
	Copper	0.2
	Zinc	0.2
	Cadmium	0.02
	Nickel	0.05
	Cobalt	
	Silver	
	Manganese	
	Mercury	
Hydride generation atomic absorption spectrometry	Arsenic	0.3
	Selenium	0.2
Cold vapour atomic absorption spectrometry	Mercury	0.005
Flameless graphite furnace atomic absorption spectrometry	Cadmium	2 pg in injected portion
Inductively coupled plasma atomic emission spectrometry	Arsenic	
	Antimony	
	Selenium	
	Cadmium	
	Arsenic	
	Boron	
	Chromium	
	Mercury	
	Molybdenum	
	Nickel	
	Lead	
	Scandium	
Gas chromatography with prior derivativisation	Mercury	
High-performance liquid chromatography with ICPAES detector	Arsenic	0.3 ng As absolute

		Detection limit ($mg\,kg^{-1}$ unless otherwise stated)
Anodic scanning voltammetry	Mercury Selenium Copper Lead Cadmium	
Neutron activation analysis	Vanadium Cobalt Chromium Selenium Silver Rubidium Nickel Zinc Aluminium Gold Bromine Calcium Chlorine Vanadium Copper Iron Iodine Potassium Magnesium Manganese Sodium Rubidium Tungsten	0.03
Spark source mass spectrometry	Copper	
Hydride generation spectrophotometric method	Arsenic	
Spectrophotometric method	Cobalt Vanadium Lead	0.4 g per injected portion

8.2   ORGANIC COMPOUNDS IN FISH

8.2.1   TOXIC EFFECTS

Polyaromatic hydrocarbons

It has been postulated that polyaromatic hydrocarbons cause liver, lip and skin tumours in brown bullhead trout (*Ictalurus nebulosus*)[47]. The concentrations of polyaromatic hydrocarbons including benzo(a)anthracene and benzo(a)pyrene found in organisms so affected were high (up to 16 μg kg^{-1} wet weight benzo (a)anthracene and up to 6.4 μg kg^{-1} wet weight benzo(a)pyrene).

Chlorobenzenes

Guppies (*Poecilia reticulata*) have been exposed to 1,2,3-trichlorobenzenes, (1.92, 3.78, 55.9 μmole l^{-1}), 1,2,3,4-tetrachlorobenzene (1.13, 1.69 μmole l^{-1}) or pentachlorobenzene (0.40, 0.54 μmole l^{-1}) in acute flow-through tests. In each experiment time of death was inversely related to toxicant concentration. Irrespective of test compound or exposure concentration, death occurred when internal toxicant concentration reached 2.0–26 μmole g^{-1} fish[48].

Carlson and Kosian[49] studied the toxicity of chlorinated benzenes to fathead minnows (*Pimephales promelas*). Compounds studied were 1,3-dichlorobenzene, 1,4-dichlorobenzene, 1,2,3,4-tetrachlorobenzene, pentachlorobenzene and hexachlorobenzene. The mean tissue residue concentrations were as follows:

	No effect concentration (*NOEC*) (mg kg^{-1})	*Lowest effect concentration* (*LOEC*) (mg kg^{-1})
1,3-dichlorobenzene	120	160
1,4-dichlorobenzene	70	103
1,2,3,4-tetrachlorobenzene	640	1100

Tissue residue concentrations in fish chronically exposed to maximal test concentrations of pentachlorobenzene and hexachlorobenzene were 380 and 97 mg kg^{-1} respectively.

Chlorophenols

As assessment of the sublethal effects on rainbow trout (*Salmo gairdneri*) of 2,4-dichlorophenol and 2,4,6-trichlorophenol has been carried out[50]. Both compounds were accumulated in fish even at the lowest concentration in water tested (5 μg l^{-1}), the greatest amount of chlorophenol being accumulated in the liver, adversely affecting liver enzyme activity. Rogers and Hall[51] determined three tetrachlorophenol isomers in starry flounder (*Platychthys stellatus*) muscle, bone and liver in polluted sites.

Polychlorobiphenyls

Reijnders[52] has reported that seals that fed on polychlorobiphenyl contaminated fish undergo reproductive failure.

Chlorinated insecticides

Kawano *et al.*[53] reported on the concentrations of chlordane compounds present in fish, seabirds, invertebrates, and mammals. The metabolite oxychordane, which is much more toxic than the parent compounds and very persistent, was found in higher concentrations in seabirds than in marine mammals.

Allyl formate

Rainbow trout (*Salmo gairdneri*) which picked up a body burden of 100 μg kg^{-1} of allyl formate developed severe liver damage[54].

8.2.2   ANALYTICAL COMPOSITION

The occurrence of organic compounds in whole fish tissue is summarised in Tables 8.4 to 8.6. The main classes of compounds so far investigated are alicyclic and aromatic hydrocarbons, and polyaromatic hydrocarbons (Table 8.4), various chlorinated compounds, including chlorinated aliphatics and aromatics, polychlorinated biphenyls and chlorinated insecticides (Table 8.5) and a compound which is causing great environmental concern, 2,3,7,8-tetrachlorodibenzo dioxin (Table 8.6).

Polyaromatic hydrocarbons are present in the exhaust gases of most vehicles operating on heavy hydrocarbon fuels. The total concentration of these found in fish is in the range of 0.1–5 mg kg^{-1} (Table 8.4) and certainly concentrations in fish at the higher end of this range gives cause for environmental concern not only in the effect on the fish but also to consumers of that fish. The maximum permitted WHO level of polyaromatic hydrocarbons in drinking water is, for example 0.2 μg l^{-1} (six, compounds fluoranthene, benzo(d)fluoranthene, benzo(k)fluoranthene, benzo(a)pyrene, benzo(ghi)perylene and indeno(1,2,3,ed)-pyrene).

Polychlorinated biphenyls

Concentrations of polychlorinated biphenyls in fish are up to 2.2 mg kg^{-1} (Table 8.6) and at this level adverse effects would be expected in birds and

Table 8.4. Concentrations of hydrocarbons (mg kg^{-1} dry weight) occurring in environmental fish samples

Compound	Tuna	Trout	White fish
Pristane	2.4 (176)		
Methylcyclohexane		0.002 (56)	
Ethylcyclohexane		0.001 (56)	
Propylcyclohexane		0.002 (56)	
Benzene		0.008 (56)	
Toluene		0.008 (56)	
m/p-xylene		0.005 (56)	
Methyl-3-methyl benzene		0.04 (56)	
1,3,5-trimethylbenzene		0.05 (56)	
1-methyl-1,4-propylbenzene		0.01 (56)	
2-ethyl-1,4-dimethylbenzene		0.01 (56)	
2-methyl cyclopentanol		0.090 (56)	
4-ethyl-1,2-dimethylbenzene		0.001 (56)	
4-methylindan		0.002 (56)	
Naphthalene		0.001 (56)	
2-Methylnaphthalene			0.001–0.006 (57)
Biphenyl			0.001–0.014 (57)
C$_2$ naphthalenes			0.017–0.18 (57)
Acetnaphthalene			0.043–0.27 (57)
Acenaphthalene			0.007–0.039 (57)
Dibenzothiophene			0.021–0.27 (57)
Phenanthacene			0.002–2.7 (57)
Methyldibenzothiophenes			0.17 (13)
Fluoranthene			0.004–1.8 (57)
Phenanthro(4,5,b,c,d) thiophane			0.016–0.078 (57)
Pyrene			0.004–1.5 (57)
Benzon(b)naphtho(2.1d)thiophene			0.006 (57)
Benz(a)anthracene			0.004–0.022 (57)
Chrysene			0.003–0.061 (57)
Benzo(e)pyrene and Benzo(a)pyrene			0.014 (57)
Perylene			0.001–0.007 (57)
Phenanthrene plus anthracene			0.008 (57)

References in parentheses.

fish and humans who consume the fish. Much the same can be said for chlorinated insecticides, between 0.05 and 20 mg kg^{-1} have been detected in fish.

2,3,7,8-tetrachlorodibenzo dioxin is one of the more toxic substances produced in the combustion of higher boiling point organic chlorine compounds. As was demonstrated in the Savisesio incident, even minute traces of the substance in the soil for many miles surrounding its release point had severe health effects. The results in Table 8.6 indicate that up to 0.48 mg kg^{-1} of this substance has been detected in edible fish and this must give rise to concern.

Table 8.5. Concentration of organic chlorine compounds ($mg\,kg^{-1}$ dry weight) occurring in environmental fish

	Perch	Trout	Pike	Cod	Salmon	White bass	Whale	Herring
Bromochloromethane		0.008 (56)						
Pentachlorobenzene				0.001–0.002 (60)				
Hexachlorobenzene				0.08–0.17 (60)				
Octachlorostyrene				1.3–4.2 (60)				
Polychlorinated biphenyl 1242			0.89 (55)					
Polychlorinated biphenyl 1254			1.01 (55)					
Polychlorinated biphenyl 1260			0.48 (55)					
Polychlorinated biphenyl				2.2 (58)			0.69–5 (62)	
DDE				2.2–20 (58)				
DDD				0.59–8.0 (58)				
DDT				0.47–7.5 (58)				
Mirex		0.05–0.36 (16)	0.05 (55)		0.09–0.33 (59)	0.06–0.43 (59)	1.25–7.4 (62)	
Polychloro-2-(chlormethyl sulphonamide) diphenylether (Eulan WA)	0.3–0.33 (61)							
Dieldrin							0.007–0.04 (62)	
Toxaphane				1.1 (63) (liver)				0.4–1 (63)

References in parentheses.

GOVERNORS STATE UNIVERSITY
UNIVERSITY PARK
IL 60466

Table 8.6. Concentrations (mg kg^{-1} dry weight) of 2,3,7,8-tetrachloro dibenzo-p-dioxin occurring in environmental fish

Lake trout	Carp	Ocean herring	Rainbow trout	Edible fish	Catfish	Buffalo fish	Predator fish	Bottom-feeding fish
< 0.004–0.014 (64)	0.001–0.094 (65)	< 0.001–< 0.01 (65)	0.031–0.038 (65)	0.48 (66)	0.04–0.05 (66)	< 0.01 (66)	0.015–0.23 (66)	0.077 (66)
0.054–0.058 (65)								

References in parentheses.

As in the case of metals, certain organic substances tend to concentrate in the organs of fish. For example, the concentration of polychlorinated biphenyl found in the liver of long nose gar (1.11–3.7 mg kg^{-1}) is appreciably higher than that found in whole fish tissue (0.5–1.0 mg kg^{-1})[55].

Rogers and Hall[51] determined polychlorobiphenyls in starry flounder (*Platichthys stellatus*) in muscle, bone and liver in polluted sites. Polychlorobiphenyls have been found in the flesh of starry flounders (*Platichthys flesus*) caught in the Elbe estuary, the German Bight[67] and San Francisco Bay[68].

O'Connor and Pizza[69] studied the pick-up by tissues and elimination routes for polychlorobiphenyls in striped bass (*Morone saxatilus*) from the Hudson River. In a single close study, measurable quantities of polychlorobiphenyls were detected in tissues 6 hours after dosing and peaked 1–2 days after dosing. Approximately 53% of the administered dose was eliminated by the fish within 120 hours. Polychlorobiphenyl burdens of the fish increased with successive doses of polychlorobiphenyls.

Organochlorine insecticides

Concentrations of αBHC, β, BHC aldrin, heptachlor, heptachlor expoxide, α-endosulphan, β-endosulphan, α-chlordane and γ-chlordane have been determined in samples of 13 commercially significant fish species caught in the North West Amercian Gulf[70]. Concentrations of all of these in fish tissue were below the analytical detection limit of 1 μg kg^{-1} wet weight. DDT was the most prevalent organochlorine pesticide with average concentrations in fish ranging from 1 to 28 μg kg^{-1}. Dieldrin was detected in about 25% of fish species examined at 1–4 μg kg^{-1} wet weight. Total DDT and Endrin residues in fish caught in an insecticide-sprayed lake were 5–72 μg kg^{-1} and 3–67 μg kg^{-1} respectively.

Levels of DDT and polychlorobiphenyl have been determined in immature cod species (*Gadus morhua*) liver and herring (*Clupea larengus*) muscle[71]. DDT levels in herring muscle between 1979 and 1986 were $0.3-2.2$ mg kg^{-1}, reducing to $0.010-0.017$ mg kg^{-1} in 1988. Polychlorobiphenyl levels in cod liver between 1979 and 1986 were 0.3 to 3.7 mg kg^{-1} reducing to $0.013-0.19$ mg kg^{-1} in 1988.

Data have been obtained on the concentrations of aldrin, endrin, endosulphan, heptachlor, heptachlor epoxide, lindane and the DDT group in black bullhead, bleak, chub, common carp, eel and tench collected in Italian rivers in 1986[72]. Aldrin, lindane, heptachlor and endosulphan were detected in less than 20% of the fish examined, whereas dieldrin was found in almost all the fish studied. Total DDT group residue concentrations in fish were between 17 and 153 μg kg^{-1}, depending on the river. Other pesticides were at lower concentrations (up to 39 μg kg^{-1}).

Fingerling rainbow trout (*Salmo gairdneri*) exposed for up to 4 days to 10 μg l^{-1} of Aminocarb in water at pH $4.6-8.2$ picked up a 9.1 mg kg^{-1} Aminocarb in fish tissue in the first 6 hours of exposure[73]. At pH 8.2 whole body Aminocarb increased to 12 mg kg^{-1} in one hour and remained elevated until the fish died in 72 hours.

Various other workers have reported on the concentrations of chlorinated insecticides found in fish and fish tissues; benzene hexachloride, hexachlorocyclhexane, heptachlor, aldrin, DDT and polychlorobiphenyl, toxaphene[75]; polychlorobiphenyls, DDT and hydrocarbons[76], polychlorobiphenyls and *p,p'*DDE[77] and lindane[78].

Concentrations of Eulan W.A. (polychloro-2-(chloromethyl sulphonamido)diphenyl ethers) in perch livers ($4.5-5.5$ mg l^{-1}) is appreciably higher than that found in the whole fish tissue ($0.30-0.33$ mg kg^{-1})[61].

Diethylhexyl phthalate

Between 0.002 and 0.02 mg kg^{-1} diethylhexyl phthalate, PVC, plasticiser has been found in fish[79].

8.2.3   ANALYTICAL METHODS

Methods for the determination of organic compounds in fish are reviewed in Table 8.7. Gas chromatographic methods of identifying the separated compounds predominate. Direct chemical ionisation mass spectrometry has been used to identify and determine polychlorobiphenyls and methyl sulphone-substituted polychlorobiphenyls and high-resolution electron impact mass spectrometry as well as chemical ionisation atmospheric pressure mass spectrometry to identify and determine 2,3,7,8-tetrachloro benzo-*p*-dioxins and other dioxins (also dibenzofurans). High-performance liquid chromatography has also been employed to examine dioxins as these, due to low volatility, are not amenable to gas chromatographic techniques.

234

Table 8.7. Methods for the determination of organic compounds in fish

Determined	Type of sample	Technique	Detection limit (mg kg^{-1} unless otherwise stated)	Ref.
Hydrocarbons aliphatic and aromatic	Fish	Gas chromatography with mass spectrometric detector	–	80, 81
Pristane	Tuna, cod, liver Fish lipids	Gas chromatography, column and thin-layer chromatography	–	82
Polyaromatic hydrocarbons	Flounder bile	Derivativisation spectro-fluorimetry	5 mg l^{-1}	83
	Fish	Capillary gas chromatography	0.5	84 85
	Fish	Gas chromatography with mass spectrometric detector	μg kg^{-1}	86
Chloroaliphatic compounds	Fish	Gas chromatography	–	87–93
Trichloroethylene tetrachloroethylene, chloroform, carbon tetrachloride	Fish		–	94, 95
1,1,1-trichloroethane, trichloroethylene, per-chloroethylene, 1,1,1,2-tetrachloroethylene, 1,1,2,2-tetrachloroethylene, pentachloroethylene, hexachloroethylene, pentachlorobutadiene, hexa-chlorobutadiene, chloroform, carbon tetrachloride	Fish		–	96–9
Haloparaffins	Fish	Gas chromatography – mass spectrometry	–	100
Organic halogen compounds	Fish	Extraction – neutron activation analysis	–	101
Pentachlorophenol	Fish Fish	Gas chromatography Conversion to silyl derivative – gas chromatography	0.15–0.3 –	102 103
	Fish	Gas chromatography with electron capture detection	–	104

Pentachlorophenol trichlorophenol, 2.4-dichlorophenol	Fish	Gas chromatography	–	105
Chlorophenols	Fish	Ion exchange chromatography conversion to methyl esters, gas chromatography	0.0001–0.001	106
Polychlorinated biphenyls	Fish, white sucker	Gas chromatography	–	107–110
	Fish	Gas chromatography with electron capture detector	–	111
	Trout muscle	Capillary gas chromatography with mass spectrometric detector	–	112
	Fish	Electron capture negative chemical ionisation mass spectrometry	–	113
Polychlorinated biphenyls	Fish	Thin-layer chromatography	$0.5–1\ \mu g$ absolute	114,
	Biota	Photoactivated luminescence	–	116
Methyl sulphone substituted polychlorobiphenyls	Adipose liver grey seals	Mass spectrometry	–	117
Polychlorinated dibenzo-p-dioxins	Fish	Gas chromatography	low ng kg^{-1}	118
	Fish	Gas chromatography with mass spectrometric detector	low ng kg^{-1}	119–122
2,3,7,8-tetrachlorodibenzo-p-dioxin	Rainbow trout	High-performance liquid chromatography	10^{-5} to 10^{-4} ($10–100$ ng kg^{-1})	123
	Fish	High-resolution electron impact mass spectrometry, also chemical ionisation atmospheric pressure mass spectrometry	ng kg^{-1} level	124–126
	Plasma blood	Collection on C$_{18}$ bonded cartridge	–	127

continued overleaf

Table 8.7. (*continued*)

Determined	Type of sample	Technique	Detection limit (mg kg^{-1} unless otherwise stated)	Ref.
Dibenzofurans	Fish	Gas chromatography mass spectrometry	–	122, 125, 126
	Fish	High-resolution electron impact mass spectrometry, also chemical ionisation atmospheric pressure mass spectrometry	ng kg^{-1} level	124
Chlorinated insecticide	Fish	Gas chromatography	0.001	128–35
	Fish	Derivativisation–gas chromatography	–	111
DDT dieldrin	Whale	Gas chromatography	–	136
DDE, DDD, DDT	Fish extracts		–	137
Toxaphene	Fish		–	138, 139
Mirex (C$_{10}$H$_{12}$)	Lake fish	Gas chromatography – mass spectrometry	–	140–44
	Trout, lamprey		500 fg absolute	145
	Fish	Gas chromatography with electron capture detector	0.005 μg kg^{-1}	146–9
Phenoxy acetic acid herbicides	Fish	Ion exchange chromatography, conversion to methyl esters, gas chromatography	0.0001–0.001	106
Pirimphos methyl 0-(2-dimethyl amino)-6-methyl-4-pyriminyl-0,0-dimethyl phosphorothioate	Fish	Gas chromatography	–	150
Dursban	Fish	Gas chromatography	0.01	151

Compound	Matrix	Method	Value	Ref.
Acephate methamidophos	Fish		–	152
Fenthion	Fish		0.01	153
Fluridone	Crayfish	Liquid chromatography	0.04–0.05	154
Atrazine, terbutryn	Fish	Assay based on inhibition of photosynthetic electron transport in spinach thylalkaloids by target compounds with spectrophotometric detection of redox dye	–	155
Triaryl phosphate	Fish	Gas chromatography	3	156
Polychloronitrobenzenes	Fish	Gas chromatography – mass spectrometry	–	157
Methylamine, trimethyl amine oxide	Fish	Gas chromatography	–	158
Hypoxanthine	oxide,			
1,1-dichloromethyl sulphone	Flounders	Gas chromatography – mass spectrometry	–	159
Polychlorostyrene	Fish		–	160–2
Hexachlorobenzene, octachlorostyrene	Fish eggs	Gas chromatography	1	163
Phthalate esters	Marine fish	Gas chromatography with electron capture detector	0.5	164
Squoxin (1,1-methylene 2-naphthol)	Fish	Derivativisation then gas chromatography	–	165
Eulan WA (polychloro-2 chloromethyl sulphamido) diphenyl esters	Fish	Methylation – gas chromatography	0.005	166
$\alpha\alpha$-tri-fluoro 4-nitro-m-cresol	Fish		0.01	167
Geosmin, 2-methyl isoborneol	Fish	Gas chromatography	0.005	168

These techniques are reviewed below together with the very satisfactory detection limits achieved.

Review of methods of detection limits applicable to the determination of organic compounds in fish

		Detection limit ($mg\,kg^{-1}$ *unless otherwise stated*)
Gas chromatography	Polyaromatic hydrocarbons	0.5
	Chloroaliphatic compounds	
	Haloforms	
	Chlorobutadienes	
	Octachlorostyrene	1
	Hexachlorostyrene	1
	di-, tri- and pentachlorophenols	
	Polychlorinated biphenyls	0.0001–0.001
	Chlorinated insecticides	0.001
	Polychloro dibenzo-*p*-dioxins	low $ng\,kg^{-1}$
	Organophosphorus insecticides	
	Triaryl phosphate	3
	Trimethyl amine	
	Trimethylamine oxide	
	Geosmin	0.005
	2-methyl iso borneol	0.005
	Fenthion	0.01
	Acephate	
	Phenthion	
With electron capture detector	Polychlorobiphenyls	
	Mirex	
	Phthalate esters	0.5
With mass spectrometric detector	Aliphatic hydrocarbons	
	Aromatic hydrocarbons	
	Polyaromatic hydrocarbons	$ng\,kg^{-1}$ level
	Haloparaffins	
	Polychlorobiphenyls	
	Polychlorodibenzo-*p*-dioxins	$ng\,kg^{-1}$
	Dibenzofurans	$ng\,kg^{-1}$
	Mirex	
	1,1-dichloromethyl sulphone	
	Polychlorostyrenes	
	Polychloronitrobenzenes	

		Detection limit (mg kg^{-1} unless otherwise stated)
With mass spectrometric detector (continued)	Pentachlorophenol as methyl ester	0.0001–0.001
Derivativisation — gas chromatography	Pentachlorophenol	0.15–0.3
	Chlorinated insecticides (as silyl derivative)	0.0001–0.001
	Phenoxyl acetic acids (as methyl esters)	
	Eulan (polychloro-2(chloromethyl) sulphamido diphenyl ethers (as methyl ester)	0.005
	Squoxin (1,1'-methylene-2-naphthol)	
High-performance liquid chromatography	2,3,7,8 tetrachloro-dibenzo-p-dioxin	10–100 ng kg^{-1}
	Pristane	
	Fluridene	0.01
Mass spectrometry	Methylsulphone substituted polychlorobiphenyls	
Chemical ionisation (CI) atmospheric pressure mass spectrometry	2,3,7,8-tetrachloro Dibenzo-p-dioxin	ng kg^{-1} level
	Dibenzofurans	ng kg^{-1} level
Electron capture negative ion chemical ionisation (CI) mass spectroscopy	Polychlorobiphenyls	
High-resolution electron impact (EI) mass spectrometry	2,3,7,8-tetrachloro dibenzo-p-dioxins, dibenzofurans	ng kg^{-1} level
Neutron activation analysis	Organohalogen hydrocarbons	
Derivativisation — spectrofluorimetry	Polyaromatic hydrocarbons	0.005

8.3   ORGANOMETALLIC COMPOUNDS IN FISH

8.3.1   TOXIC EFFECTS

Organoarsenic compounds

Juvenile rainbow trout (*Salmo gairdneiri*) fed for 8 weeks on a diet containing arsenic trioxide (180–1477 μg As g^{-1} diet), disodium arsenate (137–1054 μg As g^{-1} diet), dimethylarsinic acid (163–1497 μg As g^{-1} diet) or arsanalic acid (193–1503 μg As g^{-1} diet) all underwent adverse effects on growth, food consumption and feeding behaviour when fed with inorganic arsenic compounds, but were unaffected by diets containing the organoarsenic compounds. In all cases carcass arsenic concentrations of arsenic were related to dietary arsenic concentration[169].

8.3.2   ANALYTICAL COMPOSITION

Organic compounds of both lead and mercury have been found in fish (Tables 8.8, 8.9 and 8.12). These originate predominantly from the use of alkylead compounds in petroleum and the methylation of inorganic mercury released into the ecosystem as effluents in the chloralkali process.

Organomercury compounds

Richman *et al.*[170] have discussed the factors that might govern the uptake of mercury by fish in acid-stressed lakes. It was concluded that mercury cycling and uptake in aquatic systems were governed by a variety of interconnecting and sometimes covarying factors, the relative importance of which could differ from lake to lake.

Organotin compounds

Bailey and Davies[171] determined tributyltin concentrations in dogwhelk samples taken at various locations in Sullom Voe, Shetland. These ranged from values of 0.1 mg kg^{-1} inside Sullom Voe down to less than 0.03 mg kg^{-1} in Yell Sound. Concentrations of 0.02 to 0.03 mg kg^{-1} were found in edible tissue of queen scallops inside the Voe but tin was rarely detected in commercial shellfish outside the Voe. Only very low concentrations of tributyl tin ($\sim$2 ng l^{-1}) were found in a small proportion of seawater samples taken in the area.

Table 8.8. Concentrations ($mg\,kg^{-1}$ dry weight) of organolead compounds found in environmental fish

	Misc. fish	Carp	Bass	Small moult bass	Pike	White sucker	Range
Me_4Pb	0.43 (218)	0.14 (172)					0.14–0.43
Me_3EtPb		<0.001 (172)	<0.001 (172)				<1.001
Me_2Et_2Pb		1.43 (172)		0.057 (172)			0.057–1.43
$MeEt_3Pb$		0.14 (172)		0.19–0.25 (172)	0.10 (172)	0.29 (172)	0.14–0.29
Et_4Pb		0.78–7.5 (172)		1.20–1.83 (172)	0.15–0.17 (172)	2.95–4.38 (172)	0.78–7.5
Me_3Pb^+		0.16–2.73 (172)		<0.01 (172)	1.02–1.12 (172)	0.09–0.20 (172)	<0.01–2.73
Me_2Pb^{2+}		0.36 (172)			0.20–0.21 (172)		0.36
Et_3Pb^+		0.09–1.21 (172)		0.22–0.86 (172)	0.053 (172)	2.17–2.43 (172)	0.53–3.43
Et_2Pb^{2+}		0.71–1.31 (172)		0.09–2.75 (172)		2.2–4.3 (172)	0.09–4.3
Pb^{2+}		1.28–4.13 (172)		0.25–0.30 (172)	1.04–1.19 (172)	3.61–3.48 (172)	0.25–4.13
Total excl. Pb^{2+}		5.09–18.94	<0.01	1.76–5.55	1.52–1.65	7.70–12.60	

References in parentheses.

Source: Chau, Y.K., Wong, P.T.S., Bengent, G.A. and Dunn, J.L., *Analytical Chemistry* **56**, 271 (1984) © 1984 American Chemical Society.

Table 8.9. Concentration ($mg\,kg^{-1}$ dry weight) of organomercury compounds found in environmental fish

	Whiting	Sardine	Turbot	Halibut	Coho salmon	Salmon	Red tuna	White tuna	Rock fish
$MeHg^+$								0.93 (177)	0.1–1 (182)
$EtHg^+$								<0.01 (177)	
$PhHg^+$								<0.01 (177)	
CH_3HgCl	0.08 (173)	0.03 (173)	<0.01 (173)	5.65 (174)	0.18–0.20 (175)	0.06–0.11 (176)	1.89–8.3 (178)	0.54 (173) 0.61 (174) 0.33–0.34 (178) 0.08–0.69 (179) 0.36–0.91 (180) 0.35–0.74 (181) 0.15–0.69 (184) 0.3–11.5 (183)	

	Pike	Trout	Rainbow trout	Whale	Shark	Swordfish	Octopus	Squid
CH_3HgCl	0.11–0.88 (176)	0.06–1.46 (175)	0.05–1.97 (175)	0.56–1.09 (175)	8.41 (177) nd (177) nd (177)	0.57–1.01 (173) 0.40–3.17 (174)	<0.01 (173)	<0.01 (173)

References in parentheses.

Table 8.10. Methods for the determination of organometallic compounds in fish

Determined	Technique	Detection limit (mg kg^{-1} unless otherwise stated)	Ref.
Organomercury compounds			
Methylmercury salts	Spectrophotometric dithizone method	–	184
Organomercury compounds	Atomic absorption spectrometry	–	185–202
Organomercury compounds	Acid cuprous bromide extraction – graphite furnace atomic absorption spectrometry	–	198
Organomercury compounds in dogfish muscle, lobster hepato-pancreas	Toluene extraction, cysteine extraction then inductively coupled plasma atomic emission spectrometry using flow injection analysis	–	203
Methylmercury compounds	High-performance liquid chromatography with atomic absorption spectrometric detector	0.0037 μg absolute	204
Methylmercury compounds	High-performance liquid chromatography with electrochemical detector	0.002	205
Alkylmercury compounds	Sediment extraction, gas chromatography	0.01 0.02 0.001	206 207 208
Methylmercuric chloride in swordfish, shark, shrimp, oyster, clam, tuna	Gas chromatography with electron capture detector	0.25	209–11, 213
Organomercury compounds	Gas chromatography with helium plasma detector, ethylation with NaB(C$_2$H$_5$)$_4$, cryogenic trapping of ethylation products then gas chromatography	– 4 ng kg^{-1} as Hg (CH$_3$Hg$^+$) 75 ng kg^{-1} as Hg (labile Hg)	212
Methylmercury compounds	Neutron activation analysis	0.01	214
	Enzymic breakdown–gas chromatography	–	215
Organolead compounds			
Tetramethyl lead, methyl lead, triethyl dimethyl lead, diethyl trimethyl lead, ethyl tetraethyl lead	Gas chromatography with atomic absorption detector	0.025	216, 217
	Butylation of organolead – lead compounds – gas chromatography	–	218

continued overleaf

Table 8.10. (*continued*)

Determined	Technique	Detection limit (mg kg^{-1} unless otherwise stated)	Ref.
Alkyl lead	Differential pulse analysis anodic scanning voltrammetry	0.01	219
Organotin compounds			
Tributyltin	Flameless atomic absorption spectrometry	–	220
Tributyltin	Gas chromatography with atomic absorption detector	–	220
Organotin compounds	Gas chromatography with flame photometric detector	–	221
Tributyltin compounds	Graphite furnace atomic absorption spectrometry	–	222
Organoselenium compounds			
Organoselenium compound metabolites	Anion exchange chromatography – molecular neutron activation analysis	–	223
Organoarsenic compounds			
Mono- and dimethyl arsenic compounds	Sodium borohydride reduction, then atomic absorption spectrometry	< 1	224 225
Organoarsenic compounds in dogfish muscle	High-performance liquid chromatography with inductively coupled plasma mass spectrometric detector	0.3 μg As absolute	226
	Also, thin-layer chromatography, electron impact mass spectrometry and graphite furnace atomic absorption spectrometry		
Organoboron compounds (Et$_3$NH)$_2$ B$_{12}$H$_{12}$ CS$_2$B$_{12}$H$_{11}$SHH$_2$O C$_{15}$H$_{32}$B$_{10}$O$_5$	DC plasma atomic emission spectrometry	0.1	227
Organosilicon compounds Siloxanes	Inductively coupled plasma atomic emission spectrometry	0.01	228

8.3.3 ANALYTICAL METHODS

Methods used for the determination of fish of organometallic compounds of mercury, lead, tin, selenium, arsenic, boron and silicon are summarised in Table 8.10. Where quoted, detection limits are usually well below $0.1 \, mg \, kg^{-1}$, frequently in the $0.01 \, mg \, kg^{-1}$ range. Methods are thus available for the most important types of organometallic compounds in fish at levels of environmental concern. The techniques employed and detection limits achieved are reviewed below:

Review of methods and detection limits applicable to the determination of organo-metallic compounds in fish

		Detection limit ($mg \, kg^{-1}$ *unless otherwise stated*)
Atomic absorption spectrometry		
Hydride generation atomic absorption spectrometry	Methylmercury	< 1
Flameless atomic absorption spectrometry	Organomercury Butyltin	
Graphite furnace atomic absorption spectrometry	Organomercury Organotin	
Inductively coupled plasma atomic emission spectrometry	Organomercury Siloxanes	0.1
Solvent extraction gas chromatography	Alkylmercury	0.001–0.02
With electron capture detector	Alkylmercury	0.25
With atomic absorption detector	Alkyl lead Tributyltin	0.025
With flame photometric detector	Organotin	
Ethylation then with helium plasma detector	Organomercury	4–75
Enzyme breakdown then gas chromatography	Methylmercury	
High-performance liquid chromatography		
With atomic absorption detector	Methylmercury	0.037 ng Hg absolute

		Detection limit (mg kg^{-1} *unless otherwise stated*)
With electrochemical detector	Alkylmercury	0.002
With ICPAES detector	Organoarsenic	0.3 μg As absolute
Differential pulse anodic scanning voltammetry	Alkyl lead	0.01
Neutron activation analysis	Organomercury Organoselenium	0.01
dc Plasma atomic emission spectrometry	Organoboron	0.01
Spectrometric methods	Methylmercury salts	

8.4   METALS IN CREATURES OTHER THAN FISH

8.4.1   TOXIC EFFECTS

Samples of the barnacle (*Balanus amphitrite*) collected in the Zuan estuary, India, had copper concentrations in its tissue of 39.7–864.8 mg kg^{-1}. The copper content of the waters in the region was 1–11 μg l^{-1} [229].

Slipper limpets (*Crepidula fornicata*) exposed to 5–50 μg l^{-1} mercury for 16 weeks contained 28–75 mg kg^{-1} wet weight of mercury in their tissues[230]. The mortality of limpets was water-temperature dependent. Thus, at 13.5°C a 114-day LC_{50} value of 1100 μg l^{-1} was obtained while at 3°C no mortalities of limpet occurred upon exposure to 1600 μg l^{-1} mercury in water for up to 114 days. Mercury and selenium are both toxic to the mussels *Mytilus edulis* and both appear in the tissues[231].

8.4.2   ANALYTICAL COMPOSITION

Available information of the concentrations of metals found in these creatures is given in Table 8.11 (more detailed information and references appear in Appendix 4, Table A4.1). Again, as in the case of fish, metal concentrations vary over a wide range and certainly cover the region where adverse effects or mortalities in the creatures would occur and where the suitability of the creature as an item of human diet would be queried.

High and variable concentrations of cadmium have been reported in the tissues of the mollusc *Murex trunulus* taken in Calvi Bay, Corsica, during the tourist season[232]. Doherty *et al.*[233] have reported the occurrence of a metallothionen-like metal binding protein to the soft tissues of Asiatic clams following exposure to

Table 8.11. Metal content (mg kg⁻¹) of sea creatures other than fish

Element	Oyster	Lobster	Crab	Whelk	Canned lobster	Prawn	Scallop	Mollusc	Mussel	Clam	Shrimp	Range
Sb	0.4	0.071–0.089										0.071–0.4
As	13.4	11.9–25.5 (13.4 hepato-pancreas)	1.5 (tot) 0.06–0.1 (inorg)	3.2 (tot) 0.06–0.18 (inorg) to 26 (tot), 0.10–0.18 (inorg)	3.6 (tot) (0.06–0.08 inorg)	14 (tot) (0.02–0.04 inorg)	7.0–7.7	2–23.2				0.02–26
Bi	0.0042								0.0007–0.0025			0.0007–0.0023
Br		50.6–51.7										50.6–51.7
Cd	0.0025–25.6	0.5–1.1 3.5 (hepato-pancreas)	0.07–7.0						0.02–20.2	1.3	0.07–0.24	0.0025–20.2
Cr		0.75										0.75
Co		0.34–0.44										0.34–0.44
Cu		0.75–2.65 6.3 (hepato-pancreas)	0.75–2.65						0.75–2.65		0.75–2.65	0.75–63
Fe		212–219										212–219
Pb	0.48–0.61	0.11–12.0 12.4 (hepato-pancreas)	0.48–2.8						0.43–0.61	0.83	0.48–0.61	0.43–12.4
Mn		16.6–17.5 (hepato-pancreas)										16.6–17.5

continued overleaf

Table 8.11. (*continued*)

Element	Oyster	Lobster	Crab	Whelk	Canned lobster	Prawn	Scallop	Mollusc	Mussel	Clam	Shrimp	Range
Hg	0.14–0.16		0.02–0.05					0.02–0.05			0.02–0.05	0.02–0.16
Mo		0.16–0.31					0.1					0.16–0.31
Ni		0.98–19.4 (hepato-pancreas)				0.1						0.98–19.4
Pu									0.3–13.9			0.3–13.9
Se	1.7–2.7	2.0–6.7			4.01	0.71–1.24						0.71–6.7
Sc		0.015										0.015
Ag		0.86–0.93										0.86–0.93
Sr		11.0 / 84.9 (hepato-pancreas)										11.0–84.9
V	0.53–1.42		1.09–1.84								0.4–3.05	0.4–3.05
Zn		848–888 / 852 (hepato-pancreas)										848–888

dissolved cadmium and zinc. It was found that clams exposed to dissolved cadmium had higher concentrations of dissolved cadmium and metal binding protein in the gill, mantle and aductor muscle. Tissue concentrations increased with time of exposure.

Draback *et al.*[231] have reported concentrations of rare earth elements in the tissues of mussels (*Cyprina islandica, Mytilus edulis*) and flounder fish caught in the waste waters discharging to sea from a fertiliser production plant located at Lillebaelt, Denmark. Samples of the barnacle *Balanus amphitrite* collected in the Zuan Estuary, India, had zinc contents in tissue of $203.6-1937.5$ mg g^{-1}. The zinc content of the overlying waters was $13-46$ μg l^{-1} [229]. Gil *et al.*[235] have reported iron, zinc, manganese, copper, cadmium, lead and nickel concentrations in scallops (*Chalamys tehuelcha*) and mussels (*Aulacomya ater* and *Mylitus platensis*) from a rural uncontaminated site in San Jose Gulf and from an urban industrialized site in the Nuevo Gulf, both in Argentina. In scallops taken in both areas, iron, manganese, copper and cadmium were primarily concentrated in the liver and kidney. Zinc was concentrated in the mantle and gills. Nickel and lead were below detection limits.

Lyngby and Brix[234] determined the heavy metals in mercury, cadmium, zinc, lead and copper in mussel tissues (*Mytilus edulis*) and compared values obtained with those obtained in eelgrass (*Zostera marina*). The object of this study was to compare mussels and eelgrass as biological indicators of water pollution. Metal content of oysters taken from Darwin Harbour have been determined including lead, nickel, copper, cadmium and iron[236].

8.4.3 ANALYTICAL METHODS

Methods for the determination of metals and non-metals in sea creatures other than fish are reviewed in Table 8.12. In many cases the methods summarised below are similar to those used to analyse fish:

Review of methods and detection limits applicable to the determination of metals and non-metals in water creatures other than fish

		Detection limit (mg kg^{-1} unless otherwise stated)
Atomic absorption spectrometry		
Electrothermal atomic absorption spectrometry	Cadmium	0.3
	Lead	0.7
	Silver	
	Zinc	6
	Copper	0.7

		Detection limit (mg kg^{-1} unless otherwise stated)
Electrothermal atomic absorption spectrometry (continued)	Nickel Plantinum Uranium Iron	33
Hydride generation atomic absorption spectrometry	Arsenic Selenium Mercury	0.03 0.2 0.005
Cold vapour atomic absorption spectrometry	Mercury Arsenic Selenium	0.005 0.3 0.2
Graphite furnace atomic absorption spectrometry	Arsenic Cadmium	0.5
Zeeman atomic absorption spectrometry	Arsenic Lead Cadmium Silver Chromium Copper Manganese Selenium Nickel	1 1 1 1 5 1 15 5
Inductively coupled plasma atomic emission spectrometry	Arsenic Cadmium Lead	5
Derivativisation gas chromatography	Arsenic	10 pg absolute
Chelation-ion chromatography	Transition metals Rare earths	
Neutron activation analysis	Arsenic Cadmium Iodine Mercury Vanadium Bromine Iron Manganese Molybdenum	0.16 5 ng absolute 0.26 2.8 0.16 0.16

		Detection limit (mg kg^{-1} unless otherwise stated)
Neutron activation analysis (continued)	Nickel	0.1
	Rubidium	0.07
	Selenium	0.02
	Strontium	1.5
	Zinc	0.18
Stripping voltammetry cathodic anodic	Mercury Tin	
Non-dispersive atomic fluorescence	Mercury	
Spectrophotometric analysis	Arsenic	
Isotope dilution laser resonance ionisation mass spectrometry	Iodine	mg kg^{-1} level
Magnetron rotating dc arc plasma	44 miscellaneous elements	
Gamma-ray spectrometry	Chromium	
	Manganese	
	Iron	
	Cobalt	
	Zinc	
	Arsenic	
	Selenium	
	Cadmium	
	Antimony	
	Lead	
Photon activation analysis	Manganese	4
	Nickel	0.6
	Copper	3
	Zinc	20
	Arsenic	0.3
	Cadmium	2
	Lead	3
	Sodium	200
	Magnesium	0.4
	Chlorine	100
	Calcium	3
	Strontium	12

Table 8.12. Methods of determination of metals and non-metals in water creatures other than fish

Determined	Type of sample	Technique	Detection limit (mg kg^{-1} unless otherwise stated)	Ref.
Arsenic	Crustacae	Hydride generation atomic absorption spectrometry	0.11	9
	Oyster		0.11 ng absolute	237
	Oyster	Graphite furnace atomic absorption spectrometry	0.5	238
	Oyster	Inductively coupled plasma atomic emission spectrometry	5	238
	Oyster	Flow injection hydride generation inductively coupled plasma atomic emission spectrometry	5	238
	Crustacae Lobster hepato-pancreas	Spectrophotometric derivativisation with 2,3-dimercapto propanol–gas chromatography	– 10 pg absolute	11 239
	Oyster	Neutron activation analysis	5	238
Cadmium	Clam, mussel	Atomic absorption spectrometry	–	13, 240
	Mussel			241–3
	Crab	Inductively coupled plasma atomic emission spectrometry	–	243, 244
	Oyster	Neutron activation analysis	–	245
Gadolinium	Crab	Solvent extraction – graphite furnace atomic absorption spectrometry	–	246
Iodine	Oyster	Microwave digestion and preconcentration, neutron activation analysis	5 ng absolute	247
Iodine	Oyster	Isotope dilution laser resonance ionisation mass spectrometry	mg kg^{-1} level	248

Lead	Clam, mussel	Atomic absorption spectrometry	—	13
	Crab, lobster		—	240
	Crab	Inductively coupled plasma atomic emission spectrometry	—	244
Magnesium	Oyster		—	249
Manganese	Oyster		—	249
Mercury	Oyster	Cold vapour atomic absorption spectrometry	—	22
	Mussels, oyster		0.01	250, 251
	Crustacae	Non-dispersive atomic fluorescence	—	252
	Oyster	Neutron activation analysis	—	30
Selenium	Oyster	Cathodic stripping voltammetry	—	253, 254
Tin	Marine organisms	Anodic stripping voltammetry	—	255
Vanadium	Shrimp, crab, oyster	Neutron activation analysis	0.03	256
Pb, Cd, Hg, Zn, Cu, Ni, Pu, U	Mussel	Atomic absorption spectrometry	—	257–9
Cd, Cu, Fe, Pb, Zn	Mussel		Cd 0.3 Cu 0.7 Fe 3.3 Pb 0.7 Zn 6.0	260
As, Se, Hg	Lobster, scallop	Cold vapour atomic absorption spectrometry	As 0.3 Se 0.2 Hg 0.005	261
Misc. elements	Oyster	Magnetron relating dc arc plasma with graphite furnace sample introduction	—	262

continued overleaf

Table 8.12. (*continued*)

Determined	Type of sample	Technique	Detection limit (mg kg^{-1} unless otherwise stated)	Ref.
As, Se, Hg	Lobster, scallop	Hydride generation atomic absorption spectrometry	As 0.03 Se 0.2 Hg 0.005	261
As, Sb, Se	Oyster		As 1 μg l^{-1} Sb μg l^{-1} Se μg l^{-1}	10
As, Pb, Cd	Lobster, mussel	Zeeman atomic absorption spectrometry	–	263
Ag, Cd, Cr, Cu, Mn, Pb, Se, Ni			Ag 1 μg absolute Cd 1 Cr 1 Cu 5 Mn 1 Ni 5 Pb 1 Se 15	264
Transition metals Rare earths	Oyster	Chelation – ion chromatography	–	265
As, Br, Cd, Fe, Mn, Mo, Ni, Rb, Se, Cr, Zn	Lobster	Neutron activation analysis	As 0.16 Br 0.26 Fe 2.8 Mn 0.16 Mo 0.16 Ni 0.1 Rb 0.07 Se 0.02 Sr 1.5 Zn 0.18	266

255

Element(s)	Sample	Method	Detection limit	Ref.
44 elements	Marine bivalves	X-ray fluorescence spectrometry, neutron activation analysis, prompt γ-activation analysis	–	267
Cr, Mn, Fe, Co, Zn, As, Se, Cd, Sb, Pb	Mussels, oyster	γ-ray spectrometry	–	268, 269
Mn, Ni, Cu, Zn, As, Cd, Pb, Na, Mg, Cl, Ca, Sr	Oyster hepato-pancreas	Photon activation analysis	Mn 4, Ni 0.6, Cu 3.4, Zn 2.0, As 0.3, Sr 12, Cd 2, Pb 3, Na 200, Mg 0.4, Cl 100, Ca 3	270
Metals in biological materials				
Indium	Blood, liver, kidney, urine	Ion pair extraction – electrothermal atomic absorption spectrometry	$\mu g\,l^{-1}$	271
Selenium	Blood, liver	Inductively coupled plasma mass spectrometry using hydride generation sample introduction system	6.4 ng absolute	272
Strontium	Biological materials	Preconcentration – extraction chromatography using crown ether (bis (tertbutyl-cyclohexano) 18-crown-6) in 1-octanol	–	273
Al, Ba, Ca, Cd, Co, Fe, K, Mg, Mn, Na, P, S, Si, Se, Ti, V, Zn		Parr bomb digestion – inductively coupled plasma atomic emission spectrometry	–	274
Co, Cu, Se, Zn		Wet decomposition electrophoresis	–	275
I		Anodic stripping voltrammetry	–	276

8.5 ORGANIC COMPOUNDS IN CREATURES OTHER THAN FISH

8.5.1 TOXIC EFFECTS

Dragonfly larvae (*Odonata aeshna*) exposed to water and sediments containing di(2-ethylhexyl)phthalate $(587-623 \text{ mg kg}^{-1}$ di(2-ethylhexyl)phthalate in sediment) were shown after 40 days' exposure to contain 14.7 mg kg^{-1} di(2-ethylhexyl)phthalate in the tissue. This led to a reduction in predation efficiency of these organisms[277].

Kukkonen and Oikari[278] noted the effect of humic acid in water on the uptake by *Daphnia magna* and the toxicity of various organic pollutants. Accumulations in *Daphnia magna* were 50% less from the humic water in the case of dehydroabietic acid and benzopyrene. Consequently, the toxic effects of these to *Daphnia magna* was reduced. Humic acid did not affect pentachlorophenol uptake.

8.5.2 ANALYTICAL COMPOSITION

A variety of organic substances can occur in sea creatures other than fish, as they do in fish (Table 8.13). It is seen that in many instances the concentrations in fish and in creatures other than fish are of a similar order of magnitude.

Rice and White[279] determined the concentrations of polychlorobiphenyls in fathead minnows (*Pimephales promelas*) and fingernail clams (*Sphaerium striatinum*) before, during and up to 6 months after completion of dredging polychlorobiphenyl-contaminated sediments in the Skiawasse River, USA 1 km downstream of the pollution outfall. The concentration of polychlorobiphenyl found in fathead minnow tissue was $32.1-61.1 \text{ mg kg}^{-1}$ dry weight and the concentration in clams was $13.2-15.3 \text{ mg kg}^{-1}$ dry weight. It was concluded that dredging had increased the bioavailability of polychlorobiphenyls to these organisms.

Exposure of *Asellus aquaticus* crustacea to water containing $5 \mu\text{g l}^{-1}$ of lindane for 5 days led to a pick-up by the organism of 0.2 mg kg^{-1} lindane in the tissue[286]. Depuration was rapid, with over 40% of accumulated lindane being eliminated within 1 day.

8.5.3 ANALYTICAL METHODS

Methods used for the determination of a wide range of organic substances in creatures other than fish are shown in Table 8.14. Detection limits are usually well below 0.01 mg kg^{-1}. Gas chromatography, gas chromatography combined

Table 8.13. Occurrence of organic substances in creatures other than fish

Substance determined	Type of creature	Concentration found ($mg\,kg^{-1}$)	Ref.	See Tables 8.4 and 8.5 (concentration found in fish for comparison) ($mg\,kg^{-1}$)	Ref.
Polyaromatic hydrocarbons					
Naphthalene	Mussels	0.003–0.1	58	0.001	56
Phenanthrene and anthracene		0.008–0.032	58	0.008	51
Fluoranthrene		0.042–0.080	58	0.004–1.8	57
Pyrene		0.034–0.092	58	0.004–1.5	57
Benz(a)anthracene and chrysene		0.029–0.059	58	0.004–0.022	57
Benza(a)pyrene		0.55	280	0.007–0.083	57
Polychlorobiphenyls		0.41–0.9	58	0.48–5.0	58, 230
		0.011–0.56	281		
	Oyster	0.0002	282		
Polychloroterphenyl		0.000 15	282		–
Dibenzothiophane	Mussel	0.0001–0.8	283	–	
Diethylhexylphthalate	Shrimp, crab	0.003–0.02	284	–	
Aliphatic hydrocarbons	Mussel,	0.54	285	–	
	oyster,	0.65	285		
	clam	0.49–1.41	285		

Table 8.14. Methods for the determination of organic compounds in water creatures other than fish

Determined	Type of sample	Technique	Detection limit (mg kg^{-1} unless otherwise stated)	Ref.
Aliphatic hydrocarbons	Oyster	Gas chromatography	–	287
Hydrocarbons	Mussel, oyster, clam	Head space gas chromatography – mass spectrometry	–	289, 288
Aromatic hydrocarbons	Mussel	Atomic fluorescence spectrometry	–	290
Hydrocarbon oils	Marine organisms	Gas chromatography	0.5	291
Non-biogenic hydrocarbons	Shrimp	Gas chromatography	–	292
	Shrimp	Gas chromatography – mass spectrometry	–	292
Polyaromatic hydrocarbons	Mussel	Capillary gas chromatography	–	280, 290
Naphthalene, phenanthrene + anthracene, fluoranthrene, pyrene, benz(a)anthracene + chrysene	Marine organisms	Gas chromatography	–	137
Benzopyrene, hexachlorobenzene pentachlorophenol	Marine organisms	"	0.0002	293
Polyaromatic hydrocarbons	Lobster	Gas chromatography with ultraviolet fluorescence detector	–	294, 295
	Mussel	Atomic fluorescence spectroscopy	–	296
Phthalate esters	Shrimp, crab	Gas chromatography	0.001	164
Trichloroethylene tetrachloroethylene chloroform, carbon tetrachloride	Marine organisms	"	–	297, 298
Polychlorobiphenyls	Crab, shrimp	"	0.01	299–302

Compound	Organism	Method	Detection limit	References
Polychlorobiphenyls	Oyster, clam	Electron capture ion chemical ionisation mass spectrometry	0.0065	303, 304
Polychloroterphenyl	Oyster	Gas chromatography – mass spectrometry	–	305–7
Polychlorodibenzo-*p*-dioxins dibenzofurans	Crustacae	High-resolution mass spectrometry	–	125
Chlorinated insecticides	Oyster	Gas chromatography	0.01	308
	Mollusc	"	–	309
	Crustacae	"	–	299, 306–8, 310, 312
	Oyster	Gas chromatography – mass spectrometry	0.000 04 (lindane)	303
DDT, DDE, DDD	Scallops	"	–	306, 307
DDT	Crab, shrimp	Gas chromatography	0.001	299–302
Mirex	Crustacae	"	0.005	299–302
Dursban	Crustacae	"	0.01	313
Polychlorobenzothiophenes	Crab, lobster	Gas chromatography – mass spectrometry	–	314
Sulphur-containing hydrocarbons, e.g. dibenzothiophene	Mussel	Gas chromatography – mass spectrometry and gas chromatography with flame photometric detector	–	315, 316
Ascorbic acid-2-sulphate	Oyster, brine shrimp	Solid phase extraction	–	317
Coprostanol	Crustacae	Electron ionisation mass spectrometry	75 ng absolute	318
Miscellaneous compounds	Oyster	Magnetron rotating dc arc plasma	–	319

with mass spectrometry or various mass spectrometry methods are the most commonly employed techniques, as shown below.

Review of methods and detection limits applicable to the determination of organic compounds in creatures other than fish

		Detection limit ($mg\,kg^{-1}$ *unless otherwise stated*)
Gas chromatography	Aliphatic hydrocarbons	0.5
	Hydrocarbon oils	
	Non-biogenic hydrocarbons	
	Polyaromatic hydrocarbons,	
	e.g. benzopyrene	0.0002
	Phthalate esters	0.001
	Chloroaliphatic compounds	
	Haloforms	
	Polychlorobiphenyls	0.0065−0.01
	Chlorinated insecticides	⩽ 0.01
	e.g. DDT	0.001
	Mirex	0.005
	Dursban	0.01
With ultraviolet fluorescence detector	Polyaromatic hydrocarbons	
	Non-biogenic hydrocarbons	
Gas chromatography mass spectrometry	Chlorinated insecticides	e.g. 0.000 04 (lindane)
	Polychloroterphenyls	
	Polychlorobiphenyls	
	Polychlorobenzothiophenes	
Electron capture negative ion chemical ionization mass spectrometry	Polychlorobiphenyls	
	Coprostanol	75 ng absolute
High-resolution mass spectrometry	Polychlorodibenzo-p-dioxins, dibenzofurans	
Atomic fluorescence spectroscopy	Aromatic hydrocarbons	
	Polyaromatic hydrocarbons	

8.6 ORGANOMETALLIC COMPOUNDS IN CREATURES OTHER THAN FISH

8.6.1 TOXIC EFFECTS

Organotin compounds

The occurrence and concentrations of organotin compounds in the tissues of scallops (*Pecten maximus*), flame shells (*Lima hians*)[320], polychaetes, snails and bivalves[321] and mussels (*Mytilus edulis*) and oysters (*Crassostrea virginica*)[81] has been studied. Scallop, mussel and flame shell populations are adversely affected by organotin compounds[320]. High concentrations of tributyltin have been found in polychaetes, snails and bivalves living in marinas containing 2–646 ng l^{-1} tributyltin[321], i.e. above the Environmental Quality Target for tributyltin of 20 ng l^{-1}. San Diego Bay mussels exposed to 0.7 μg l^{-1} organotin for 60 days sustained a 50% mortality in the case of mussels and a decline in condition in the case of oysters[322]. Various tissues in these organisms showed tin uptake within 0–30 days.

8.6.2 ANALYTICAL COMPOSITION

The only compound that has been investigated is that of lead. From the limited data available concentrations of organolead in creatures other than fish are appreciably lower than occur in fish (Table 8.15).

Data have been presented on the concentrations of ionic alkylead compounds in saltmarsh periwinkles (*Littorina irrorata*) collected in Maryland, Virginia. Male periwinkles accumulated higher concentrations of several alkyl lead species than females.

8.6.3 ANALYTICAL METHODS

Methods used for the determination of organometallic compounds in water creatures other than fish are summarised overleaf (see also Table 8.16).

Table 8.15. Comparison of organolead levels (mg kg^{-1} dry weight) in crustacea and fish

	In creatures other than fish[a]	Range in fish (from Table 8.8)
R_4Pb	< 0.02	0.92–7.93[b]
R_3Pb^+	< 0.01–0.5	0.54–6.16[b]
R_2Pb^{2+}	< 0.01	0.54–0.79[b]
Pb^{+2}	1.1–1.8	0.25–4.13

[a]From reference 429.
[b]Ethyl plus methyl compounds.

Table 8.16. Methods for the determination of organometallic compounds in water creatures other than fish

Determined	Type of sample	Technique	Detection limit ($mg\,kg^{-1}$ unless otherwise stated)	Ref.
Organotin compounds				
Butyltin	Oyster	Hydride generation atomic absorption spectrometry	1.1–2.5 ng absolute	324
Butyltin, $Me_3SnC_5H_{11}$ $Me_2Sn(C_5H_{11})_2$ $MeSn(C_5H_{11})_3$ Pr_4Sn	Mussels	Gas chromatography with flame photometric detector	Pr_4Sn 0.3 pg Sn absolute $Me_3SnC_5H_{11}$ $Me_2Sn(C_5H_{11})_2$ $MeSn(C_5H_{11})_3$ 2-3pg as Sn absolute	325
Methyltin and butyltin	Oyster	Hydride generation atomic absorption spectrometry	MeSn 0.000 023 BuSn 0.000 025	326
Organolead compounds				
Eb_3Pb^+ Me_3Pb^+ Et_2Pb^{2+} Me_2Pb^{2+}	Oyster macuma	Differential pulse anodic scanning voltrammetry	–	327
Organomercury compounds				
Methyl mercury	Crustacae	Extraction gas chromatography	0.000 01	328
Organoarsenic compounds				
Arsenobetaine	Crab	Inductively coupled plasma atomic emission spectrometry	–	329

Review of methods and detection limits applicable to the determination of organometallic compounds in water creatures other than fish

		Detection limit ($mg\,kg^{-1}$ *unless otherwise stated*)
Hydride generation atomic absorption spectrometry	Butyltin compounds	$1-2.5$ ng as Sn absolute or 0.000 025
	Methyltin compounds	0.000 023
Gas chromatography	Methylmercury compounds	0.000 01
With flame photometric detector	Butyltin compounds Propyltin compounds $Me_3Sn(C_5H_{11})$	3 pg as Sn absolute
	$Me_2Sn(C_5H_{11})_2$ $MeSn(C_5H_{11})_3$	$2-3$ pg as Sn absolute
Inductively coupled plasma atomic emission spectrometry	Organoarsenic compounds	
Differential pulse anodic scanning voltammetry	Et_3Pb^+ Me_3Pb^+ Et_2Pb^{2+} Me_2Pb^{2+}	

8.7 METALS IN PHYTOPLANKTON, ALGAE AND WEEDS

8.7.1 TOXIC EFFECTS

Phytoplankton and algae

Dallakyan *et al.*[330] studied the combined effect of zinc $(100-1000\,\mu g\,l^{-1})$ chromium $(100-1000\,\mu g\,l^{-1})$ and cadmium $(100-500\,\mu g\,l^{-1})$ on the phytoplankton in a reservoir. Zinc and cadmium additions maximally inhibited phytoplankton production at the beginning of a blue-green bloom. Chromium had no such inhibiting effect.

Brand *et al.*[331] investigated the effect of copper and cadmium on the reproduction rates of 38 clones of marine phytoplankton. Cyanobacteria were the most sensitive to the copper toxicity and diatoms were the least sensitive.

Reproduction rates of cyanobacteria were reduced at copper-11 activities above 10PM (picomole), whereas eukaryotic algae still maintained maximal reproductive rate at 10 nM. Trends for cadmium-11 were the same as for copper. Concentrations of cadmium in natural seawater were not of significance in unpolluted water, but copper concentrations in upwelling water might affect cyanobacteria.

In a study of the toxic effect of total aluminium and copper concentrations on the green alga (*Scenedesmus*) it was observed that toxicity effects mainly on growth rate was due almost entirely to an increase in cupric ion activity as a result of indirect competition of aluminium in the growth media that displaced copper from chelators[332].

Claesson and Tornqvist[333] studied toxicity of aluminium to two acido-tolerant green algae, chlorophycae (*Monoraphidium dybowskii* and *Stichoccus sp*). Exposure to $100-800$ μg l^{-1} aluminium at pH 5 to 6 led to cell decomposition even at 100 μg l^{-1} aluminium. Growth was also affected.

Growth of pure cultures of phytoplankton *Scenedesmus bijugatus* and *Nitzchia palea* in $10-50$ μg l^{-1} and $20-40$ μg l^{-1} cadmium, respectively, showed that the physiology of the algae was affected during the experimental growth phase and the ratio of carbohydrate, protein and lipid was affected by cadmium[334].

Increase in cyanide concentration in the range of $100-700$ μg l^{-1} inhibited the growth of the Nile water algae *Scenedesmus* but had no effect on the growth of *Anabaena*[335].

The aquatic macrophytes *Elchornia crassipes*, *Hydrilla verticillata* and the alga *Oedogonium aerolatum* upon exposure to solutions of mercuric chloride and sodium chromate for 28 days accumulated more chromium than mercury. Exposure did not produce any significant changes in Hill activity, chlorophyll, protein, free amino acid, inorganic phosphorus, RNA, DNA, dry weight permeability or protease activity[336].

In studies of chronic exposure of algal-periphyton the communities were exposed to $50-1000$ μg l^{-1} zinc for up to 30 days. Treatments as low as 50 μg l^{-1} zinc significantly changed algal community composition from diatoms to green or blue-green algae. A zinc concentration of 47 μg l^{-1} is the criterion of the US Environmental Protection Agency for the 24-hour average to total recoverable zinc. Starodub et al.[338] carried out short- and long-term studies on the individual and combined toxicities of copper, zinc and lead to *Scenedesmus quadricanda* freshwater green alga. The short-term exposure to the effect of combinations of $0-200$ μg l^{-1} copper, $0-500$ μg l^{-1} zinc and $0-6000$ μg l^{-1} lead and the long-term effects of the single and combined metals on primary productivity were studied[338]. Low concentrations of single metals had the greatest effect on primary productivity, copper was the most toxic and lead the least toxic in short- and long-term studies. The combined metals exhibited an antagonistic effect in short-term exposure and both synergistic and antagonistic effects in long-term experiments. Kuwabara[339] has reviewed the physico-chemical processes affecting copper, tin and zinc toxicity to algae.

Weeds

Nickel at the 100 μg l^{-1} level would reduce the growth rate of common duck-weed (*Lemna minor*) by 30% in most surface waters and by 70% in very soft water[340]. The angiosperm *Cuseuta reflexa* undergoes a reduction in chlorophyll and protein content and percentage dry matter in biomass as well as an increase in tissue permeability in the presence of certain metals in the overlying water[341]. It is most sensitive to arsenic, then to cadmium, lead, mercury and chromium in that order. Pick-up of aluminium, copper-11 and lead-11 from water by duck-weed in amounts above a certain concentration causes the plant to die[342]. Toxicity is believed to be due to the replacement of magnesium in chlorophyll and hence loss of its normal activity.

Samples of coral *Pocillopora damicornis* in a coral reef adjacent to a tin smelter contained significantly higher concentrations of calcium, strontium, zinc, chromium, cobalt, molybdenum, nickel, magnesium, sodium and potassium than those found in coral from an uncontamination site[343]. There was distinct evidence that these contaminants caused reduced coral growth rate and a low number of branching coral species.

8.7.2 ANALYTICAL COMPOSITION

The available limited information on the occurrence of metals in phytoplankton, algae and weed is reviewed in Table 8.17 (see also Appendix 4, Table A4.1). Significant correlations have been found between the concentrations of mercury, lead, cadmium and zinc in the Limfjord, Denmark, and the concentrations of these elements found in eelgrass leaves and root rhizomes (*Zostera marino*) present in the fjord[234].

Quite enhanced levels of metals can occur in algae and weeds in areas of high pollution, making them useful indicators of pollution.

8.7.3 ANALYTICAL METHODS

Methods used for the determination of metal and non-metals in algae, plankton and weeds are reviewed in Table 8.18. As seen below, methods of analysis that have been employed are based mainly on atomic absorption spectrometry, neutron activation analysis and X-ray based methods.

Review of methods and detection limits for the determination of metals and non-metals in algae, plankton and weeds

Detection limit

Atomic absorption spectrometry   Cadmium
                                                  Chromium

		Detection limit
Atomic absorption spectrometry (*continued*)	Copper Iron Manganese Mercury Tin Zinc	
Graphite furnace atomic absorption spectrometry	Tin	0.3 $\mu g\,l^{-1}$
Zeeman atomic absorption spectrometry	Cadmium Copper Lead Zinc	
Flameless atomic absorption spectrometry	Bismuth	3 pg absolute
Electrothermal vaporisation DC argon plasma emission spectrometry	Mercury Iodine	
X-ray fluorescence spectrometry	20 elements	
Energy-dispersive X-ray spectrometry	Aluminium Copper Iron Manganese	
Neutron activation analysis spectrometry	Mercury Plutonium Americium	$31-81$ mg kg^{-1}
Spectrophotometric methods	Arsenic Cobalt Zinc	

8.8 ORGANIC COMPOUNDS IN PHYTOPLANKTON, ALGAE AND WEEDS

8.8.1 TOXIC EFFECTS

Phytoplankton

Di-*n*-butyl phthalate has a distinct adverse effect on the distribution and survival of marine phytoplankton. It also markedly affects growth and/or aggregation behaviour of algae and diatoms[367].

Table 8.17. Metal and non-metal contents of algae, phytoplankton and weeds

Determined	Type of sample	Concentration ($mg\,kg^{-1}$ unless otherwise stated)	Ref.
		(a) Algae	
Aluminium	Plankton, algae	$12\ \mu g\,l^{-1}$	344
Arsenic	Algae	20.0–56.1	345
Bismuth	Marine algae	3 pg absolute	
Bromine	Plankton, algae	$0.3\ \mu g\,l^{-1}$	344
Chromium	Algae	40–630	
Copper	Algae	50–660	
Iron	Algae	340–9720	
Mercury	Plankton, algae (Lake Eyrie)	31–81	346
Manganese	Algae	230–4170	
Silver	Algae	0.03–1.06	
Tin	Macro algae	0.03 (inner tissue)	347
	Narrangansett Bay California	0.83 (algal blade)	
Zinc	Algae	20–700	
Iron	Freshwater	340–9720	348
Manganese	plankton,	87–4170	
Copper	*Platihypnidium*	40–690	
Chromium	*riparoides,*	40–630	
	Olea europa,	20–700	
Zinc	*Lagarosiphon major*		
		(b) Plankton	
Chromium	Plankton	60–70	
Copper		40	
Iron		3700–3800	
Manganese		230–250	
Mercury		31.2–81.0	
Rare earths			
Lanthanum		0.15	
Cerium		0.24	
Neodynium		0.03	
Samarium		0.006	
Gadolinium		0.019	
Europium		0.003	
Dysprosium		0.008	
Ytterbium		0.001	
		(c) Weeds	
Bismuth	Kelp	0.005	345
	Macrocystis	0.009	345
Thorium	Seaweed (Baltic)	0.01–0.06	349

continued overleaf

Table 8.17. (*continued*)

Determined	Type of sample	Concentration (mg kg^{-1} unless otherwise stated)	Ref.
Uranium		0.07–0.41	349
Copper		*See plant Copod*	350
		198 μmole 121 μmol kg^{-1}	
Zinc	Sea plant, copod	979 kg^{-1} 367 "	
Mercury		1.7 1.4 "	
Cadmium		6.2 6.7 "	
Cadmium		198–541	351
	Seagrass, *Posidonia*		
Manganese	*australis* (near lead smelter)	112–537	
Lead		116–379	
Zinc		728–4241	
Potassium		0.83–26.8 mg g^{-1},	352
		dry weight	
Sodium		0.39–11.7	
Magnesium		1.5–2.85	
Nitrogen	Macrophyte, *Juncus*	19.4–25.5	
	bulbosus		
Phosphorus		0.14–0.29	
Manganese		3.40–17.17	
Iron		4.47–35.81	
Zinc		0.12–0.42	

Rhee *et al.*[368] studied the long-term responses of phytoplankton (*Selanastrum capricornutum*) to 2,5,2^1,5^1-tetrachlorobiphenyl in water. This compound caused a reduction in the percentage of fixed carbon incorporated into the cells and this carbon was probably excreted.

Concentrations of permethrin between 0.75 and 1.5 μg l^{-1} in pond water caused a decline in populations of *Daphnia rosea* and at 10 μg l^{-1} a complete elimination of this species. *Acanthodiaptomus pacificus* behaved similarly. *Tropocylops praciuus* was slightly more tolerant[369].

Zooplankton

Ali *et al.*[370] obtained no evidence that very low concentrations of diflubenzuron had any adverse effects on zooplankton and benthic invertabrates in ponds which had been contaminated by this insect growth regulator present as an air drift from a nearby citrus grove.

Day and Kaushik[371] studied the effect of short-term exposure to the synthetic pyrethroid fenvalerate in water on the rate of filtration and the rate of assimilation of *Chlamydomonas reinhardii* by three species of freshwater zooplankton, namely *Daphnia galeata mendotae*, *Ceriodaphnia cacustris* and *Diaptomus*

Table 8.18. Methods for the determination of metals and non-metals in algae, plankton and weeds

(a) *Algae*

Determined	Technique	Detection limit ($mg\,kg^{-1}$ unless otherwise stated)	Ref.
Arsenic	Conversion to $AsCl_3$, spectrophotometry	–	353
	Arsine generation spectrophotometry	–	354
Bismuth	Flameless atomic absorption spectrometry	3 pg absolute	345
Halogens	Ion-selective electrodes	–	355
Mercury	Electrothermal vaporisation dc argon plasma emission spectrography	–	356
	Neutron activation analysis	0.2–0.79	346
Phosphorus	Biological availability	–	357
Tin	Graphite furnace atomic absorption spectrometry	$0.3\ \mu g\,l^{-1}$	358, 359
	Atomic absorption spectrometry	–	347
Fe, Mn, Cu, Cr, Zn	Atomic absorption spectrometry	–	348
Cd, Pb, Cu, Zn	Zeeman atomic absorption spectrometry	–	360
20 Elements	X-ray fluorescence spectrometry	from $0.3\ \mu g\,l^{-1}$ (Br) to $12\ \mu g\,l^{-1}$ (Al)	344
Mn, Cu, Fe, Al	Energy-dispersive X-ray spectrometry	–	361

(b) *Weeds*

Determined	Sample	Technique	Detection limit	Ref.
Actinides incl. plutonium and americium, Bismuth	Seaweed Kelp	Alpha spectrometry,	–	362
	Seaweed	Flameless atomic absorption spectrometry	3 pg absolute	345
Cobalt	Seaweed	Ashing spectrophotometric method	–	363

continued overleaf

Table 8.18. (*continued*)

(b) *Weeds* (*continued*)

Determined	Sample	Technique	Detection limit	Ref.
Iodine	Seaweed	Cold vapour atomic absorption spectrometry	–	364
Mercury	Seawater weeds		–	365
Tin	Aquatic plants	Graphite furnace atomic absorption spectrometry	–	358
Zinc	Seaweed	Ashing spectrophotometric method	–	363
Misc. radioactive elements	Bladderwrack	γ-ray spectrometry	–	366
Cu, Zn, Hg, Cd		Atomic absorption spectrometry	–	350

oregonensis. Rates of filltration of *Chlamydomonas reinhardii* by all three species were decreased significantly at sublethal concentrations ($\leqslant 0.05\ \mu g\,l^{-1}$) of fenvalerate in water after 24 hours' exposure. Rates of assimilation of algae by the three species were decreased at lethal concentration of more than 0.05 μg fenvalerate. Changes in rates of filtration and assimilation can be used to monitor the effects of sublethal levels of toxicants.

Applications of 1000 $\mu g\,l^{-1}$ of carbaryl insecticide to pond water killed off all zooplankton but had no effect on phytoplankton, though changes in zooplankton densities affected phytoplankton community structures[372]. Lindane (γBHC) has no significant effect on natural zooplankton populations, but the population density of zooplankton was reduced even at concentrations of lindane as low as 20 $\mu g\,l^{-1}$ [373]. Rotifers and nauplii were particularly adversely affected.

Arthur[374] has studied the effects of pollution by diazinon in water (0.3–3.0 $\mu g\,l^{-1}$), chlorpyrifos (0.2–11 $\mu g\,l^{-1}$), pentachlorophenol (48–432 $\mu g\,l^{-1}$) on plankton and invertabrate communities and on survival, growth and reproduction.

Algae

Exposure of periphyton communities from brackish water mesocosinus to 1–10 $\mu g\,l^{-1}$ 4,5,6-trichloroguiacol produced no evidence for adverse effects[427]. Exposure of natural periphyton communities to atrazine, alalchlor, metolchlor and metribuzin reduced growth rate and rate of uptake of nutrients at least temporarily[375].

Minimum concentrations of terbutryn, diuron, monouron and atrazine for inhibiting the growth of microalgae have been reported as $1100-2800$ $\mu g\,l^{-1}$ (terbutryn and diuron) and $1100-17\,100$ $\mu g\,l^{-1}$ (monouron and atrazine)[376].

Hamilton et al.[377] have studied the effect of up to 2 years' exposure of lake periphyton communities to concentrations of atrazine in the range of $80-1500$ $\mu g\,l^{-1}$. Chlorophyll-a, freshwater biomass, ash-free weight, cell numbers, species diversity, community carbon uptake and species-specific carbon uptake were measured. There was a shift from chlorophyte to a diatom-dominated community over the two-year period but Cylindrospernum stagnate and Tetraspora cyclindrica showed evidence of resistance to atrazine at 1560 $\mu g\,l^{-1}$.

Community productivity was reduced by 21% and 82% in the low and high exposure, respectively, returning to control levels in 21 days. The productivities of the larger algae were most affected. Reduced growth rates were obtained after exposure to the herbicide. Other workers have reported on a growth rate depression when green algae are exposed to atrazine[378].

Marine unicellular algae Skeletonema costatum, Thalassiosira pseudonana and Chlorella sp have been exposed to water containing the brominated organic compounds decabromobiphenyloxide, pentabromomethyl benzene and pentabromomethyl benzene. The corresponding LC_{50} values were greater than 1.1, 1.0 and 0.5 mg l^{-1}, respectively, the highest exposure concentrations tested[379].

The effect has been studied of atrazine $(50-30.000$ $\mu g\,l^{-1})$ combined with either ethanol $(0.1-3\%$ v/v) or acetone $(0.1-5\%)$ on the growth of the green alga Chlorella pyrenoidosa[380]. Acetone and atrazine interacted antagonistically, but only at solvent concentrations exceeding $4-5\%$ with both solvents. Atrazine EC_{50}'s (calculated using growth data in the additive solvent range) were between 50 and 80 $\mu g\,l^{-1}$.

The effect of $0-100$ mg l^{-1} concentrations of lindane (γ-BHC) in freshwaters on the alga S. obliquus have been studied[381]. Daily samples were examined for algal growth, pigment content and accumulation and degradation. Above 50 mg l^{-1} lindane in water the algal pigment content was affected. Accumulation was enhanced by exposure time and by vibration.

Walsh et al.[382] evaluated the effect of 21 pesticides in water on five different algal species by determining EC_{50} values.

The effects have been examined of the organophosphorus insecticide Phosalone on the sexual life cycle of the algal Chlamydomonas reinhardii[383].

The formation of gametes, young, mature zygotes and the meiotic division of mature zygotes were examined following 2 hours' exposure to 36.7 mg l^{-1} phosalone. The formation of gametes, young zygotes mature zygotes and the meiotic division of mature zygotes were examined following 2 hours' exposure to 36.7 mg l^{-1} phosalone. The formation of gametes and young zygotes was not affected by the treatment. Unlike control groups, the mature zygotes thus formed did not exhibit the ability of meiotic division in the first days of light exposure but remained in the same state for 5 days then underwent meiotic division on the sixth day of exposure. Stratten[384] has studied the inhibitory effect of from 0.1%

to 14% of six organic solvents (methanol, acetone, hexane, ethanol, dimethyl sulphoxide and N,N-dimethyl formamide) in water towards five species of blue-green algae (*Anabaena sp*, *Anabaena cylindrica*, *Anabaena variablilis*, *Nos-toc cp* and *Anabaena inaequalis*. Acetone and dimethyl-sulphoxide were of intermediate toxicity as regards growth inhibition (EC_{50} values 0.36% and 4.4%, respectively). Dimethyl sulphoxide and ethanol were highly toxic.

In 10–14-day growth experiments methyl-formamide and ethanol have been confirmed as the most toxic organic solvents towards the green algae *Chlorella pyrenoidosa*[380] (EC_{50}'s of 0.84 and 1.18% v/v respectively), followed by dimethyl sulphoxide, hexane, methanol and acetone (EC_{50}'s of 2.01%, 2.66%, 3.02% and 3.60% v/v respectively).

Chlorella vulgaris cultures exposed to *p*-nitrophenol or *m*-nitrophenol in water at concentrations between 5 and 20 mg l^{-1} for 20–30 days exhibited inhibited growth in the case of *p*-nitrophenol at 10 mg l^{-1} and stimulated growth in the case of *m*-nitrophenol at 5 μg l^{-1} during 20–30 days' exposure but inhibited growth at 15 mg l^{-1} during 15 days' exposure[385].

Weeds

Thorhang and Marins[386] studied the effect of three oil dispersants (Corexit 9527, Arcochem D609 and Canco K(K)) on the subtropical/tropical sea grasses *Thalassia testudinum*, *Halodule wrightii* and *Syringodium filiforme*. At concentrations below 1 ml dispersant per 10 ml oil in 100 litres of seawater, mortality rates were low even for long exposure times. At 10 ml dispersal per 100 ml oil in 100 litres of seawater *Syringodium filiforme* and *Halodule wrightii* died. Conco K(K) was far more toxic than either of the other two dispersants.

Diatoms

Goutex *et al.*[387] studied the effects of 50 mg l^{-1} 9,10-dihydroanthracene and its biodegradation products on the marine diatom *Phaeodactylum tricornatum*. Growth of the diatom was inhibited. Synergistic effects between 9,10-dihydroanthracene and its biodegration products increased the toxicity of the hydrocarbon. Resistance to polychlorobiphenyls and cross-resistance to DDT were induced in a polychlorobiphenyl-resistant clone of *Ditylum brightwelli* by 30 days' exposure to 10 μg polychlorobiphenyl^{-1} or polychlorobiphenyl concentrations which increased progressively from 10 to 30 μg l^{-1} over the 30 days[428].

Polychlorobiphenyl resistance persisted for 2 years. The polychlorobiphenyl-resistant *Ditylum brightwelli* exhibited greater tolerance to polychlorobiphenyl than did the sensitive strain under all environmental conditions which permitted its growth, even when the conditions of salinity, temperature and nitrogen availability were very different from those maintained during induction. Polychlorobiphenyl resistance decreased the tolerance of the strain to lower salinities and nitrogen limitation but increased its tolerance to lower temperatures.

8.8.2 ANALYTICAL COMPOSITION

Organic compounds

A wide variety of organic compounds have been determined in algae and plant tissues (see Appendix 4, Table A4.3). These include hydrocarbons, phenols, acrylic acid, carbohydrates, chlorinated phenols, chlorinated insecticides, polychlorobiphenyls, chlorophylls and other plant pigments, humic acids, organosulphur compounds, adenosine triphosphate and Anatoxin a.

8.8.3 ANALYTICAL METHODS

Methods for the determination of organic compounds in phytoplankton, algae, weeds and plant tissues are reviewed in Table 8.19.

Table 8.19. Methods for the determination of organic compounds in algae, weeds and plant tissues

Determined	Sample	Technique	Detection limit ($mg\,kg^{-1}$ unless otherwise stated)	Ref.
	(a) *Algae*			
Hydrocarbons	Algae	Dynamic head space analysis gas chromatography	–	81
	Lichens	Gas chromatography – mass spectrometry	–	388
Acrylic acid	Algal cultures (*Hymenomonas carterae*, *Skeletonema costatum*)	Capillary gas chromatography with electron capture detector	–	389
Phenols and substances containing sulphydryl groups	Microalgal secretions	Chemiluminescent oxidation of luminol with KI	–	390
Carbohydrates	Plankton algae	Conversion to trimethylsilyl ether–gas chromatography	–	391
Chlorinated insecticides	Algae	Hexane extraction–gas chromatography	–	392
DDT	Freshwater algae	Liquid scintillation counting of [^{4}C] DDT	–	393

continued overleaf

Table 8.19. (*continued*)

Determined	Sample	Technique	Detection limit (mg kg^{-1} unless otherwise stated)	Ref.
Chlorinated insecticides, polychlorobiphenyls	Algae, e.g. *Chlorella pyrensidosa*	Capillary gas chromatography with electron capture and flame ionisation detector	–	394
Chlorophylls	Diatom material and algae	Spectrophotometric	–	395
	Algae		–	396
Chlorophylls, Coprostanol	Algae (*Oocystis* sp, *Oscillatoria* sp)		–	397
Chlorophylls a,b,c Carotene, Xanthophylls	Algae	Spectrophotometric and thin-layer chromatography	12 μg absolute	398
Phaeophytins, a,b,c, and chlorophyll a,b,c	Algae	Thin-layer chromatography	–	
Chlorophylls a,b,c fucoxanthin, diatinoxanthin lutein, violxanthin, echineenore, myxoxanthophyll	Algae	High-performance liquid chromatography	–	400, 401
Chlorophyll a,b,c				400–5
Carotenoids			–	406
Xanthophylls			–	407
Phaeophytins a,b,c			–	402, 408
Humic substances	Aquatic algae Macrophytes	Pyrolysis – gas chromatography	–	409
Dimethyl sulphide	Algae	Gas chromatography with flame photometric detector	ng level	410, 411
Adenosine triphosphate	Algae	Solvent extraction – luciferin–luciferase assay	–	412
		High-performance liquid chromatography and ^{31}PNMR analysis	–	413
Anatoxin a		High-performance liquid chromatography with ultraviolet detection	–	414
	(b) *Plankton*			
Uronic acids aldoses	Plankton	Capillary gas chromatography	–	415

continued overleaf

Table 8.19. (*continued*)

Determined	Sample	Technique	Detection limit (mg kg⁻¹ unless otherwise stated)	Ref.
	(c) *Water weeds and plant tissues*			
Uronic acids aldoses	Plant tissues	Capillary gas chromatography	–	415
2,4-dichlorophenol		Supercritical fluid chromatography	–	416
Ethylene thiourea		Particle beam liquid chromatography – mass spectrometry	5 μg absolute	417
BHC isomers (DDE, DDT, hexachlorobenzene	Water weeds	Solvent extraction gas chromatography and thin-layer chromatography	α BHC 0.15 β BHC 0.75 γ BHC 0.2 δ BHC 0.15 pp DDE 0.45 op DDT 0.5 pp DDD 1.5 pp DDT 1.5	418, 419
Hydrocarbons (oil spills)	Marine biota	Spectrofluorimetry	–	420

8.9 ORGANO METALLIC COMPOUNDS IN PHYTOPLANKTON, ALGAE AND WEEDS

8.9.1 ANALYTICAL COMPOSITION

The results in Table 8.20 indicate that total organolead compounds in phytoplankton can occur at similar levels to the maximum total level found in fish (see Table 8.8).

8.9.2 ANALYTICAL METHODS

Methods for the determination of organometallic compounds are reviewed in Table 8.21 (see also Appendix 4, Table A4.2).

References to further information on the presence of metals, organometallic compounds and organic compounds in phytoplankton and weeds is reviewed in Appendix 4, Tables A4.1–A4.3.

Table 8.20. Concentration of organometallic compounds found in algae and plankton

Compound	Type of sample	Concentration (mg kg^{-1} unless otherwise stated)		Ref.
	(a) *Algae*			
Organoarsenic compounds	Macroalgae	0.2–0.6 monomethyl arsenic and 7.6–15.6 dimethyl arsenic acid		421
	Marine brown algae (*Haminariaceae*)	40.3–89.7		422
	(b) *Plankton*			
Organolead compounds	Macrophytes	4 m deep samples	Surface water samples	423
		Me$_3$EtPb 0.038	–	
		Me$_2$Et$_2$Pb 1.5	–	
		MeEt$_3$Pb 3.61	–	
		Et$_4$Pb 16.5	0.07	
		Et$_3$Pb$^+$ 0.59	0.13	
		Et$_2$Pb^{2+} 0.11	–	
		Total Pb 59.2	4.3	

Table 8.21. Methods for the determination of organometallic compounds in algae, plankton and weeds

Determined	Type of sample	Technique	Ref.
	(a) *Algae*		
Organoarsenic compounds	Marine brown algae	Arsine generation – silver diethyl dithiocarbamate spectrophotometric method	422
Monomethyl arsenic acid, dimethyl arsinic acid	Macro algae	Arsine generation – atomic absorption spectrometry	421
Organotin compounds	Algae		424
	(b) *Plankton*		
Organolead compounds Me$_3$EtPb, Me$_2$Et$_2$Pb MeEt$_3$Pb, Et$_4$Pb, Et$_3$Pb$^+$, Et$_2$Pb^{2+}	Macrophytes	Gas chromatography with atomic absorption spectrometric detector	423
	(c) *Weeds*		
Organotin compounds methyltin, butyltin	Eelgrass (*Zostera marina L*)	Hydride generation atomic absorption spectrometry	425

REFERENCES

1. Papontsoglou, S.E. and Abel, D., *Bulletin of Environmental Contamination and Toxicology* **41**, 404 (1988)
2. Segner, H., *Journal of Fish Biology* **30**, 423 (1987)
3. Lauren, D.J. and McDonald, D.G., *Canadian Journal of Fisheries and Aquatic Sciences* **44**, 105 (1987)
4. Presser, T.S. and Ohlendorf, H.M., *Environmental Management* **11**, 805 (1987)
5. Behar, J.V., Schuck, E.A., Stanley, R.E. and Morgan, G.B., *Environmental Science and Technology* **13**, 49 (1979)
6. Welz, B. and Melcher, M., *Analytical Chemistry* **57**, 427 (1985)
7. Agemian, H. and Thomas, M., *Analytical Chemistry* **105**, 902 (1980)
8. Heggie, D.T., *Study of Reservoirs, Fluxes and Pathways in an Alaskan Fjord*, PhD dissertation, University of Alaska (1977)
9. Brooke, P.J. and Evans, W.H., *Analyst (London)* **106**, 574 (1981)
10. de Oliveria, E., McLaren, J.W. and Berman, S.S., *Analytical Chemistry* **55**, 2047 (1983)
11. Maher, W.A., *Analyst (London)* **108**, 939 (1983)
12. Beauchemin, D., Bednas, M.E., Berman, S.S., McLaren, J.W., Siu, K.W.M. and Sturgeon, R.E., *Analytical Chemistry* **60**, 2209 (1988)
13. Poldoski, J.W., *Analytical Chemistry* **52**, 1147 (1980)
14. Sperling, K.R., *Fresenius Zeitschrift für Analytische Chemie* **287**, 23 (1977)
15. Sperling, K.R., *Fresenius Zeitschrift für Analytische Chemie* **301**, 294 (1980)
16. Sperling, K.R., *Fresenius Zeitschrift für Analytische Chemie* **310**, 254 (1982)
17. Sperling, K.R., *Fresenius Zeitschrift für Analytische Chemie* **332**, 565 (1988)
18. Borg, H., Edin, A., Holm, K. and Skold, E., *Water Research* **15**, 1291 (1981)
19. Harvey, B.R., *Analytical Chemistry* **50**, 1866 (1978)
20. Pagenkopf, G.K., Neuman, D.K. and Woodruff, R., *Analytical Chemistry* **44**, 2248 (1972)
21. May, T.W. and Brumbaugh, W.G., *Analytical Chemistry* **54**, 1032 (1982)
22. Louie, H.W., *Analyst (London)* **108**, 1313 (1983)
23. Davidson, J.W., *Analyst (London)* **104**, 683 (1979)
24. Tong, S.L. and Leow, W.K., *Analytical Chemistry* **52**, 581 (1980)
25. Konishe, T. and Takahashi, H., *Analyst (London)* **108**, 827 (1983)
26. Thomas, R.J., Hogstrom, R.A. and Kuchar, E.J., *Analytical Chemistry* **44**, 512 (1972)
27. Jones, P. and Nickless, O., *Journal of Chromatography* **89**, 201 (1974)
28. Gustarsson, I. and Giolimowski, J., *Science of the Total Environment* **22**, 85 (1981)
29. Svasankara-Pillay, K.K., Thomas, C.C., Sondel, J.A. and Hyche, C.M., *Analytical Chemistry* **43**, 1419 (1971)
30. Lo, J.M., Wei, J.C., Young, M.H. and Yeh, S.J., *Journal of Radioanalytical Chemistry* **72**, 571 (1982)
31. Goulden, P.D., Anthony, D.H.J. and Austen, K.D., *Analytical Chemistry* **53**, 2027 (1981)
32. Dogan, S. and Haerdi, W., *International Journal of Environmental Analytical Chemistry* **8**, 249 (1980)
33. Bloteky, A.J., Medira, V.A., Falcone, C. and Rack, E.P., *Analytical Chemistry* **51**, 178 (1979)
34. Kuriyama, T. and Kuroda, R., *Analyst (London)* **107**, 505 (1982)
35. Agemian, H., Sturtevant, D.P. and Austen, K.D., *Analyst (London)* **105**, 125 (1980)
36. Armannsson, H., *Analytica Chimica Acta* **110**, 21 (1979)
37. Ramelow, G., Ozkan, M.A., Tuncel, G., Saydam, C. and Balkas, T.I., *International Journal of Environmental Analytical Chemistry* **5**, 125 (1978)

38. Sakar, M.K. and May, W., *Science of the Total Environment* **74**, 199 (1988)
39. Adeljau, S.B., Bond, A.M. and Hughes, H.C., *Analytica Chimica Acta* **148**, 59 (1983)
40. Greig, R.A. and Jones, J., *Archives of Environmental Contamination and Toxicology* **4**, 420 (1976)
41. Awadallah, R.M., Mohamed, A.E. and Gabr, S.A., *Journal of Radioanalytical and Nuclear Chemistry Letters* **95**, 145 (1985)
42. Linde, G., Gether, J. and Steinnes, E., *Ambio* **5**, 180 (1976)
43. Analytical Methods Committee, Society for Analytical Chemistry (London), Recommended General Methods for the Examination of Fish and Fish Products. *Analyst (London)* **104**, 434 (1979)
44. Kirkpatrick, D.S. and Bishop, S.H., *Analytical Chemistry* **43**, 1707 (1971)
45. White, R.H., Thesis, Illinois University, University Microfilms Ltd, Tylers Green Penn, Bueke (1968)
46. Addison, R.I. and Ackmann, R.G., *Journal of Chromatography* **47**, 421 (1970)
47. Banmann, P.C., Smith, W.D. and Parkland, W.K., *Transactions of the American Fisheries Society* **116**, 79 (1987)
48. Hoogen, G.V. and Opperhuizen, A., *Environmental Toxicology and Chemistry* **7**, 213 (1988)
49. Carlson, A.R. and Kosian, P.A., *Archives of Environmental Contamination and Toxicology* **16**, 129 (1987)
50. Tana, A., *Water, Science and Technology* **20**, 77 (1988)
51. Rogers, I.H. and Hall, K.J., *Water Pollution Research Journal of Canada* **22**, 197 (1987)
52. Reijnders, P., Research Institute for Nature Management, Environmental Protection of the North Sea, London, 24–27 March 1987 Paper 5 (1987)
53. Kawano, M., Ingue, T., Wada, T., Hidaka, H. and Tasukawa, R., *Environmental Science and Technology* **22**, 792 (1988)
54. Droy, B.F. and Hinton, D.E., *Marine Environmental Research* **24**, 259 (1988)
55. Kaiser, K.L.E., *Science* **185**, 523 (1974)
56. Hiatt, M.H., *Analytical Chemistry* **55**, 506 (1983)
57. Vassilaros, D.L., Stoker, P.W., Booth, G.M. and Lee, M.L., *Analytical Chemistry* **54**, 106 (1982)
58. Ozretich, R.J. and Schroeder, W.P., *Analytical Chemistry* **58**, 2041 (1986)
59. Laseter, J.L., De Lean, R. and Remele, P.C., *Analytical Chemistry* **50**, 1169 (1978)
60. Norheim, G. and Okland, E.O., *Analyst (London)* **105**, 990 (1980)
61. Ichinose, N., Adachi, K. and Schwedt, G., *Analyst (London)* **110**, 1505 (1985)
62. Gaskin, D.E., Smith, G.J.D., Arnold, P.W., Louisy, M.V., Frank, R., Moldrinet, M. and McWade, J.W., *J. Fisheries Research Board, Canada* **31**, 1235 (1974)
63. Musial, C.J. and Uthe, J.F., *International Journal of Environmental Analytical Chemistry* **14**, 117 (1983)
64. Lamparski, L.C., Nestrick, J.J. and Stehl, R.L., *Analytical Chemistry* **51**, 1453 (1979)
65. Smith, L.M., Stalling, D.L. and Johnson, J.L., *Analytical Chemistry* **56**, 1830 (1984)
66. Mitchum, R.K., Moier, G.F. and Korfmacker, W.A., *Analytical Chemistry* **52**, 2278 (1980)
67. Luckas, B. and Harms, U., *International Journal of Environmental Analytical Chemistry* **29**, 215 (1987)
68. Spies, R.B. and Rice, D.W., *Marine Biology* **98**, 191 (1988)
69. O'Connor, J.M. and Pizza, J.C., *Estuaries* **10**, 68 (1987)
70. Douabul, A.A.Z., Al-Saad, H.T., Al-Obaidy, Z. and Al-Rekabi, H.N., *Water, Air and Soil Pollution* **35**, 187 (1987)

71. Haahti, H. and Perttila, M., *Marine Pollution Bulletin* **19**, 29 (1988)
72. Amodio-Coecchiero, R. and Arnese, A., *Bulletin of Environmental Contamination and Toxicology* **40**, 233 (1988)
73. Doe, K.G., Ernst, W.R., Parker, W.R., Julien, G.R.J. and Hennigar, P.A., *Canadian Journal of Fisheries and Aquatic Sciences* **45**, 287 (1988)
74. Venant, A. and Cumont, G., *Environmental Pollution* **43**, 163 (1987)
75. Gooch, J.W. and Matsumara, F., *Archives of Environmental Contamination and Toxicology* **16**, 349 (1987)
76. Albaiges, J., Farran, A., Soler, M. and Gallifer, A., *Marine Environmental Research* **22**, 1 (1987)
77. Devaux, A. and Monod, G., *Environmental Monitoring and Assessment* **9**, 105 (1987)
78. Cossarini-Dunier, M., Monod, G., Demael, A. and Lepot, D., *Ecotoxicology and Environmental Safety* **13**, 339 (1987)
79. Giam, C.S., Chau, H.S. and Nelt, G.S., *Analytical Chemistry* **47**, 2225 (1975)
80. Law, R.J., *Analytical Proceedings (London)* **19**, 248 (1982)
81. Chesler, S.N., Gump, B.H., Hertz, H.S., May, W.E. and Wise, S.A., *Analytical Chemistry* **50**, 805 (1978)
82. Farrington, J.W., Teal, J.M., Quinn, J.G., Wade, T. and Burns, K.A., *Bulletin of Environmental Contamination and Toxicology* **10**, 129 (1973)
83. Ariese, I., Kok, S.J., Verback, M., Py, G., Hoorning, C.D., Govifer, C., Velthorst, N.H. and Hafstraat, J.W., *Analytical Chemistry* **65**, 1100 (1993)
84. Vassilaros, D.L., Stoker, P.W., Booth, G.M. and Lee, M.L., *Analytical Chemistry* **54**, 106 (1982)
85. Birkholz, D.A., Coutts, A.T. and Hrudy, S.E., *Journal of Chromatography* **449**, 251 (1988)
86. Johnson, J.L., Stalling, D.L. and Hogen, J., *Bulletin of Environmental Contamination and Toxicology* **11**, 393 (1974)
87. Itoh, K., Chikumo, M. and Tanaka, H., *Fresenius Zeitschrift für Analytische Chemie* **330**, 600 (1988)
88. Cantillo, A.Y., Sinex, S.A. and Helz, G.R., *Analytical Chemistry* **56**, 33 (1984)
89. Suhr, N.H. and Ingasmello, C.O., *Analytical Chemistry* **38**, 730 (1968)
90. Sturgeon, R.E., Desauliniers, J.A.H., Berman, S.S. and Russell, D.S., *Analytica Chimica Acta* **134**, 283 (1982)
91. McQuaker, N.R., Kluckner, P.D. and Chang, G.N., *Analytical Chemistry* **51**, 888 (1979)
92. Walsh, J.N. and Howie, E.A., *Min. Management* **43**, 967 (1980)
93. Sinex, S.A., Cantillo, A.Y. and Helz, G.R., *Analytical Chemistry* **52**, 2342 (1980)
94. Murray, A.J. and Riley, J.P., *Analytica Chimica Acta* **65**, 261 (1973)
95. Murray, A.J. and Riley, J.P., *Nature (London)* **242**, 37 (1973)
96. Deetman, A.A., Demeulemeester, P., Garcia, M., Hauck, G., Hauck, G., Prigge H., Palin D.E., Krochenberger, L., Hollies, J.I., Rohrschneider, L. and Schmidthammer, L., *Analytica Chimica Acta* **82**, 1 (1976)
97. Parejko, R. and Keller, R., *Bulletin of Environmental Contamination and Toxicology* **14**, 480 (1975)
98. Solomon, J., *Analytical Chemistry* **51**, 186 (1979)
99. De Leon, I.R., Maberry, M.A., Overton, E.B., Roschke, C.K., Remele, P.C., Steele, C.F., Warren, V.L. and Leister, J.L., *J. Chromatographic Science* **18**, 85 (1980)
100. Hiatt, M.H., *Analytical Chemistry* **55**, 506 (1983)
101. Linde, G., Gether, J. and Steinnes, E., *Ambio* **5**, 180 (1976)
102. Hobien, H.J., Ching, S.A., Casarett, J.J. and Young, R.A., *Bulletin of Environmental Contamination and Toxicology* **15**, 78 (1976)

103. Stark, A., *J. Agriculture and Food Chemistry* **17**, 871 (1969)
104. Rudling, L., *Water Research* **4**, 533 (1970)
105. Sackmasserova-Vennigerova, M. and Uhnak, J., *Vodni Hospodarstvi Series B* **31**, 133 (1981)
106. Renberg, L., *Analytical Chemistry* **46**, 459 (1974)
107. Tausch, H., Stelik, G. and Widlidal, H., *Chromatographia* **41**, 403 (1981)
108. Neeley, W.B., *Science of the Total Environment* **7**, 117 (1977)
109. Frederick, L.L., *J. Fisheries Research Board, Ottawa, Canada* **32**, 1705 (1975)
110. Jan, J. and Malseric, S., *Bulletin of Environmental Contamination and Toxicology* **6**, 772 (1978)
111. Luckas, B., Pscheidl, H. and Haberland, P., *J. Chromatography* **147**, 41 (1978)
112. Laramee, J.A. and Deinzer, M.L., *Analytical Chemistry* **66**, 719 (1994)
113. Yu Ma, C. and Bayne, C.K., *Analytical Chemistry* **65**, 772 (1993)
114. Sackmauer, O.M., Pal'Usova, O. and Szokolay, A., *Water Research* **11**, 551 (1977)
115. Szelewski, M.J., Hill, D.R., Spiegel, S.J. and Tifft, E.C., *Analytical Chemistry* **51**, 2405 (1979)
116. Jones, A., *Analytical Chemistry* **66**, 1264 (1994)
117. Buser, H.R., Zook, D.R. and Rappe, C., *Analytical Chemistry* **64**, 1176 (1992)
118. Phillipson, D.W. and Puma, B.J., *Analytical Chemistry* **52**, 2328 (1980)
119. Mitchum, R.K., Moier, G.F. and Korfmacher, W.A., *Analytical Chemistry* **52**, 2278 (1980)
120. Lawrence, J., Onuska, F., Wilkinson, R. and Afghan, B.K., *Chemosphere* **15**, 1085 (1986)
121. Harless, R.L., Oswald, E.O., Lewis, R.G., Dupey, A.E., McDaniel, D.D. and Tai, H., **11**, 193 (1982)
122. Clement, R.E., Babbie, B. and Taguchi, V., *Chemosphere* **15**, 1147 (1986)
123. Lamparski, L.L., Nestrich, T.J. and Shehl, R.H., *Analytical Chemistry* **51**, 1453 (1979)
124. Smith, L.M., Stalling, D.L. and Johnson, J.L., *Analytical Chemistry* **56**, 1830 (1984)
125. Taguchi, V.Y., Reiner, E.J., Wang, D.T., Meresz, O. and Hallas, B., *Analytical Chemistry* **60**, 1429 (1988)
126. Crummelt, W.B., *Chemosphere* **12**, 429 (1983)
127. Chang, R.D., Jarman, W.M. and Hennings, J.A., *Analytical Chemistry* **65**, 2420 (1993)
128. Norheim, G. and Okland, M.O., *Analyst (London)* **105**, 990 (1980)
129. Markin, G.P., Hawthorne, J.C., Collins, H.L. and Ford, J.H., *Pesticides Monitoring Journal* **7**, 139 (1974)
130. Frank, R., Armstrong, A.F., Boeleus, R.G., Braun, H.H. and Douglas, C.N., *Pesticides Monitoring Journal* **7**, 165 (1974)
131. Hesselberg, R.J. and Johnson, J.L. *Bulletin of Environmental Contamination and Toxicology* **7**, 115 (1972)
132. Simal, J., Crous Vidal, D., Maria-Chareo, A. Boado, M.A., Diaz, R. and Vilas, D., *An. Bromat. (Spain)* **23**, 1 (1971)
133. Chau, A.S.Y., *Journal of the Association of Official Analytical Chemists* **55**, 519 (1972)
134. Kuem, D.W., *Analytical Chemistry* **49**, 521 (1977)
135. Sackmauerova, H., Pal'Usova, O. Szokolay, A., *Water Research* **11**, 537 (1977)
136. Gaskin, D.E., Smith, G.J.D., Arnold, A.W., Louisy, M.V., Frank, R., Moldrinet, M. and McWade, J.W., *J. Fisheries Research Board, Ottawa, Canada* **31**, 1235 (1974)
137. Ozretich, R.J. and Schroeder, W.P., *Analytical Chemistry* **58**, 2041 (1986)

138. Musial, C.J. and Uthe, J.F., *International Journal of Environmental Analytical Chemistry* **14**, 117 (1983)
139. Hughes, R.A. and Lee, G.F., *Environmental Science and Technology* **7**, 934 (1973)
140. Kaiser, K.L.E., *Science* **185**, 523 (1974)
141. Reynolds, L.M., *Research Review* **34**, 27 (1971)
142. Chau, A.S.Y. and Wilkinson, W.J., Personal communications. Pesticide Analytical Manual (Department of Health, Education and Welfare), Food and Drug Administration, Washington, DC Volumes 1 and 2 (1971)
143. Bonelli, E.J., *Analytical Chemistry* **44**, 603 (1972)
144. Laseter, J.L., De Leon, I.R. and Remele, P.C., *Analytical Chemistry* **50**, 1169 (1980)
145. Onuska, F.I., Comba, M.E. and Coburn, J.C., *Analytical Chemistry* **52**, 2272 (1980)
146. Markin, G.P., Hawthorne, J.C., Collins, H.L. and Ford, J.H., *Pesticides Journal* **7**, 139 (1974)
147. Armour, J.A. and Burke, J.A., *Journal of Association of Official Analytical Chemists* **53**, 761 (1970)
148. Gaul, J. and Cruz-La Grange, P., Separation of Mirex and PCB's in fish. *Laboratory Information Bulletin*, Food and Drug Administration, New Orleans District (1971)
149. Markin, G.P., Ford, J.H., Spence, J.H., Davies, J. and Loftis, C.D., *Environmental Monitoring for the Pesticide Mirex*, USDA APHIS 81–83, November (1972)
150. Zakitis, L.H. and McCray, E.M., *Bulletin of Environmental Contamination and Toxicology* **28**, 334 (1982)
151. Deutsch, M.E., Westlake, W.E. and Gunther, F.A., *J. Agriculture and Food Chemistry* **18**, 178 (1970)
152. Szeto, S.Y., Yee, J., Brown, M.J. and Oloffs, P.C., *Journal of Chromatography* **240**, 526 (1982)
153. Lores, E.M., Moore, J.C., Knight, J., Forester, J., Clark, J. and Moody, P., *Journal of Chromatographic Science* **23**, 124 (1985)
154. West, S.D. and Day, W., *Journal of the Association of Official Analytical Chemists* **69**, 856 (1986)
155. Brewster, J.D. and Lightfield, A.R., *Analytical Chemistry* **65**, 2415 (1993)
156. Murray, D.A.J., *J. Fisheries Research Board, Ottawa, Canada* **32**, 457 (1975)
157. Steinwandter, H., *Fresenius Zeitschrift für Analytische Chemie* **326**, 139 (1987)
158. Chau, Y.K., Wong, P.T.S., Bengert, G.A. and Kramer, O., *Analytical Chemistry* **51**, 186 (1979)
159. Lindstrom, K. and Schubert, R., *Journal of High Resolution Chromatography and Chromatography Communications* **7**, 68 (1984)
160. Kuchl, D.W., Kopperman, H.L., Veith, G.D. and Glass, G.E., *Bulletin of Environmental Contamination and Toxicology* **16**, 127 (1976)
161. Steinwandter, H. and Zimmer, L., *Fresenius Zeitschrift für Analytische Chemie* **316**, 705 (1983)
162. Ramdahl, T., Carlberg, G.E. and Kolsker, P., *Science of the Total Environment* **48**, 147 (1986)
163. Bush, B. and Bernard, E.L., *Analytical Letters* **15**, 1643 (1982)
164. Giam, C.S., Chan, H.S. and Nett, G.S., *Analytical Chemistry* **47**, 2225 (1975)
165. Kiigemagi, U., Burnard, J. and Terriere, L.C., *J. Agriculture and Food Chemistry* **23**, 717 (1975)
166. Wells, D.E. and Cowan, A.A., *Analyst (London)* **106**, 862 (1981)
167. Allen, J.L. and Sills, J.B., *Journal of Association of Official Analytical Chemists* **57**, 387 (1974)

168. Martin, J.F., McCoy, C.P., Greenleaf, W. and Bennett, L., *Canadian Journal of Fisheries and Aquatic Sciences* **44**, 909 (1987)
169. Cockell, K.A. and Hilton, J.W., *Aquatic Toxicology* **12**, 73 (1988)
170. Richman, L.A., Wren, C.D. and Stokes, P.M., *Water, Air and Soil Pollution* **37**, 465 (1988)
171. Bailey, S.K. and Davies, I.M., *Environmental Pollution* **55**, 161 (1988)
172. Chau, Y.K., Wong, P.T.S., Bengent, G.A. and Dunn, J.L., *Analytical Chemistry* **56**, 271 (1984)
173. Holak W., *Analyst (London)* **107**, 1457 (1982)
174. Shum, G.T.C., Freeman, H.C. and Uthe, J.F., *Analytical Chemistry* **51**, 414 (1979)
175. Uthe, J., *J. Fisheries Research Board, Ottawa, Canada*, private communication
176. Farrington, J.W., Teal, J.M. Quinn, J.G., Wade, T. and Burns, K.A., *Bulletin of Environmental Contamination and Toxicology* **10**, 129 (1973)
177. McCrehan, W.A. and Durst, R.A., *Analytical Chemistry* **50**, 2108 (1978)
178. Capelli, R., Fezia, C. and Franchi, A., *Analyst (London)* **104**, 1197 (1979)
179. Westoo, F., *Anal. Chem. Scan.* **21**, 1790 (1967)
180. Callum, G.I., Ferguson, M.M. and Lenchan, J.M.A., *Analyst (London)* **106**, 1009 (1981)
181. Analytical Methods Committee Society for Analytical Chemistry, London, *Analyst (London)* **102**, 769 (1977)
182. Matsunaga, K. and Takahashi, S., *Analytica Chimica Acta* **87**, 487 (1976)
183. Uthe, J.F., Solomon, J. and Griff, B., *Journal of Association of Official Analytical Chemists* **55**, 583 (1972)
184. Jones, P. and Nickless, G., *Analyst (London)* **103**, 1121 (1978)
185. Hendzel, M.R. and Jamieson, D.M., *Analytical Chemistry* **48**, 926 (1976)
186. Schulz, C.D., Clear, D., Pearson, J.E., Rivers, J.B. and Hylin, J.W., *Bulletin of Environmental Contamination and Toxicology* **15**, 230 (1976)
187. Analytical Methods Committee, Chemical Society, London, *Analyst (London)* **92**, 403 (1976)
188. Analytical Methods Committee, *Analyst (London)* **101**, 62 (1976)
189. Friend, M.T., Smith, C.A. and Wishart, D., *Atomic Absorption Newsletter* **16**, 46 (1977)
190. Agemian, H. and Chau, A.S.Y., *Analytical Chimica Acta* **75**, 297 (1975)
191. Matsunaga, K. and Takahashi, S., *Analytical Chimica Acta* **87**, 487 (1976)
192. Analytical Methods Committee, Chemical Society, London, *Analyst (London)* **102**, 769 (1977)
193. Stuart, D.C., *Analytical Chemistry* **96**, 83 (1978)
194. Agemian, H. and Cheam, V., *Analytica Chimica Acta* **101**, 193 (1978)
195. Davies, I.M., *Analytical Chemistry* **102**, 189 (1978)
196. Jones, P. and Nickless, G., *Analyst (London)* **103**, 1120 (1978)
197. Aspila, K.I. and Carron, J.M., Interlaboratory Quality Control Study No. 1, Total Mercury in Sediments, Report Series, Inland Waters, Directorate Water Quality Branch, Special Services Section, Inland Water Directorate Water Quality Branch, Department of Fisheries and Environment, Burlington, Ontario, Canada
198. Shum, G.T.C., Freeman, H.C. and Uthe, J.F., *Analytical Chemistry* **51**, 414 (1979)
199. Capelli, R., Fezia, C., Franchi, A. and Zanicchi, *Analyst (London)* **104**, 1197 (1979)
200. Collett, D.L., Fleming, D.E. and Taylor, G.E., *Analyst (London)* **105**, 897 (1980)
201. Abo-Rady, M.D.K., *Fresenius Zeitschrift für Analytische Chemie* **299**, 187 (1989)
202. Holden, A.V., *Pesticide Science* **4**, 399 (1973)
203. Beauchemin, D., Siu, K.W. and Berman, S.S., *Analytical Chemistry* **60**, 2587 (1988)
204. Holak, W., *Analyst (London)* **107**, 1457 (1982)

205. MacCrehan, W.A. and Durst, R.A., *Analytical Chemistry* **50**, 2108 (1978)
206. Uthe, J.F., Solomon, J. and Grift, B.J., *Journal of Official Analytical Chemists* **55**, 583 (1972)
207. Kamps, L.R. and McMahom, I., *Journal of Association of Official Analytical Chemists* **18**, 351 (1970)
208. Cappon, C.J. and Crispin-Smith, V., *Analytical Chemistry* **49**, 365 (1977)
209. Analytical Methods Committee, Society for Analytical Chemistry London, *Analyst (London)* **102**, 769 (1977)
210. Hight, S.C. and Corcoran, M.T., *Journal of Association of Official Analytical Chemists* **70**, 24 (1987)
211. Hight, S.C., *Journal of Association of Official Analytical Chemists* **70**, 667 (1987)
212. Bache, C.A. and Lisk, D.J., *Analytical Chemistry* **43**, 950 (1971)
213. Fischer, R., Rapsomankiz, S. and Andreae, M.O., *Analytical Chemistry* **65**, 763 (1993)
214. Pillay, K.K.S., Thomas, C.C., Sondel, J.A. and Hyche, C.M., *Analytical Chemistry* **43**, 1419 (1971)
215. Callum, G.I., Ferguson, M.M. and Lenihan, J.M.A., *Analyst (London)* **106**, 1009 (1981)
216. Chau, Y.K., Wong, P.T.S., Bengert, G.A. and Kramer, O., *Analytical Chemistry* **51**, 186 (1979)
217. Chau, Y.K., Wong, P.T.S. and Goulden, P.D., *Analytica Chimica Acta* **421**, 85 (1976)
218. Chau, Y.K., Wong, P.T.S., Bengert, G.A. and Dunn, J.L., *Analytical Chemistry* **56**, 271 (1984)
219. Birnie, S.E. and Hodges, D.J., *Environmental Technology Letters* **2**, 433 (1981)
220. Short, J.W., *Bulletin of Environmental Contamination and Toxicology* **39**, 412 (1987)
221. Sasaki, K., Ishizaku, T., Suzuki, T. and Saito, Y., *Journal of Association of Official Analytical Chemists* **71**, 360 (1988)
222. Smith, J.D., *Nature (London)* **225**, 103 (1970)
223. Blotcky, A.J., Ebrahim, A. and Rack, E.P., *Analytical Chemistry* **60**, 2734 (1988)
224. Maher, W.A., *Analytica Chimica Acta* **126**, 157 (1981)
225. Agemian, H. and Cheam, V., *Analytica Chimica Acta* **101**, 193 (1978)
226. Beauchemin, D., Bednas, M.E., Berman, S.S., McLaren, J.W., Siu, K.W.M. and Sturgeon, R.E., *Analytical Chemistry* **60**, 2209 (1988)
227. Barth, R.J., Adams, D.M., Soloway, A.H., Mechetner, E.B., Alam, F. and Anisuzzamen, A.K.M., *Analytical Chemistry* **63**, 890 (1991)
228. Wanatake, N., Yasuda, Y., Kato, K., Nakamura, T., Funasaka, R., Shimokawa, K., Sato, E. and Ose, Y., *Science of the Total Environment* **34**, 169 (1988)
229. Anil, A.C. and Wagh, A.B., *Marine Pollution Bulletin* **19**, 177 (1988)
230. Micallef, S. and Tyler, P.A., *Marine Pollution Bulletin* **18**, 180 (1987)
231. Draback, I., Eichner, P. and Rasmussen, L., *Journal of Radioanalytical and Nuclear Chemistry Articles* **114**, 29 (1987)
232. Bunguegneau, J.M. and Martoja, M., *Bulletin of Environmental Contamination and Toxicology* **39**, 69 (1987)
233. Doherty, F.G., Failla, M.L. and Cherry, D.S., *Water Research* **22**, 927 (1988)
234. Lyngby, J.E. and Brix, H., *Science of the Total Environment* **64**, 239 (1987)
235. Gil, M.N., Harvey, M.A. and Esteves, J.L., *Marine Pollution Bulletin* **19**, 181 (1981)
236. Peerzada, N. and Dickinson, C., *Marine Pollution Bulletin* **19**, 182 (1988)
237. Uthus, E.O., Collings, M.E., Cornatzer, W.E. and Nielson, F.H., *Analytical Chemistry* **53**, 2221 (1981)
238. Brzezinska-Pandyn, A., Vanhoon, J. and Hancock, R., *Atomic Spectroscopy* **7**, 72 (1986)

239. Siu, K.W.M., Roberts, S.Y. and Berman, S.S., *Chromatographia* **19**, 398 (1984)
240. Topping, G., Report on the 6th ICES Trace Metal Intercomparison Exercise for Cadmium and Lead in Biological Tissue. ICES Loop Research Report No. 111 (1982).
241. Gabrielli, L.F., Marletta, G.P. and Favretto, L., *Atomic Spectroscopy* **1**, 35 (1980)
242. Ashworth, M.J. and Farthing, R.H., *International Journal of Analytical Chemistry* **10**, 35 (1981)
243. Mazzucotelli, A., Viarengo, A., Martino, G. and Frache, R., *Marine Environmental Research* **24**, 129 (1988)
244. McLaren, J.W. and Berman, S.S., *Applied Spectroscopy* **35**, 403 (1981)
245. Greenberg, R.R., *Analytical Chemistry* **52**, 676 (1980)
246. Liang, L., D'Haese, P.C., Lamberts, L.V., Van der Vyver, F.L. and De Broe, M.E., *Analytical Chemistry* **63**, 423 (1991)
247. Rae, R.R. and Chatt, A., *Analytical Chemistry* **63**, 1298 (1991)
248. Fassett, J.D. and Murphy, T.J., *Analytical Chemistry* **62**, 386 (1990)
249. Brewer, S.W. and Sacks, R.D., *Analytical Chemistry* **60**, 1769 (1988)
250. Zhe-Ming, N., Xiao-Chun, L. and Heng-Bin, H., *Analytica Chimica Acta* **186**, 147 (1986)
251. Kunkel, E., *Fresenius Zeitschrift für Analytische Chemie* **258**, 337 (1972)
252. Hutton, R.C. and Preston, B., *Analyst (London)* **105**, 981 (1980)
253. Ahmed, R.B., Hill, J.O. and Magee, R.J., *Analyst (London)* **108**, 835 (1983)
254. Maher, W.A., *Marine Pollution Bulletin* **16**, 33 (1985)
255. Florence, T.M. and Farrer, Y.I., *J. Electroanalytical Chemistry* **51**, 191 (1974)
256. Blotcky, A.J., Falcone, C., Medina, V.A. and Rack, E.P., *Analytical Chemistry* **51**, 178 (1979)
257. *The International Mussel Watch*, US National Academy of Science XVI (1980)
258. Goldberg, E.D., Bowen, J.W., Farrington, G., Harvey, J.H., Martin, P.L., Parker, R.W., Risebrough, W., Robertson, E., Schneider, O. and Gamble, E., The Mussel Watch. *Environmental Conservation* **5**, 101 (1978)
259. Patterson, D., Settle, B., Schaule, O. and Burnett, M., Transport of lead to the oceans and within ocean ecosystems, in Windam, H.L. and Duce, R.H. (eds), *Marine Pollution Transfer* Lexington Books, Lexington, MA (1976)
260. Solchaga, M. and De La Guardia, M., *Journal of Association of Official Analytical Chemists* **69**, 874 (1986)
261. Welz, B. and Melcher, M., *Analytical Chemistry* **57**, 427 (1985)
262. Slinkman, D. and Sacks, R., *Analytical Chemistry* **63**, 343 (1991)
263. Schlemmer, G. and Welz, B., *Fresenius Zeitschrift für Analytische Chemie* **320**, 648 (1985)
264. Amiard, J.C., Pineau, A., Boiteau, H.C., Metayer, C. and Amiard-Triquet, C., *Water Research* **21**, 693 (1987)
265. Siriaks, A. and Kingston, H.M., *Analytical Chemistry* **62**, 1185 (1990)
266. Chisela, F., Gawlik, D. and Bratter, P., *Analyst (London)* **111**, 405 (1986)
267. Zeisler, R., Stone, S.F. and Sanders, D.W., *Analytical Chemistry* **60**, 2760 (1988)
268. Dutton, J.W.R., Technical Report FRL4 Fisheries Radiological Laboratory Hamilton, Lowestoft, Suffolk, UK (1969)
269. Fourie, H.O. and Peisach, M., *Analyst (London)* **102**, 193 (1977)
270. Lansberger, S. and Davidson, W.F., *Analytical Chemistry* **57**, 197 (1985)
271. Zheng, W., Sipes, G. and Carter, D.E., *Analytical Chemistry* **65**, 2174 (1993)
272. Buckley, W.T., Budac, J.J. and Godfrey, D.V., *Analytical Chemistry* **64**, 724 (1992)
273. Horowicz, E.P. and Fisher, D.E., *Analytical Chemistry* **63**, 522 (1991)
274. Que Hee, D.G. and Boyle, J.R., *Analytical Chemistry* **60**, 1033 (1988)

275. Yang, J.Y. and Liu, S.M., *Analytical Chemistry* **62**, 146 (1990)
276. Yang, S., Fu, S. and Wang, M., *Analytical Chemistry* **63**, 2970 (1991)
277. Woin, P. and Larssen, P., *Bulletin of Environmental Contamination and Toxicology* **38**, 220 (1987)
278. Kukkonen, J. and Oikari, A., *Science of the Total Environment* **62**, 399 (1987)
279. Rice, C.P. and White, D.S., *Environmental Toxicology and Chemistry* **6**, 259 (1987)
280. Dunn, B.P. and Stich, H.F.J., *J. Fisheries Research Board, Ottawa, Canada* **33**, 2040 (1976)
281. Tanabe, S., Tatsukawa, R. and Phillips, D.J.H., *Environmental Pollution* **47**, 41 (1987)
282. Freudenthal, J. and Greve, P.A., *Bulletin of Environmental Contamination and Toxicology* **10**, 108 (1973)
283. Kira, S., Izumi, T. and Ogata, M., *Bulletin of Environmental Contamination and Toxicology* **31**, 518 (1983)
284. Giam, C.S., Chan, H.S. and Nett, G.S., *Analytical Chemistry* **47**, 2225 (1975)
285. Chesler, S.N., Gump, B.H., Hertz, H.S., May, W.E. and Wise, S.E., *Analytical Chemistry* **50**, 865 (1978)
286. Thybaud, E. and Le Bras, S., *Bulletin of Environmental Contamination and Toxicology* **40**, 731 (1988)
287. Benthon, E., Gourmelun, Y., Dreano, Y. and Friocourt, M.C., *Journal of Chromatography* **203**, 279 (1981)
288. May, W.E., Chester, S.N., Cram, S.P., Gump, B.H., Hertz, D.S. and Enagonio, D.P., *Analytical Chemistry* **50**, 867 (1978)
289. Chester, S.N., Gump, B.H., Hertz, H.S., May, W.E., Dryzel, S.M. and Enagonio, D.P., National Bureau of Standards (US) Technical Note No. 889, Washington, DC (1976)
290. Mason, P.R., *Marine Pollution Bulletin* **18**, 528 (1987)
291. Morgan, N.L., *Bulletin of Environmental Contamination and Toxicology* **14**, 309 (1975)
292. Meyers, P.A., *Chemosphere* **7**, 385 (1978)
293. Giam, C.S., Trujillo, D.A., Kira, S. and Hrung, Y., *Bulletin of Environmental Contamination and Toxicology* **25**, 824 (1980)
294. Bjorseth, A., Knutsen, J. and Skei, J., *Science of Total Environment* **13**, 71 (1979)
295. Kunte, H., *Archives Hyg. Bakt.* **151**, 193 (1967)
296. Uthe, J.F. and Musal, C.J., *Journal of Association of Official Analytical Chemists* **71**, 363 (1988)
297. Murray, A.J. and Riley, J.P., *Analytica Chimica Acta* **65**, 261 (1973)
298. Murray, A.J. and Riley, J.P., *Nature (London)* **242**, 37 (1973)
299. Markin, G.P., Hawthorne, J.C., Collins, H.L. and Ford, J.H., *Pesticide Monitoring Journal* **7**, 139 (1974)
300. Armour, J.A. and Burke, J.A., *Journal of Association of Official Analytical Chemists* **53**, 761 (1970)
301. Gaul, J. and Cruz-La Grange, P., Separation of Mirex and PCB's in fish. *Laboratory Information Bulletin*, Food and Drug Administration, New Orleans District (1983)
302. Markin, G.P., Ford, J.H., Hawthorne, J.C., Spence, J.H., Davies, J. and Loftis, C.D., Environmental monitoring for the insecticide Mirex. USDA, APHIS 81–83, November (1972)
303. Teichman, J., Bevenue, A. and Hylin, J.W., *J. Chromatography* **151**, 155 (1978)
304. Ya Ma, C. and Bayne, C.K., *Analytical Chemistry* **65**, 772 (1993)
305. Freudeythal, J. and Greve, P.A., *Bulletin of Environmental Contamination and Toxicology* **10**, 108 (1973)

306. Ernst, W., Goerke, H., Eder, G. and Schaefer, R.C., *Bulletin of Environmental Contamination and Toxicology* **15**, 55 (1976)
307. Ernst, W., Schaefer, H., Goerke, H. and Eder, G.Z., *J. Analytical Chemistry* **227**, 378 (1974)
308. Mills, P.A., Caley, J.F. and Grithen, R.A., *Journal of Association of Official Analytical Chemists* **46**, 106 (1963)
309. Arias, C., Vidal, A. and Maria, J., *An. Bromat. (Spain)* **22**, 273 (1980)
310. Kouyonmjiam, H.H. and Uglow, R.F., *Environmental Pollution* **7**, 103 (1974)
311. US Environment Protection Agency Report No. EPA-600-4-74 1974T, 108 (1974)
312. Wilson, A.J., Forester, J. and Knight, J., US Wildlife Service, Circular 335 18–20 Centre for Estuaries and Research, Gulf Breeze, Florida, USA (1969)
313. Deutsch, M.E., Westlake, W.E. and Gunther, F.A., *J. Agriculture and Food Chemistry* **18**, 178 (1970)
314. Buzer, R.H. and Rappe, C., *Analytical Chemistry* **63**, 1210 (1991)
315. Kira, S., Izumi, T. and Ogata, M., *Bulletin of Environmental Contamination and Toxicology* **31**, 518 (1983)
316. Otaga, M., Mujake, Y. and Yamazaki, Y., *Water Research* **13**, 1179 (1979)
317. Nels, H.J. and Merchie, G., *Analytical Chemistry* **66**, 1330 (1994)
318. Matusik, J.E., Moskin, G.P. and Splon, J.A. *Journal of Association of Official Analytical Chemists* **71**, 994 (1988)
319. Slinkman, D. and Sacks, R., *Analytical Chemistry* **63**, 343 (1991)
320. Minchin, D., Duggan, C.B. and King, W., *Marine Pollution Bulletin* **18**, 604 (1987)
321. Langston, W.J., Burt, G.R. and Mingjiang, Z., *Marine Pollution Bulletin* **18**, 634 (1987)
322. Pickwell, G.V. and Steinert, S.A., *Marine Environmental Research* **24**, 215 (1988)
323. Krishnan, K., Marshall, W.D. and Hatch, W.I., *Environmental Science and Technology* **22**, 806 (1988)
324. Jones, A., *Analytical Chemistry* **60**, 316 (1988)
325. Giang, G.B., Maxwell, B., Siu, K.W.M., Luong, V.T. and Berman, S.S., *Analytical Chemistry* **63**, 1506 (1991)
326. Han, J.S. and Weber, J.H., *Analytical Chemistry* **60**, 316 (1988)
327. Birnie, S.E. and Hodges, D.E., *Environmental Technology Letters* **2**, 433 (1981)
328. Uthe, J.F., Solomon, J. and Grift, B., *Journal of Association of Official Analytical Chemists* **55**, 583 (1972)
329. Francesconi, K.A., Hicks, P., Stockton, R.A. and Irgolic, K.J., *Chemosphere* **14**, 1443 (1985)
330. Dallakyan, G.A., Korsak, M.N. and Nikiforova, E.P., *Water Resources* **15**, 53 (1988)
331. Brand, L.E., Sunda, W.G. and Guillard, P.R.L., *Journal of Experimental Marine Biology and Ecology* **96**, 225 (1986)
332. Rueter, J.G., O'Reilly, K.T. and Peterson, R.R., *Environmental Science and Technology* **21**, 435 (1987)
333. Claeesson, A. and Tornquist, L., *Water Research* **22**, 977 (1988)
334. Sathya, K.S. and Balakrishnan, K.P., *Water, Air and Soil Pollution* **36**, 283 (1988)
335. Shogata, S.A., Aboelela, S.I. and Ali, G.H., *Environmental Technology Letters* **9**, 1137 (1988)
336. Jana, S., *Water, Air and Soil Pollution* **38**, 105 (1988)
337. Genter, R.B., Cherry, D.S., Smith, E.P. and Cairns, J., *Hydrobiologia* **153**, 261 (1987)
338. Starodub, M.E., Wong, P.T.S. and Mayfield, C.I., *Science of the Total Environment* **63**, 101 (1987)
339. Kuwabara, J.S., *Studies in Environmental Science* **28**, 129 (1986)

340. Wang, W., *Environmental Toxicology and Chemistry* **6**, 961 (1987)
341. Jana, S., Dalal, T. and Barua, B., *Water, Air and Soil Pollution* **33**, 23 (1987)
342. Mo, S.C., Choi, D.S. and Robinson, J.W., *Journal of Environmental Science and Health* **A23**, 139 (1988)
343. Howard, L.S. and Brown, B.E., *Marine Pollution Bulletin* **18**, 451 (1987)
344. Pietilainen, K., Adams, F., Nullens, H. and Van Espen, P., *X-ray Spectrometry* **10**, 31 (1981)
345. Lee, D.S., *Analytical Chemistry* **54**, 1682 (1982)
346. Svasankara-Pillay, K.K., Thomas, C.C., Sondel, J.A. and Hyche, C.M., *Analytical Chemistry* **43**, 1419 (1971)
347. Hodge, V.F., Seider, G.L. and Goldberg, D., *Analytical Chemistry* **51**, 1256 (1979)
348. Bando, R., Galanti, G. and Varinci, P.G., *Analyst (London)* **108**, 722 (1983)
349. Szefer, P., *Marine Pollution Bulletin* **18**, 439 (1987)
350. IAEA International Atomic Energy Agency, Monaco, Intercalibration of analytical methods in marine environment samples. Progress Report No. 19, November (1978)
351. Ward, T.J., *Marine Biology* **95**, 315 (1987)
352. Aylid, K., *Environmental Pollution* **44**, 1 (1987)
353. Whyte, J.N.C. and Englar, J.R., *Botanic Marina* **26**, 159 (1983)
354. Maher, W.A., *Analyst (London)* **108**, 939 (1983)
355. Whyte, J.N.C. and Englar, J.R., *Analyst (London)* **101**, 815 (1976)
356. Mitchell, P.G., Greene, B. and Sneddon, *J. Mikrochimica Acta* No. 314, 249 (1986)
357. Darich, R.A., Nelson, D.W. and Sommers, L.E., *Journal of Environmental Quality* **14**, 400 (1985)
358. Dogan, S. and Haerdi, W., *International Journal of Environmental Analytical Chemistry* **8**, 249 (1980)
359. Sheugjun, M. and Holcombe, J.A., *Analytical Chemistry* **62**, 1994 (1990)
360. Flechenstein, J., *Fresenius Zeitschrift für Analytische Chemie* **328**, 396 (1987)
361. Bistricki, T. and Munawar, M., *Canadian Journal of Fisheries and Aquatic Sciences* **39**, 506 (1982)
362. Hampson, B.L. and Tennant, D., *Analyst (London)* **98**, 873 (1973)
363. Van Raaphorst, J.G., Van Weers, A.W. and Haremaker, H.M., *Analyst (London)* **99**, 523 (1974)
364. Kuldver, A., *Analyst (London)* **107**, 1343 (1982)
365. Kuldver, A. and Andreassen, B.T., *Atomic Spectroscopy* **18**, 106 (1979)
366. Dutton, J.W.R., Technical Report No. FR14, Gamma Spectrometric Analysis of Environmental Materials. Fisheries Radiobiological Laboratory, Hamilton Dock, Lowestoft, Suffolk, UK (1969)
367. Acey, R., Healy, P., Unger, T.F., Ford, C.E. and Hudson, R.A., *Bulletin of Environmental Contamination and Toxicology* **39**, 1 (1987)
368. Rhee, G.Y., Shane, L. and Denucci, A., *Applied and Environmental Microbiology* **54**, 1394 (1988)
369. Yasuno, M., Hanazato, T., Iwakuma, T., Takamura, K., Ueno, R. and Takamura, T., *Hydrobiologia* **159**, 247 (1988)
370. Ali, A., Nigg, N.H., Stamper, J.H., Kok-Yokami, H.L. and Weaver, M., *Bulletin of Environmental Contamination and Toxicology* **41**, 781 (1988)
371. Day, K. and Kaushik, N.K., *Archives of Environmental Contamination and Toxicology* **16**, 423 (1987)
372. Hanazato, T. and Yasuno, M., *Environmental Pollution* **48**, 145 (1987)
373. Lay, J.P., Muller, A., Peichl, L., Lang, R. and Korte, F., *Chemosphere* **16**, 1527 (1987)

374. Arthur, J.W., *International Journal of Environmental Studies* **32**, 97 (1988)
375. Krieger, K.A., Baker, D.B. and Kramer, J.W., *Archives of Contamination and Toxicology* **17**, 299 (1988)
376. Paterson, D.M. and Wright, S.J.L., *Letters in Applied Microbiology* **7**, 87 (1988)
377. Hamilton, P.B., Jackson, G.S., Kaushik, N.K. and Solomon, K.R., *Environmental Protection* **46**, 83 (1987)
378. Hersh, C.M. and Crumpton, W.G., *Bulletin of Environmental Contamination and Toxicology* **39**, 1041 (1987)
379. Walsh, G.E., Yoder, M.J., McLaughlin, L.L. and Lores, L.M., *Ecotoxicology and Environmental Safety* **14**, 215 (1987)
380. Stratton, G.W. and Smith, T.M., *Bulletin of Environmental Contamination and Toxicology* **40**, 736 (1988)
381. Yi-Xiong, L. and Bo-zen, S., *Hydrobiologia* **153**, 249 (1987)
382. Walsh, G.E., Deans, C.H. and McLaughlin, L.L., *Environmental Contamination and Toxicology* **6**, 767 (1987)
383. Pednekar, M.D., Gandhi, S. and Netrawate, M.S., *Environment International* **13**, 219 (1987)
384. Stratten, G.W., *Bulletin of Environmental Contamination and Toxicology* **38**, 1012 (1987)
385. Megharaj, M., Venkateswarlu, K. and Rao, A.S., *Ecotoxicology and Environmental Safety* **15**, 320 (1988)
386. Thorhang, A. and Marins, J., *Marine Pollution Bulletin* **18**, 124 (1987)
387. Goutex, M.M., Al-Mallah, M. and Bertrand, J.C., *Marine Biology* **94**, 111 (1987)
388. Smith, D.H., *Analytical Chemistry* **44**, 536 (1972)
389. Vairovamurthy, A., Andreae, M.O. and Brooks, J.M., *Analytical Chemistry* **58**, 2684 (1986)
390. Dallakyan, G.A., Veselovski, V.A., Tarusou, B.N. and Peogosyan, S.I. *J. Hydrobiology* **14**, 90 (1978)
391. Cowie, G.L. and Hedges, J.I., *Analytical Chemistry* **56**, 497 (1984)
392. Sodergren, A., *Bulletin of Environmental Contamination and Toxicology* **10**, 116 (1984)
393. Picer, N., Picer, M. and Strohal, P., *Bulletin of Environmental Contamination and Toxicology* **14**, 565 (1975)
394. Sodergren, A., *Journal of Chromatography* **160**, 271 (1978)
395. Jensen, K.S., *Vatten* **32**, 337 (1976)
396. Youngman, R.E., Report TR 82. Water Research Centre, Medmenham, Marlow, Bucks, UK (1978)
397. Wun, C.K., Rho, J., Walker, R.W. and Litsky, W., *Water, Air and Soil Pollution* **11**, 173 (1979)
398. Garside, C. and Riley, J.P., *Analytica Chimica Acta* **46**, 179 (1969)
399. Shoaf, W.I. and Lium, B.W., *J. Research US Geological Survey* **5**, 263 (1977)
400. Abayashi, J.K. and Riley, J.P., *Analytica Chimica Acta* **107**, 1 (1979)
401. UNESCO Monographs on Oceanographic Methodology No. 1 (1966)
402. Sartory, D.P., *Water Research* **19**, 605 (1985)
403. Jeffrey, N.W., *Biochimica Biophys. Acta* **162**, 271 (1968)
404. Strickland, D.H., in Riley, J.P. and Skirrow, G. (eds), *Chemical Oceanography*, Volume 1, p. 494, Academic Press, New York (1965)
405. Shoaf, W.T., *Journal of Chromatography* **152**, 247 (1978)
406. Davies, B.H., in Goodwin, T.W. (ed.), *Chemistry and Biochemistry of Plant Pigments*, Volume 2, p. 108, Academic Press, New York (1976)
407. Garside, C. and Riley, J. P., *Analytica Chimica Acta* **46**, 179 (1969)

408. Evans, N., Games, D.E., Jackson, A.H. and Marlin, S.A., *Journal of Chromatography* **115**, 325 (1975)
409. Gadel, F. and Brucket, A., *Water Research* **21**, 1195 (1987)
410. Andreae, M.O., *Analytical Chemistry* **52**, 150 (1980)
411. Andreae, M.O., *Analytical Chemistry* **49**, 820 (1977)
412. Shoaf, W.T. and Lium, B.W., *J. Research US Geological Survey* **4**, 241 (1976)
413. Martin, G., *Revue Français des Sciences de l'Eau* **2**, 407 (1983)
414. Wong, S.H. and Hindin, E., *Journal of American Water Works Association* **74**, 528 (1982)
415. Walters, J.S. and Hedges, J.J., *Analytical Chemistry* **60**, 988 (1988)
416. Thomson, C.A. and Chesney, D.J., *Analytical Chemistry* **64**, 848 (1992)
417. Doerge, D.R. and Miles, J., *Analytical Chemistry* **63**, 1999 (1991)
418. Sackmauerova, M., Pal'usova, O. and Szokolay, A., *Water Research* **11**, 537 (1977)
419. Sackmauerova, M., Pal'usova, O. and Hluchan, E., *Vodni Hospodarstvi* **10**, 267 (1972)
420. Law, R.J., Fileman, T.W. and Partman, J.E., Ministry of Agriculture, Fisheries and Food, Lowestoft, Suffolk UK. Report No. 2. Methods of analysis for hydrocarbons in marine and other samples (1988)
421. Maher, W.A., *Analytica Chimica Acta* **126**, 157 (1981)
422. White, J.N.C. and Englar, J.R., *Botanica Marina* **26**, 159 (1983)
423. Chau, Y.K., Wong, P.T.S., Bengert, G.A. and Dunn, J.L., *Analytical Chemistry* **56**, 271 (1984)
424. Hodge, V.J., Seider, S.L. and Goldberg, D., *Analytical Chemistry* **51**, 1256 (1979)
425. François, R. and Weber, J.H., *Marine Chemistry* **5**, 279 (1988)
426. Hilmy, A.L., El-Domiatry, N.A., Daabees, A.Y. and Latik, H.A.A., *Comparative Biochemistry and Physiology* **86C**, 263 (1987)
427. Molander, S. and Blanck, H., *Water Science and Technology* **20**, 193 (1988)
428. Casper, E.M., Snyder, B.J., Arnold, L.M., Zaikowski, A. and Wurster, C.F., *Marine Environmental Science* **23**, 207 (1987)
429. Birnie, S.S. and Hodges, D.F., *Environmental Technology Letters* **2**, 433 (1981)

CHAPTER 9

Pollution of Potable Water

9.1 METALS

Concentrations of metals encountered in potable waters are shown in Table 9.1 (further details are given in Appendix 5) together with the concentrations of metals found in environmental river waters for comparison. It is seen that the concentrations of those metals in potable water controlled by the World Health Organisation (cadmium, chromium, cobalt, copper, mercury and zinc) are all less than the stipulated limits, whatever the source of the potable water[1-3]. Generally, due to the treatment it receives, metal contents in potable water are lower than those in riverwater, as might be expected.

Mercury and cadmium are two of the more common environmental pollutants encountered in the raw water intake to water treatment plants. Even at very low levels, these elements have been suspected of causing detrimental health effects to humans and other creatures. Both mercury and cadmium appear on the US Environmental Protection Agency's Priority List of Toxic Substances[4]. The US Safe Water Drinking Act allows a maximum concentration of mercury and cadmium, respectively, of 2 and 10 μg l^{-1} and it is seen that all the potable waters quoted in Table 9.1 fall within these limits. The World Health Organisation limits for mercury and cadmium are lower than those of the US authorities, respectively 0.5 and 5 μg l^{-1}. Potable water containing 0.05 μg l^{-1} mercury will contribute 0.1 μg l^{-1} mercury per day to the human intake compared to 1–20 μg mercury per day from food and 1 μg per day from air containing 0.05 μg m^3 of mercury. Thus the contribution of mercury by potable water is a very small proportion of total daily human intake even when the mercury content of the water approaches the US Safe Water Drinking Act limit of 2 μg l^{-1}.

9.2 ORGANOMETALLIC COMPOUNDS

Very low concentrations of organomercury (0.7–3.4 ng l^{-1}) and organotin compounds have been found in potable water. The World Health Organisation has suggested a maximum daily intake of organomercury by humans via potable water of 29 μg (43 μg total mercury).

Table 9.1. Concentration of trace elements in potable water

| Element | Concentration range (μg l^{-1}), metals controlled by World Health Organisation limits | | |
	Potable water[a]	WHO Limit	Rivers[a]
Cadmium[b]	< 1	5[d]	(0.03−5)
Chromium[b]	< 1−18.1	50	(0.05−23)
Cobalt[b]	< 1	50	(0.2−10)
Copper[b]	< 3−945	50	(0.11−200)
Mercury[b]	0.001−0.010	0.5[c]	(0.009−1.3)
Zinc[b]	< 2−2043	5000	(0.9−630)
Additional elements included in EU directive			
Antimony[b]	< 13	−	(14−520)
Arsenic[b]	5.4−11.4	−	(0.4−490)
Iron[b]	4.5−1960	−	(1−3925)
Manganese[b]	1.9−20.5	−	(1−1835)
Nickel[b]	0.19−6.1	−	(1.5−40)
Silver[b]	< 3	−	(0.3−32)
Aluminium (total)	34−38	−	(73−360)
Lead	< 0.6−565	−	(0.13−60)
Other elements not subject to regulations			
Titanium	< 0.7		(3−31)
Barium	1−47	−	(10−23)
Beryllium	< 0.01−0.17		(0.4)
Bismuth	< 30		(0.005)
Molybdenum	< 7		(0.7−4.1)
Selenium	< 10		(< 0.0002−750)
Uranium	0.4−0.7		(0.4−1.4)
Vanadium	7.4−14.8		(0.1−0.2)
Zirconium	50−86		−
Tellurium	< 30		−
Thallium	< 60		−
Thorium	< 4		−
Tin	< 13		−
Yttrium	< 0.3		−
Non-metallic elements			
Bromine	17.3−743		(78)
Iodine (free)	0.7−30		(10)
Iodine (total)	0.9−28		−
Boron	< 13		−

Concentration ranges obtained from river quoted in parentheses for comparison.
[a] Samples taken at a variety of locations (mainly European and USA) over a period of time (years).
[b] Metals in EU drinking water limits.
[c] US Environmental Protection Agency Drinking Water Act limit (2 μg l^{-1} mercury).
[d] US Environmental Protection Agency Drinking Water Act limit (10 μg l^{-1} cadmium).

9.3 ORGANIC COMPOUNDS

The occurrence of organic compounds in potable water is summarised in Tables 9.2 and 9.3 (further details are given in Appendix 5, Table A5.2).

9.3.1 HALOFORMS

Numerous articles have appeared referring to the presence of carcinogens in water and questioning the safety of chlorine when used as a disinfectant of water supplies[5-7]. The impetus for this relatively sudden development was the release of a study by the Environmental Protection Agency (EPA) of the presence of

Table 9.2. Haloform concentrations ($\mu g\,l^{-1}$) in potable water

Haloforms	Chlorinated	Chlorinated and treated with carbon
$CHCl_3$	0.57–182	0.5–4.06
$BrCl_2CH$	0.2–74.7	0.08
Br_2ClCH	< 0.01–36.8	0.16–15.0
Br_3CH	< 0.1–8.0	0.13–0.37
CCl_4	0.017–1.2	0.02–0.15
CH_2ClCH_2Cl	21.9–24.5	0.06–24.5
CH_2ClCH_2Cl	0.06–24.5	–
Cl_2CCCl_2	0.008–0.9	–
CH_3CCl_3	0.06–0.5	–
$CHClCCl_2$	0.015–0.9	–
$CHCl_2I$	0.007	–
CH_2Cl_2	0.3	–
$CH_2ClCHClCH_3$	0.8	–
Total haloforms	2.4–259	0.95–44.2

Table 9.3. Organic substances in potable water

Compound	Concentration ($\mu g\,l^{-1}$)
Polychlorinated biphenyls	< 0.0001 (as Aroclor 1016)
Pentachlorophenol	< 140–340
2,4,6-trichlorophenol	< 1
Di- and trichloracetic acid	30–160
Total nitrosamine	< 0.001
%-Chlorouracil	0.1–14.1
5-Chlorouridine	0.7–26.7
4-chlororesorcinol	1.60–4.7
5-chloro salicylic acid	2.3–12.5
p-isopropyl phenylamine	10–15
2-hydroxybenzothiazole	5–25

potentially toxic organic substances in the New Orleans Water supply[8] and an epidemiological study of the implications of cancer-causing substances in the Mississippi River water by the Environmental Defence Fund[9]. The latter document is of some importance in that it suggests a relationship between the above-average incidence of cancer in certain communities and the Mississippi-derived water supply. Subsequent to the passage of the Safe Drinking Act (PL93-523) in 1974, the US Government Environmental Protection Agency sponsored two finished water studies, the National Organics Reconnaissance Survey (NORS) and the National Organics Monitoring Survey (NOMS). These studies confirmed the widespread occurrence of chloroform as well as other trihalomethanes, bromodichloromethane ($CHBr_4Cl_2$), chlorodibromomethane ($CHBr_2Cl$) and bromoform ($CHBr_3$). Since this work, there have been numerous other reports of the presence of chlorinated and brominated haloforms in river and potable water[10-28].

As well as chlorohydrocarbons a wide range of bromohydrocarbons and bromochlorohydrocarbons such as bromoform, bromodichloromethane, and di-bromochloromethane have been identified in water supplies. Luong et al.[11] have recently drawn attention to the role of bromide in water supplies in the formation of brominated trihalomethanes with reference to its interaction during the chlorination process with humic material present in natural waters. Changes during water treatment were examined and subsequent trihalomethane formation on chlorination of those waters evaluated. For bromide levels in lowland waters of up to 120 $\mu g\,l^{-1}$ brominated trihalomethanes were shown to account for up to 54% of the total trihalomethanes formed on treatment.

Chloroform apparently results from reaction between hypochlorite and any of several types of organic precursors in the chlorinated raw water. The brominated and mixed brominated/chlorinated trihalomethanes are presumed to be formed from the reaction of hypobromite and hypochlorite with the same precursors; the hypobromite is formed from the oxidation of bromide by hypochlorite. If iodide salts are also present in the water being chlorinated, an analogous reaction with hypochlorite results in trihalomethanes containing iodine. Bunn et al.[12] detected all ten possible mixed and single halogen-containing trihalomethanes of chlorine, bromine and iodine when salts containing fluoride, bromide and iodide were added to a riverwater sample before chlorination (no fluorinated trihalomethanes were detected). Glaze et al.[13] identified seven trihalomethanes containing chloride, bromine, and iodine in potable water samples. Also found in potable water but not necessarily formed during the chlorination process are compounds such as methylene chloride, dichlorobenzene, hexachlorobutadiene[14], tetra-chloroethylene[15], trichloroethylene[14,15], carbon tetrachloride[16], 1,2-dichloro-ethane[15,16] and ethylene dibromide[17].

The formation of volatile organohalogen compounds by the chlorination of waters containing organic contaminants[18] has received wide attention[18,20,21]. Investigation carried out by Tardiff and Dunzer[5] confirmed the presence of six halogenated compounds (i.e. chloroform, bromoform, bromodichloromethane, dibromochloromethane, tetrachloromethane and 1,2-dichloroethane) in potable waters with concentrations varying up to 100 μg l^{-1} while Tabor[22] identified 3-(2-chloroethoxy) 1,2-dichloropropene.

As a consequence of the concern regarding possible adverse effects of minute quantities of trihalomethanes in potable water, the US Environmental Protection Agency[23,24] in 1978 drew up an amendment to US National Interim Primary Drinking Water Regulations designed to protect the public from exposure to undesirable amounts of trihalomethanes (including chloroform) in potable water. A maximum contaminant level of 100 μg l^{-1} has been prescribed for total trihalomethanes applicable initially to community water supplies for populations as part of the treatment process. It is seen in Table 9.2 that only after treatment with active carbon can it be assumed that chlorinated potable water will meet this specification.

9.3.2 POLYAROMATIC HYDROCARBONS

It is seen in Appendix 5 (Table A5.2) that the total polyaromatic hydrocarbons content of potable waters are in the range of 0.05–0.13 μg l^{-1}. This compares favourably with the required maximum of 0.2 μg l^{-1} in potable water by the World Health Organisation and 0.25 μg l^{-1} by the Federal Republic of Germany standards.

9.3.3 OTHER ORGANICS

Information on concentrations of other organics found in potable water is presented in Table 9.3 (further details are given in Appendix 5, Table A5.2).

Toxicity is of prime importance when considering the effects of organic chemicals on humans, and both chronic and acute toxicities must be taken into account. Compounds having low acute toxicities but which accumulate in the body may have much lower long-term acceptable intake levels than those of much higher acute toxicity which do accumulate in the body. These materials may be taken into the body from air, food or water by ingestion or adsorption through the skin, and in many cases residues in water may not be the prime source of the compound.

In assessing the effects of organics on water supplies it is important to consider not only the active ingredient but also its degradation products in soil, water and plants. Acceptable intake of toxic compounds is assessed from long-term feeding and dermal application experiments on those animals found to be most sensitive to the toxicant. The transference of these data to acceptable daily intakes to humans is naturally a task for expert toxicologists and involves the setting of a suitable safety factor. This is necessary because humans may be more sensitive to the toxicant than the most sensitive test animal, and because individuals may vary in their reactions. The size of the safety factor, which is determined by the nature of the compound, may be of the order of 1000. In setting acceptable levels of materials in potable water not only must humans be considered but also animals, plants, etc. which might use the supply.

Many organic chemicals (e.g. chlorophenols) can give rise to unpleasant tastes and odours in water when present at levels considerably less than 1 mg l^{-1}. Water contaminated in this way will be unacceptable to consumers even though it is perfectly safe from biological and toxicological aspects. Even pleasant tastes and odours will be rejected, as the water is obviously 'contaminated'.

REFERENCES

1. Council of European Communities Directive of 15/7/80—80/778/EEC OJL 229 30/8/80 Drinking Water for Human Consumption (1980)
2. Behar, J.V., Schuck, E.A., Stanley, R.E. and Morgan, G.B., *Environmental Science and Technology* **13**, 34 (1979)
3. Minagawa, K., Takizawa, Y. and Kifune, I. *Analytica Chimica Acta* **15**, 103 (1980)
4. Oda, C.E. and Ingle, J.D., *Analytical Chemistry* **53**, 2205 (1981)
5. Tardiff, R.G. and Dunzer, M., *Toxicity of Organic Compounds in Drinking Water*, Environmental Protection Agency, Cincinnati, Ohio, USA (1974)
6. Report on the Carcinogenesis Assay of Chloroform, US Natl. Cancer Inst., Bethesda, MD (1976)
7. Cantor, R.H., Nason, T.J. and McCabe, L.J., Association of cancer mortality rates and trihalomethane levels in municipal drinking water supplies. *Proc. 10th Ann. Mtg. Soc. Epidemo Res.* (1977)
8. Analytical Report—New Orleans Area Water Supply Study, EPA 906/10-74-002. Surveillance and Analysis Division, USEPA Region VI, Dallas, Texas, mimeo (1974)
9. *The Implications of Cancer Causing Substances in Mississippi River Water*. Environmental Defence Fund, Washington, DC (1974)
10. Glase, W.H. and Rawley, R., *Journal of the American Water Works Association* **71**, 509 (1979)
11. Luong, T., Peters, C.J., Young, R.J. and Perry, R. *Environmental Technology Letters* **1**, 299 (1980)
12. Bunn, W.W., Haas, B.B., Deane, E.R. and Kleopfer, R.D., *Environmental Letters* **10**, 205 (1975)
13. Glaze, W.H., Henderson, J.E. and Smith, G., private communication
14. Rook, J.J., *Water Treatment Exam.* **21**, 259 (1972)
15. Dowty, B.K., Carlisle, D.R. and Laseter, J.L., *Environmental Science and Technology* **9**, 762 (1975)
16. New Orleans Area Water Supply Study (Draft Analytical Report), Lower Mississippi River Facility USEPA (1974)
17. Libbey, A.J., *Analyst (London)* **111**, 1221 (1986)
18. Stevens, A.A., Slocum, C.J., Seiger, D.R. and Robeck, G.G. Conf. Environ. Impact of Water Chlorination, Oak Ridge National Lab., Oak Ridge, Tennessee (1975)
19. Fielding, M., McLoughlin, K. and Steel, C., Water Research Centre Enquiry Report ER 532, August, 1977. Water Research Centre, Stevenage Laboratory, Elden Way, Stevenage, Herts, UK (1977)

20. Symons, J.M., Bellar, T.A., Carswell, J.K., Demarco, J., Kropp, K.L., Roebeck, G.G., Seegar, D.R., Slocum, C.J., Smith, B.L. and Stevens, A.A. National Organics Reconnaissance Survey for Halogenated Organics in Drinking Water, EPA, Cincinnati, Ohio (1975)
21. Nicholson, A.A. and Mersz, O., *Bulletin of Environmental Contamination and Toxicology* **14**, 453 (1975)
22. Tabor, M.W., *Environmental Science and Technology* **17**, 324 (1983)
23. US Environmental Protection Agency Federal Register No. 28, **43**, 5756 (1978)
24. USEPA Control of Organic Chemical Contaminants in Drinking Water, Federation Register, **44**, 5755 *et seq.* (9 February 1978)
25. Quimby, B.R., Delaney, M.F., Uden, P.C. and Barnes, R.M., *Analytical Chemistry* **52**, 259 (1980)
26. Harris, L.E., Budde, W.L. and Eichelberger, J.W., *Analyst* (*London*) **46**, 1912 (1974)
27. Nicholson, A.A. and Merscz, O. *Bulletin of Environmental Contamination and Toxicology* **14**, 4 (1975)
28. Kissinger, L.D. and Fritz, J.S., *J. Am. Water Works Ass.* **68**, 435 (1976)

CHAPTER 10
Radioactivity in the Environment

The measurement of radioactivity in the environment is not the subject matter of this book. However, for completeness and to aid the reader in obtaining further information a detailed list of published literature is included in Appendix 2 covering the determination of radioactive isotopes in freshwater, seawaters and the sediments occurring in these. Obviously, as far as the consumer of river or sea fish is concerned, in the case of radioactive elements in many cases the concentration of the element in the sea creature required to produce a hazard due

Table 10.1. Radioactivity measurements in natural waters

Isotope	Type of sample	Disintegration min^{-1}/100 l sample	pCi l^{-1}	Ref.
Radium-224	River	904–2567	–	1
Radium-226		1346–1845	–	1
Radium-228		1563–2823	–	1
Caesium-137	Lake Ontario	–	0.015–0.06	2
Radium-226		–	0.042–0.055	2
Radium-228		–	0.015–0.038	2
Beryllium-7		–	0.020–0.037	2
Zirconium-95		–	0.004–0.017	2
Niobium-95		–	0.006–0.039	2
Ruthenium-108		–	< 0.001–0.021	2
Antimony-125		–	< 0.001–0.003	2
Cerium-141		–	< 0.001–0.005	2
Cerium-144		–	0.013–0.089	2
Thorium-228		–	0.022–0.030	2
Caesium-137	Rain (following Chinese atomic bomb test, 24 October 1980)	–	27	3
Cerium-144		–	2	3
Barium-140		–	320	3
Lanthanum-237		–	314	3
Polonium-210	Seawater	–	0.025–0.115	4
Radium-226		15.4–30.5	–	5
Radium-228		5.3–22.3	–	6
Plutonium-238		–	0.015–0.018	7
Plutonium-239, 240		–	0.096–0.119	7

Table 10.2. Radioactivity measurements on river and marine sediments

Isotope	Type of sample	Disintegrations min^{-1} g^{-1}, dry sediment						fCi g^{-1} dry sediment	Ref.
	depth of core (m)	Nil	389	900	2778	4280	5000		
Actinium-237	marine sediment	—	< 0.03	< 0.01	< 0.03	< 0.20	< 0.03–0.52	—	8
Uranium-238		—	0.67–3.78	0.62–0.76	0.41	0.20	0.44–0.83	—	8
Thorium-230		—	0.36–0.38	1.22–1.93	10.4	16.6	5.7–27.4	—	8
Palladium-231		—	< 0.03–0.026	0.087–0.105	0.34	0.48	0.20–0.91	—	8
Thorium-232		—	0.06–0.21	0.07–0.6	0.11	0.12	0.23–1.26	—	8
Thorium-228		—	0.7–13.9	5.5–13.8	5.9	5.4	10.0–21.7	—	8
Plutonium-238, 239		—	0.18–0.28	0.14–1.61	2.0	2.4	0.12–2.2	—	8
Plutonium-239, 240		—	—	—	—	—	—	501–629	9
Plutonium-238		—	—	—	—	—	—	14–31	9
Thorium-228	River sediment	2.27	0.038	—	—	—	—	—	10
Thorium-230		1.47	0.024	—	—	—	—	—	10
Thorium-232		2.36	0.039	—	—	—	—	—	10

to the ingestion of radioactivity is much lower than that required to produce a hazard due to the ingestion of non-radioactive toxic elements. This, however, is not always the case: thus the ingestion level of a low α-emitting element may be less harmful than the ingestion of 2,3,7,8-tetrachlorobenzo-p-dioxin.

Releases of radioactive isotopes into the ecosystem will cause perceptible increases in the levels of radioactivity in environmental water as is evidenced by the approximately two hundredfold increase in the level of cerium-144 in water following a Chinese atomic bomb test carried out in 1980, and by the high levels of radioactivity found in the environment following the Chernobyl nuclear accident.

Some levels of radioactivity found in water and sediments are illustrated in Tables 10.1 and 10.2 (see also Appendix 2).

Many radioactive elements concentrate from water into sediment. This is demonstrated in Tables 10.1 and 10.2, for the data shown for plutonium-239 and plutonium-240. Its concentration in marine sediments ($501-629$ fCi g^{-1}, i.e. $501-629$ pCi kg^{-1}) when divided by the concentration of these isotopes found in seawater $0.096-0.119$ Ci l^{-1} gives a concentration fraction of (pCi kg^{-1}/pCi l^{-1}) of about 5200.

REFERENCES

1. Elsinger, R.J., King, P.T. and Moore, W.S. *Analytica Chimica Acta* **144**, 277 (1982)
2. Durham, R.W. and Joshi, S.R., *Water Research* **15**, 83 (1981)
3. Anderson, G., *Trends in Analytical Chemistry* **1**, 281 (1982)
4. Cowen, J.P., Hodge, V.F. and Folson, T.R., *Analytical Chemistry* **49**, 494 (1977)
5. Michel, J., Moore, W.S. and King, P.T., *Analytical Chemistry* **53**, 1885 (1981)
6. Rey, R.M., Brewer, R.L., Stockwell, J.H., Guinasso, N.L. and Schuck, R.D., *Marine Chemistry* **7**, 251 (1979)
7. Delle Site, A., Marchionni, V. and Testa, C., *Analytica Chimica Acta* **117**, 217 (1980)
8. Anderson, R.F. and Fleer, A.I., *Analytical Chemistry* **54**, 1142 (1982)
9. Delle Site, A., Marchionni, V. and Testa, C., *Analytica Chimica Acta* **117**, 219 (1980)
10. Joshi, S.R., *Analytical Chemistry* **57**, 1023 (1985)
11. Friedman, L., Amann, W. and Lux, D., *Gas-u-Wassfach Wasser-Abwasser* **127**, 604 (1986)
12. Haberer, K., *GWF Wasser/Abwasser* **127**, 597 (1986)
13. Fowler, S.W., Buat Menard, P., Yokoyama, Y., Ballastera, S., Holm, E. and Van Nguyen, H., *Nature* **329**, 56 (1987)

APPENDIX 1
Ranges of Metal Concentrations Found in Environmental Waters

Table A1.1. Ranges of metal concentration in freshwaters

Element	Type of water	Concentration (μg l^{-1})		Ref.
Aluminium	River	Total	1300 (pH 7.7)	1
			3600 (pH 4.9)	
		Labile	15 (pH 7.7)	
			520 (pH 4.9)	
		Total	200 (pH 4.6)	
			94–103 (pH 8.5)	
			73 (pH 8.1)	
		Labile	200 (pH 4.6)	
			39–42 (pH 8.5)	
			14.0–16.4 (pH 8.1)	
	Surface water		20–1430	2
	North Florida		210–260	2
Antimony	River		1	3
	River Arve, Germany		0.32	4
	River Arne, Germany		0.066–0.14	4
	Lake		0.08–0.42	5
	Groundwater		0.77	6, 7
Arsenic	River		258–490	8
			210–240	9
			1.1–275	10
			2	3
	River Arve, Germany		2.8	4
	River Arne, Germany		0.42–0.69	4
	Groundwater		2.3	6, 7
Barium	River		10	3
			10–30	4
	River Arne, Germany		23	4
	Surface water		100–103	11
	Groundwater		41	6, 7
Beryllium	River		0.4	3
	Surface water		< 0.01–0.31	11
			1	2

continued overleaf

Table A1.1. (*continued*)

Element	Type of water	Concentration (μg l^{-1})	Ref.
Bismuth	River	0.005	3
	Lake water	< 0.000 15	42
		(total and	
		dissolved)	
Caesium	Groundwater	0.006	6, 7
Cadmium	River	0.013–0.29	12
		0.07–0.13	4
		0.03	3
	River Arne, Germany	5	4
	River Arne, Germany	0.88	2
	Surface water	4–130	2
	Groundwater	100–2600	13
Chromium	River	1	3
		16–23	14
	River Arne, Germany	0.05–0.25	15
	River Arve, Germany	1.44	4
	River Rhine, Germany	10	4
	Surface water	180	2
	North Florida	0.2–0.3	15
	Groundwater	1.0	6, 7
Cobalt	River	0.2	3
	River Arne, Germany	0.013–0.09	4
	River Arve, Germany	0.12	4
	River Rhine, Germany	10	4
	Lakewater	54	16
	Groundwater	0.11	6, 7
Copper	River	7	3
		0.51–6.5	12
		123–178	14
		0.48	17
	River Arve, Germany	14.8	4
	River Rhine, Germany	30	4
	River Thames, UK	30–200	18
	River Arne, Germany	0.53–2.35	4
	Surface water	14–15	2
		110	2
	Groundwater	3.7	6, 7
Europium	River Arve, Germany	0.018	4
	River Arne, Germany	0.000 08–0.0011	4
Gold	River (western USA and Alaska)	< 0.001–0.036	19
Iron	River	200–2950	20
		50–3925	14
		100	3
	River Arne, Germany	1–12.4	4
	River Arve, Germany	57	4
	Surface water (North Florida)	220–350	15
		150–5000	2
	Groundwater	0.15	6, 7

Table A1.1. (*continued*)

Element	Type of water	Concentration (μg l^{-1})	Ref.
Lead	River	0.9–1.0	12
		6–37	14
		2.1–34.8	21
	River Thames, UK	40–60	18
		0.13–0.15	22
	Surface water (North Florida)	17–42	15
Manganese	River	11–1835	14
		5.8–19.9	23, 24
		7	3
	River Arne, Germany	0.97–8.9	4
	River Arve, Germany	7.9	4
	River Thames, UK	200–1720	18
	Surface water	70–500	2
	Groundwater	3.2	6, 7
Mercury	River	0.51–1.3	25, 26
		0.07	3
	River Arne, Germany	0.017	4
	River Arve, Germany	0.009–0.047	4
	River Rhine, Germany	0.5	4
Molybdenum	River	1	3
	River Arne, Germany	0.74–1.12	4
	River Arve, Germany	4.08	4
Nickel	River	1.5	3
	River Thames, UK	20–40	18
	Surface water	10–40	2
	North Florida	8–10	15
Scandium	Groundwater	0.009	6, 7
Selenium	River	0.2	3
		0.2–0.9	10
	River Arne, Germany	0.0006–0.0023	4
	River Arve, Germany	0.031	4
		< 0.0002–> 50 (as selenate and selenite)	27
	Japan	0.005–0.012 (as elemental selenium)	28
		0.008–0.012 (as SeIV)	29
		< 0.002–0.016 (as SeIV)	–
		0.036–0.052 (as SeVI)	29
		0.003–0.020 (as SeVI)	–
		0.022–0.023 (as total Se)	29
		0.016–0.023 (as total Se)	–

continued overleaf

Table A1.1. (*continued*)

Element	Type of water	Concentration (μg l^{-1})	Ref.
Selenium	Groundwater	0.4	6, 7
(*continued*)		0.02–0.7	30
Silver	River	0.3	3
		24–32	14
Titanium	River	3	3
	Surface water (North Florida)	24–31	15
Uranium	River Arve, Germany	1.36	4
	River Arne, Germany	0.37–0.49	4
Vanadium	River	0.9	3
		0.1–1	32
		24 (as V^{4+})	31
		23 (as V^{5+})	31
		21 (as total V)	31
	Surface water	4.5–5.2	11
	North Florida	3.9–24	15
	Groundwater	0.63	6, 7
	Lake	0.1–1.5	32
Zinc	River	20	3
		14–202	14
	River Arne, Germany	0.86–5.13	4
	River Arve, Germany	630	4
	River Rhine, Germany	250	4
	Groundwater	8.9	6, 7
	Surface water	10–250	2
	North Florida	2.5–48	15

Total non-metallic elements

Element	Type of water	Concentration (μg l^{-1})	Ref.
Bromine	Groundwater	78	6, 7
Iodine	Groundwater	10	6, 7
Nitrogen	Lake water	1060–2940	33
	Surface water	1500–91 000	34
Phosphorus	River	250–800	35
		20	3
Silicon		3000–5800	35
Sulphur	Lake water	20	36
Anions			
Borate	River	0.12–0.25	37
	Groundwater	44	6, 7
Bromide	River	1.5–109.8	38
		< 500	39
	River Arne, Germany	0.7–4.7	4
	River Arve, Germany	4.7	4
	Surface water (North Florida)	40–140	15
	Groundwater	2000–280 000	40
Fluoride	River	100–180	41
	Well water	600	39
Phosphate	River	160–550	41

Table A1.2. Range of metal concentrations in open ocean seawater

Element	Location	Concentration (μg l^{-1})	Consensus value (μg l^{-1})	Ref.
Aluminium	Open seawater surface	0.1		43
	Open seawater 3 km depth	0.6		43
Bismuth	Pacific, surface	< 0.000 05	–	44
	Pacific, 2500 m depth	< 0.000 003	–	45
Cadmium	Open ocean, Salinity 35%	0.03	–	46
	Arctic Sea	0.010–0.045	–	47
	Arctic Sea, surface	0.0127	–	48
	Arctic Sea, 2000 m depth	0.023	–	48
	Arctic Sea	0.018	–	49
	Pacific	0.02–0.04	–	50
	Kattergat/Skaggerat	0.022	–	51
	Norwegian Sea	0.02–0.025, (surface)	–	52
		0.02–0.025, (3000 m)	–	
	Sargasso Sea	0.035–0.042, (216 m)	–	53
		0.109–0.126, (4926 m)	–	53
	Baltic Sea	0.03–0.06	–	54
	Open sea	0.03	–	46
	Open sea	0.079	–	55
	Open sea	0.12–0.30	–	56
	Open sea	0.03–0.17	–	57
Chromium	Pacific	CrIII 0.005–0.52	–	58, 59
		CrVI 0.03–0.96	–	58, 59
		Organic Cr 0.07–0.32	–	58, 59
		Total Cr 0.06–1.26	0.03	58, 59
	Mediterranean	CrIII 0.02–0.05	–	60
		CrVI 0.05–0.38	–	60
	Open ocean	Total 0.07–0.97	0.03	61
		CrIII 0.08–0.22	–	62
		CrVI 0.13–0.68	–	62
		Total 0.18–0.19	0.03	46
Cobalt	North Sea	0.07–0.16	0.005	63
	Open ocean (salinity 35%)	0.003		46
	Open sea	0.04	–	64
	Open sea	0.003	–	46
	Open sea	0.15–0.16	–	56
Copper	Pacific	0.3–2.8	0.05	50
	Open ocean (salinity 35%)	0.121		46
	Good quality seawater	0.36–8.6		65
	Sargasso Sea	0.072–0.081 (216 m)		53
	Sargasso Sea	0.26–0.33 (4926 m)		53
	Baltic Sea	0.59–0.99		51
	Baltic Sea	0.6–1.0		54
	Baltic Sea	0.0063–0.0252 (organic)		67
		0.6–0.751 (total)		67
	North Sea	0.208		51

continued overleaf

Table A1.2. (*continued*)

Element	Location	Concentration (μg l^{-1})	Consensus value (μg l^{-1})	Ref.
Copper	Norwegian Sea	0.08–0.10 (surface)		52
(*continued*)	Norwegian Sea	0.08–0.10 (3000 m)		52
	Danish Sound	0.48		51
	Arctic Sea	0.097		67
	Open sea	0.341		55
	Open sea	0.48–1.51		56
Iron	Pacific	140–320	0.2	50
	Open ocean (salinity 35%)	0.2		46
	Open seawater	2.1		64
	Open seawater	3.25		55
	Pacific	< 0.01–0.7		68
Lead	Pacific	0.6–0.8	–	50
	Arctic Sea	0.01–28	–	47
	Arctic Sea	0.019–0.021	–	49
	Open ocean (salinity 35%)	0.095	–	46
	Sargasso Sea	0.000 041 (surface)	–	69
		0.0083–0.012 (4800 m)		69
	West North Atlantic	0.000 17–0.0003		70
	Norwegian Sea	< 0.0002, (3000 m)		52
		0.025–0.065 (surface)		
	Open sea	0.095		46
	Arctic Sea	0.015		
	Open sea	0.0083		
	Open sea	0.03–9.0		
	Open sea	< 0.04–0.28		
Manganese	Open ocean (salinity 35%)	0.018	0.02	46
Mercury	Atlantic, open sea	0.021–0.078	< 0.2	71
	Open ocean	0.002–0.011		71
	Off Iceland	0.04		72
Molybdenum	Pacific	11.2–12.0		50
	Non-central Pacific	3.2		73
	Seawater, Japan	11.5		74
	Open sea	5.3		64
Nickel	Pacific, 4000 m depth	0.45–0.84	0.17	75
	Pacific	0.15–0.93		50
	Pacific, surface	0.16–0.29		75
	Open ocean (salinity 35%)	0.341–0.608		76
	Open ocean	0.38–0.46		77
	Open ocean	0.27		46
	Norwegian Sea	0.175–0.20 (surface)		52
		0.175–0.20 (3000 m)		52
	Sargasso Sea	0.26–0.27 (216 m)		53
		0.45–0.47 (4926 m)		53
	Baltic Sea	0.6–0.9		54
	Arctic Sea	0.099		67
	Open sea	0.545		55

Table A1.2. (*continued*)

Element	Location	Concentration (μg l^{-1})	Consensus value (μg l^{-1})	Ref.
Nickel	Open sea	0.76–1.58		56
Rare earths	North Atlantic, below mixed layer	La 13.0 × 10^{-12} mole kg^{-1}		78
		Ce 16.8	–	
		Nd 12.8	–	
		Sm 2.67	–	
		Eu 0.644	–	
		Gd 3.4	–	
		Dy 4.78	–	
		Er 4.07	–	
		Yb 3.55	–	
Rhenium	Atlantic	6–8	–	79
Selenium	Seawater	0.021–0.029	–	80
	Open ocean	0.000 95	–	64
Silver	Open sea	0.08		56
Thorium	Open sea	< 0.0002		64
Tin	Open sea	0.02 (SnIV)		81
		0.05 (SnII)		81
Uranium	Seawater	1.9	–	82
		2.6	–	77
Vanadium	Pacific	1.73–2.00	2.5	83
		1.29–1.87		50
	Adriatic	1.64–1.73		83
	Open sea	0.45		64
Zinc	Pacific	1.9–3.0	0.49	50
	Arctic Sea	0.125–0.16		49
		0.05–0.34		47
	Open sea (salinity 35%)	0.28		46
		4.9		64
	Norwegian sea	0.08–0.30 (surface)		52
		0.10–0.18 (3000 m)		52
	Open sea	0.074		55
		0.3–10.9		57
		2.6–10.1		56

Table A1.3. Ranges of metal concentrations in coastal, bay and estuary waters

Element	Location	Concentration (μg l^{-1})	Ref.
Aluminium	Seto Upland Sea, Japan and Pacific Ocean	6.4–63	84
Antimony	North Sea	0.3–0.82	85
Arsenic	North sea (soluble metals)	1.0	86

continued overleaf

Table A1.3. (*continued*)

Element	Location	Concentration ($\mu g\ l^{-1}$)	Ref.
Arsenic	North Sea Coastal water	1.04	87
Barium	Kwangyana Bay, Korea	4.8	88
Bismuth	Seawater	0.02–0.11	89
	Kattegat	0.0015–0.003	90
	San Diego Bay	0.000 05–0.000 06 (dissolved)	45
		0.000 13–0.002 (total)	
	North sea	0.2–0.68	85
Cadmium	Near shore seawater (salinity 29‰)	0.02–0.0025	46
	Sandy Cove, USA	0.04–0.05	91
		0.24–0.28	92
	Bermuda	0.029	91
	Kwangyana Bay, Korea	0.20	88
	Coastal seawater	0.05–0.2	93
		0.020–0.28	93
	Seawater	0.056–0.08	76
		0.053	94
		0.053–0.07	95
		0.2	96
	North Sea (soluble metals)	0.02	86
	Straits of Gibralter	< 2.8	97
	Heligoland Bight	0.02–0.07	98, 99
	Sea off California	0.015–0.016 (surface)	100
		0.94–0.099 (2950 m)	100
	German Bight	0.024–0.768	101
	Danish coastal water	0.06–0.80	102
	Clyde coastal water	0.11–0.25	57
	Chesapeake Bay	0.05	103
	Canadian coastal water	0.035–0.048	104
	Danish coastal water	0.2–5.0	102
	Mediterranean	< 5.4	97
	Southampton Water	< 0.1–0.35	56
	Cape San Blay	0.013 (5 m)	105
		0.0045 (70 m)	105
	North Sea	0.2–0.4	90
	Near shore water	0.02–0.025	46
	Coastal water	0.3–1.0	106
	Estuary water (salinity 10‰)	0.5	107
	Estuary water (salinity 24.1‰)	2.1	107
	Coastal water	0.1	97
	Coastal water	0.05–0.07	108
	Gota River Estuary (salinity 0.5‰)	0.02	109
	Gota River Estuary (salinity 32‰)	0.02	109
Cerium	Kwangyana Bay, Korea	16.7	110

Table A1.3. (*continued*)

Element	Location	Concentration (μg l^{-1})				Ref.
		CrIII	CrVI	Organic Cr	Total Cr	
Chromium	North Sea, soluble metals	–	–	–	0.4	86
	Sea of Japan	0.057–0.093	0.088–0.15	0.18–0.32	0.37–0.50	111
	Japan Coast	0.04–0.06	–	–	–	112
	Sandy Cove, USA	–	–	–	0.84	92
	Kwangyana Bay, Korea	–	–	–	2.33	88
	Port Hacking, Australia	0.27	0.49	0.56	1.32	113
	Drummoyne Bay, Australia	0.32	0.95	0.69	1.96	113
	Botany Bay, Australia	0.45	1.26	0.71	2.41	113
	Coastal seawater	–	–	0.095–0.100		61
	UK	0.46	0.60	–	–	114
		–	–	–	3.3	82
	Kwangyana Bay, Korea	–	–	–	2.3	111
	Canadian coastal water	–	–	–	0.15–0.5	104
	Coastal water	–	–	–	0.25	97
	Coastal water	–	–	–	0.25–0.29	108
	Estuary water (salinity 10‰)	–	–	–	0.9	107
	Estuary water (salinity 24.1‰)	–	–	–	0.5	107
Cobalt	Sandy Cove, USA			0.02		92
	Coastal seawater			0.018–0.02		93
				0.044		82
	(salinity 29.5‰)			0.017–0.018		46
				0.015–0.028		95
	Shitukawa Bay, Japan			0.07–0.16		73
	North Central Pacific			0.24		73
	Botany Bay, Australia			0.25		73
	North West Coast, USA			0.13		73
	Port Hacking, Australia			0.25		73
	Cronhulla Beach, Australia			0.21		103
	Chesapeake Bay			< 0.1		103

continued overleaf

Table A1.3. (*continued*)

Element	Location	Concentration (μg l^{-1})	Ref.
Cobalt	Southampton Water	< 0.1–0.16	56
(*continued*)	Menai Straits	0.07	73
	Shore seawater	0.017–0.018	46
	Canadian coastal water	0.01	104
	Coastal water	< 0.1	97
	Coastal water	0.015–0.028	108
Copper	North Sea, soluble metals	0.2	86
	Kwangyana Bay, Korea	1.1	88
	Sandy Cove, USA	0.6–0.7	92
	Chirihaua, Japan	20	115
	Gironde Estuary	3.7	116
	Estuary water	2.0–2.01	46
	Near shore water (salinity 29.5‰)	0.17–1.03	46
	Coastal seawater	0.6–0.7	93
	Sunlace Water, North Pacific	0.64	117
	Seawater	0.66–0.72	118
		0.50–0.73	77
		0.16–0.34	76
		0.2	117
	Poor-quality seawater	6.8–15.8	60
	Chesapeake Bay	2.0	103
	Osaka Bay, Japan	0.89–2.66	119
	Delaware Bay	0.83–2.18 (surface)	120
		0.73–0.91 (16 m)	120
	North Sea	2.82–9.7	85
	Canadian Coast	1.1–1.2	104
	Cape San Blas	0.123 (5 m)	105
		0.065 (70 m)	105
	Southampton Water	0.48–2.6	56
	Heligoland Bight	0.3–2.04	98, 99
	Sandy Cove, USA	0.6–0.7	121
	Sea off California	0.069–0.105 (surface)	100
		0.098–0.24 (2950 m)	100
	Kwangyana Bay, Korea	1.1	110
	Coastal water	0.5–0.73	108
	Coastal water	1.0	97
	Coastal water	0.6–3.4	106
	Gota River Estuary (salinity 0.5‰)	1.2	109
	Gota River Estuary (salinity 32‰)	0.3	109
Iron	Sandy Cove, USA	1.4–1.5	92
	Kwangyana Bay, Korea	250	88
	Coastal seawater	1.4–1.6	93
	Salinity 29.5‰	1.0–7.2	46
	Estuary water	2.1	82

Table A1.3. (*continued*)

Element	Location	Concentration (μg l^{-1})	Ref.
Iron	Seawater	3.2–3.4	77
(*continued*)	Delaware Bay	2.46–35.1 (surface)	120
		2.1–5.2 (10 m)	120
	Heligoland Bight	1.13	99
	Osaka Bay, Japan	15.4–65.5	119
	Canadian Coastal Water	3.5–4.2	104
	Chesapeake Bay	2.1	103
	Coastal water	5.0	97
	Coastal water	3.2–3.7	108
	Gota River Estuary (salinity 0.5‰)	170	109
	Gota River Estuary (salinity 32‰)	16	109
Lanthanium	Kwangyana Bay, Korea	0.72	110
Lead	Guanalana Bay	0.07–0.55	122
	Sandy Cove, USA	0.22–0.35	92
	Near shore seawater (salinity 29.5‰)	0.14–0.22	46
	Coastal seawater	0.22–0.35	93
	Seawater	0.2–0.3	118
		7.1	123
		0.038–0.29	76
		0.06–0.11	77
		0.51–0.65	124, 125
	North Sea soluble metals	0.05	86
	Canadian coastal water	0.34–0.36	104
	North Sea	1.8–7.44	85
	Chesapeake Bay	0.3	103
	Danish Coastal water	0.8–80	102
	Clyde water	0.02–0.36	57
	Heligoland Bight	0.07	99
	Southampton Water	< 0.1–0.6	56
	Danish coastal water	4.5–200	56
	Coastal water	0.5–2.4	106
	Coastal water	0.06–0.11	108
	Coastal water	3.1–12	126
	Coastal water	0.25	97
	Near shore water	0.22	46
	Gota River Estuary (salinity 0.5%)	0.30–0.36	109
	Gota River Estuary (salinity 32‰)	0.06–0.07	109
Manganese	Sandy Cove, USA	1.4–1.8	92
	Kwangyana Bay, Korea	1.5	88
	Chirihama Bay, Japan	60	115
	Near shore seawater (salinity 29.5‰)	0.71–1.06	46
	Tamar Estuary, UK	20–250	127
	Coastal seawater	1.4–1.6	92

continued overleaf

Table A1.3. (*continued*)

Element	Location	Concentration (μg l^{-1})	Ref.
Manganese	Estuary and seawater	1.89–2.0	82
(*continued*)	South-west Bermuda	1.4–1.8	128
	Canadian coastal water	0.78–0.95	104
	Osaka Bay, Japan	11.1–30.6	119
	Chesapeake Bay	2.0	103
	Heligoland Bight	0.35	99
	Near shore seawater	0.7–1.06	46
	Coastal water	1.4–75.1	72
	Coastal water	4.0	97
	Coastal water	1.9–2.5	108
Mercury	Seawater	0.000 018–0.000 026	118
	Seawater	0.01	44
	Coastal samples	0.05	44
	River Loire Estuary (salinity 20–30‰)	0.6–1.1	129
	River Loire 0–10 km upstream of estuary (salinity 10–20‰)	1.4–11.6	129
	River Loire 10–15 km upstream of estuary (salinity 1–10‰)	1.0–7.0	129
	River Loire 15–30 km upstream of estuary (salinity < 1‰)	1–15.1	129
	North Sea soluble metals	0.002	86
Molyb-denum	Kagoshima Bay, Japan	8.16–9.7	130
	Estuary water	5.3	82
	Seawater	2.1–18.8	131
	Coastal water	7–200	126
	Coastal water	10.1–10.3	87
Nickel	Sandy Cove, USA	0.33–0.40	92
	Profile to 1200 m in Santa Catalina	0.3–0.6	132
	Near shore seawater (salinity 29.5‰)	0.33–0.39	46
	Coastal seawater	0.33–0.4	92
	Estuary water	1.2–1.3	82
	Seawater	0.341–0.608	76, 77
	North Sea soluble metals	0.25	86
	Southampton Water	0.50–1.58	56
	Cronhulla Beach, Australia	2.5	73
	Chesapeake Bay	1.2	103
	Heligoland Bight	0.2–1.2	98, 108
	Osaka Bay, Japan	2.41–5.33	119
	North West coast, USA	1.1	72
	Canadian coastal waters	0.37–0.43	104
	Port Hastings, Australia	2.9	73
	Botany Bay, Australia	3.8	73
	Sea off California	0.22–0.3 (surface)	100
		0.60–0.67 (2950 m)	100
	Menai Straits	1.9	72

Table A1.3. (*continued*)

Element	Location	Concentration (μg l^{-1})	Ref.
Nickel	Coastal water	0.58	97
(*continued*)	Gota River Estuary (salinity 0.5‰)	1.2–1.3	109
	Gota River Estuary (salinity 32‰)	0.4	109
Thorium	Estuary and seawater	$\leq$ 0.0002	82
Uranium	Estuary and seawater	1.90	82
	Kwangyana Bay, Korea	1.36–1.86	88, 110
	Coastal water	3.08–3.1	87
Rare earths	Kwangyana Bay, Korea	Ce 16.7	88
		La 0.72	88
Scandium	Kwangyana Bay, Korea	0.098	88
	Estuary water	0.000 95	82
Selenium	Seawater	0.4	133
Silver	Southampton Water	< 0.01–0.08	56
Vanadium	Kwangyana Bay, Korea	2.14	88
	Estuary and seawater	0.45	82
	Osaka Bay, Japan	0.23–0.88	119
	Coastal water	1.22–1.23	87
	Coastal water	< 0.01–5.1	126
Zinc	North Sea soluble metals	1.0	92
	Sandy Cove, USA	1.5–1.9	92
	Kwangyana Bay, Korea	45.9	88
	Near shore seawater (salinity 29.5‰)	0.29–0.44	46
	Estuary and seawater	4.5–4.9	82
	Cape San Blas	0.055 (5 m)	105
		0.030 (70 m)	105
	Heligoland Bight	1.3–6.6	98, 99
	Osaka Bay, Japan	5.3–29.1	120
	Chesapeake Bay	4.8	103
	Danish coastal water	0.5–250	102
	Clyde water	2.0–23.0	57
	North Sea	7.0–22.0	85
	Southampton Water	1.9–13.2	56
	Sea off California	0.007 (surface)	100
		0.60–0.65	100
		(2950 m)	
	Coastal	0.29–0.44	46
	Coastal	3.28	97
	Coastal	1.6–2.0	108
	Coastal seawater	1.5–1.9	93
	Seawater	1.6–1.9	77
		4.1	123
		0.72–0.84	75
	Gota River Estuary (salinity 0.5‰)	7.6–8.4	109
	Gota River Estuary (salinity 32‰)	0.5	109

continued overleaf

Table A1.3. (*continued*)

Element	Location	Concentration (μg l^{-1})	Ref.
	Anions in seawater		
	Phosphate		
	Japanese inland waters	12.8–46.0 μg l^{-1}	134, 135
	Fluoride		
	Port Lonsdale, Victoria Australia	1280–1430	136
	(salinity 35%)		
	Iodate	30–60	137
	Iodide	0–20	137
	Organic iodine	< 5	137

Note: The North Sea and the Mediterranean are included in this list as these are both subject to a high degree of metal contamination originating from surrounding coastal areas.

REFERENCES

1. Zoltzer, D. and Schwedt, G., *Fresenius Zeitschift für Analytische Chemie* **317**, 422 (1984)
2. Janssens, E., Schutyser, P. and Dams, R., *Environmental Technology Letters* **3**, 35 (1982)
3. Thompson, M., Ramsey, M.H. and Pahlavanpour, B., *Analyst (London)* **107**, 1330 (1982)
4. Bart, G. and Von Gunter, H.R., *International Journal of Environmental Analytical Chemistry* **6**, 25 (1979)
5. Abu Hilal, A.H. and Riley, J.P., *Analytica Chimica Acta* **131**, 175 (1981)
6. Kasuka, Y., Tsuji, H., Fujimoto, Y., Ishida, K., Fukai, Y., Mamuro, T., Matsumani, I., Mizuhata, A. and Hurai, S., *Journal of Radioanalytical Chemistry* **71**, 7 (1982)
7. Kusaka, J.Y., Tsuji, H., Fujimoto, Y., Ishida, K., Mamuro, I., Matsumami, I., Mizohata, A. and Hirai, S., *Bull. Inst. Chem. Res. Kyoto Univ.* **58**, 171 (1980)
8. Chakrabarthi, D. and Irgolic, K.J., *International Journal of Environmental Analytical Chemistry* **17**, 241 (1984)
9. Subramamian, K., Leung, P.C. and Meranger, J.C., *International Journal of Environmental Analytical Chemistry* **11**, 121 (1982)
10. Thompson, M., Paklavanpour, B., Walton, S.J. and Kirkbright, G.F., *Analyst (London)* **103**, 568 (1978)
11. Logos, P., *Analytica Chimica Acta* **98**, 201 (1978)
12. Podoski, T.E. and Glass, G.E., *Analytica Chimica Acta* **101**, 79 (1978)
13. Hasan, M.Z. and Kumar, A., *Indian Journal of Environmental Health* **25**, 161 (1983)
14. West, M.M., Molina, J.F., Yuan, C.L. and Davis, D.G. *Analytical Chemistry* **51**, 2370 (1979)
15. Tanaka, S., Darzi, M. and Winchester, J.W., *Environmental Science and Technology* **15**, 354 (1981)
16. McMahon, J.W., Docherty, K.E. and Judd, J.M., *Hydrobiologia* **126**, 103 (1985)
17. Yoshimura, K., Nigo, S. and Tarutani, T., *Talanta* **29**, 173 (1982)
18. Inami, E.B., *Journal of Chromatography* **256**, 253 (1983)
19. McHugh, J.B., *Journal of Geochemical Exploration* **20**, 303 (1984)
20. Give, M.F., Bergamin, F.H., Zagatto, E.A.C. and Peis, B.F., *Analytica Chimica Acta* **114**, 191 (1980)

21. Bemis, P., Koe, J. and Stulik, K., *Water Research* **13**, 967 (1979)
22. Batley, G.E., Thompson, J.H. and Jenne, E.A., *Analytica Chimica Acta* **98**, 67 (1978)
23. Hadjiiannon, T.P., Hadjiiannon, S.I., Avery, J. and Malinstadt, H.V., *Analytica Chimica Acta* **89** 231 (1977)
24. Hadjiiannon, T.P., Hadjiiannon, S.J., Avery, J. and Malinstadt, H.V., *Clinical Chemistry* **22**, 802 (1976)
25. Mercury Analysis Working Party of the Bureau International Technique du Chlore, *Analytica Chimica Acta* **109**, 209 (1979)
26. Kopp, J.F., Longbottom, H.C. and Lohring, L.B., *Journal of the American Water Works Association* **64**, 20 (1972)
27. Chau, V. and Agemian, H., *Analytica Chimica Acta* **113**, 237 (1988)
28. Ushida, H., Shimoishi, V. and Toei, K., *Environmental Science and Technology* **14**, 541 (1980)
29. Shimoishi, Y. and Toei, K., *Analytica Chimica Acta* **100**, 65 (1978)
30. Radau, D.R. and Tallman, D.E., *Analytical Chemistry* **54**, 307 (1982)
31. Orvini, E., Ladola, L., Sabbioni, E., Pietra, R. and Goetz, L., *Science of the Total Environment* **13**, 195 (1979)
32. Meineche, G., Zur Geochemie des Vanadiums, Clausthaler, Hefte zur Lagerstattenkunde Und Geochemie der Mininalischen Rohstoffe 2 (1973)
33. Simeonov, V., Andrew, G. and Stoianov, A., *Fresenius Zeitschrift für Analytische Chemie* **297**, 418 (1979)
34. Johansen, H.S. and Middeboe, V., *International Journal of Applied Radiation and Isotopes* **27**, 591 (1976)
35. Urasa, I.T., *Analytical Chemistry* **56**, 904 (1984)
36. Landers, D.H., David, M.B. and Mitchell, M.J., *International Journal of Environmental Analytical Chemistry* **14**, 245 (1983)
37. Aznarez, J., Bouilla, A. and Vidal, J.C., *Analyst (London)* **108**, 368 (1983)
38. Morrow, C.M. and Minear, R.A., *Water Research* **18**, 1165 (1984)
39. Mosko, J., *Analytical Chemistry* **56**, 629 (1984)
40. Maaschelein, W.J. and Denis, M., *Water Research* **15**, 857 (1981)
41. Zelensky, I., Zelenska, V., Kamiansky, D., Havassi, P. and Lednarova, V., *Journal of Chromatography* **294**, 317 (1984)
42. Pillay, K.K.S., Thomas, C.C., Sondel, J.A. and Hyche, C.M., *Analytical Chemistry* **43**, 1419 (1971)
43. Moore, R.M., in Wong, C.S. *et al.* (eds), *Trace Metals in Seawater*, Proceedings of a NATO Advanced Research Institute on Trace Metals in Seawater. 30 March to 3 April 1981, Sicily, Italy. Plenum Press, New York (1981)
44. Fitzgerald, W.F., Mercury analysis in seawater using cold trap preconcentration and gas phase detection, in Gibb, T.R.P. (ed.), *Analytical Methods in Oceanography*, American Chemical Society, Washington, DC (1975)
45. Pillay, K.K.S., Thomas, C.C., Sondel, I.A. and Hyche, C.M., *Analytical Chemistry* **43**, 1419 (1971)
46. Sturgeon, R.E., Berman, S.S., Willie, S.N. and De Sanliniers, J.A.U., *Analytical Chemistry* **53**, 2337 (1981)
47. Jagner, D., Josefson, M. and Westerlund, S., *Analytica Chimica Acta* **129**, 153 (1981)
48. Danielson, L.G., Magnusson, R. and Westerlund, S., *Analytica Chimica Acta* **98**, 47 (1978)
49. Jagner, D., Josefson, M. and Westerlund, S., *Analytica Chimica Acta* **129**, 153 (1981)
50. Mujazaki, A., Kimura, A., Bansho, K. and Amezaki, Y., *Analytica Chimica Acta* **144**, 213 (1981)

51. Magnusson, B. and Westerlund, A., in Wong, C.S. *et al.* (eds), *Trace Metals in Seawater*, Proceedings of a NATO Advanced Research Institute on Trace Metals in Seawater, 30 March to 3 April 1981, Sicily, Italy, Plenum Press, New York (1981)
52. Brugmann, L., Danielsson, L.G., Magnusson, B. and Westerlund, S., *Marine Chemistry* **13**, 327 (1983)
53. Boyle, E.A. and Edmund, J.M., *Analytica Chimica Acta* **91**, 189 (1977)
54. Magnusson, B. and Westerlund, S., *Marine Chemistry* **8**, 231 (1980)
55. Stukas, V.J. and Wong, C.S., in Wong, C.S. *et al.* (eds), *Trace Metals in Seawater*, Proceedings of a NATO Advanced Research Institute on Trace Metals in Seawater, 30 March to 3 April 1981, Sicily, Italy, Plenum Press, New York (1981)
56. Armannsson, H., *Analytica Chimica Acta* **110**, 71 (1979)
57. Campbell, W.C. and Ottaway, J.M., *Analyst (London)* **102**, 495 (1977)
58. Grimaud, D. and Michord, G., *Marine Chemistry* **2**, 229 (1974)
59. Kuwamoto, T. and Murai, S., Preliminary Report of the Hakuho-Maru Cruise K11-68-4, Ocean Research Institute University, Tokyo, p. 72 (1970)
60. Tukai, R., *Nature (London)* **213**, 901 (1967)
61. Willie, S.N., Sturgeon, R.E. and Berman, S.S., *Analytical Chemistry* **55**, 981 (1983)
62. Batley, G.E. and Matonsik, J.P., *Analytical Chemistry* **52**, 1570 (1980)
63. Motomizu, S., *Analytica Chimica Acta* **64**, 217 (1973)
64. Greenberg, R.R. and Kingston, H.M., *Journal of Radioanalytical Chemistry* **71**, 147 (1982)
65. Sheffrin, N. and Williams, E.E., *Analytical Proceedings* **19**, 483 (1982)
66. Krembing, K.P., in Wong, C.S. *et al.* (eds), *Trace Metals in Seawater*, Proceedings of a NATO Advanced Research Institute on Trace Metals in Seawater, 30 March to 3 April 1981, Sicily, Italy, Plenum Press, New York (1981)
67. Mort, L., Nurnberg, H.W. and Dyrssen, D., in Wong, C.S. *et al.* (eds), *Trace Metals in Seawater*, Proceedings of a NATO Advanced Research Institute on Trace Metals in Seawater, 30 March to 3 April 1981, Sicily, Italy, Plenum Press, New York (1981)
68. Miyazaki, A., Kimura, A., Bansho, K. and Umezaki, Y.Z., *Coastal Analytica Chimica Acta* **144**, 213 (1981)
69. Schaube, B.K. and Patterson, C.C., in Wong, C.S. *et al.* (eds), *Trace Metals in Seawater*, Proceedings of a NATO Advanced Research Institute on Trace Metals in Seawater, 30 March to 3 April 1981, Sicily, Italy, Plenum Press, New York (1981)
70. Leyte, E. and Huested, S.W., in Wong, C.S. *et al.* (eds), *Trace Metals in Seawater*, Proceedings of a NATO Advanced Research Institute on Trace Metals in Seawater, 30 March to 3 April 1981, Sicily, Italy, Plenum Press, New York (1981)
71. Fitzgerald, W.F., Lyons, W.B. and Hunt, C.D., *Analytical Chemistry* **46**, 1882 (1974)
72. Olafsson, J., in Wong, C.S. *et al.* (eds), *Trace Metals in Seawater*, Proceedings of a NATO Advanced Research Institute on Trace Metals in Seawater, 30 March to 3 April 1981, Sicily, Italy, Plenum Press, New York (1981)
73. Batley, G.E. and Matonsck, J.P., *Analytical Chemistry* **49**, 2031 (1977)
74. Sheriadah, H.M.A., Katoaka, M. and Ohzeki, K., *Analyst (London)* **110**, 125 (1985)
75. Pihlar, B., Valenta, P. and Nurnberg, H.W., *Fresenius Zeitschrift für Analytische Chemie* **307**, 337 (1981)
76. Stuckas, V.J. and Wong, C.S., in Wong, C.S. *et al.* (eds), *Trace Metals in Seawater*, Proceedings of a NATO Advanced Research Institute on Trace Metals in Seawater, 30 March to 3 April 1981, Sicily, Italy, Plenum Press, New York (1981)
77. Mykytiuk, A.P., Russell, D.S. and Sturgeon, R.E., *Analytical Chemistry* **52**, 1281 (1980)
78. Elderfield, H. and Greaves, H.J., in Wong, C.S. *et al.* (eds), *Trace Metals in Seawater*,

Proceedings of a NATO Advanced Research Institute on Trace Metals in Seawater, 30 March to 3 April 1981, Sicily, Italy, Plenum Press, New York (1981)

79. Matthews, A.D. and Riley, J.P., *Analytica Chimica Acta* **51**, 445 (1970)
80. Willie, S.R., Sturgeon, R.E. and Berman, S.S., *Analytical Chemistry* **58**, 1140 (1986)
81. Brinkmann, F.E., in Wong, C.S. *et al.* (eds), *Trace Metals in Seawater*, Proceedings of a NATO Advanced Research Institute on Trace Metals in Seawater, 30 March to 3 April 1981, Sicily, Italy, Plenum Press, New York (1981)
82. Collela, M.B., Siggia, S. and Barnes, R.M., *Analytical Chemistry* **52**, 2347 (1980)
83. Van Der Berg, C.M.G. and Huang, O., *Analytical Chemistry* **56**, 2383 (1984)
84. Korenaga, T., Motomizu, S. and Toei, K., *Analyst (London)* **105**, 328 (1980)
85. Gillam, G., Duyckaents, G. and Disteche, A., *Analytica Chimica Acta* **106**, 23 (1979)
86. Hill, J.M., O'Donnell, A.R. and Mance, G., The quantities of some heavy metals entering the North Sea. Technical Report TR205, Water Research Centre, Stevenage, UK (1984)
87. Schreadhara Murthy, R.S. and Ryan, D.E., *Analytical Chemistry* **55**, 682 (1983)
88. Lee, C., Kim, N.B., Lee, I.C. and Chung, K.S., *Talanta* **24**, 241 (1977)
89. Florence, T.M., *Journal of Electroanalytical Chemistry* **49**, 255 (1974)
90. Eskillson, H. and Jaguer, D., *Analytica Chimica Acta* **138**, 27 (1982)
91. Peuszkowska, E., Carnrick, G.R. and Slavin, W., *Analytical Chemistry* **55**, 182 (1983)
92. Berman, S.S., McLaren, J.W. and Willie, S.N., *Analytical Chemistry* **52**, 488 (1980)
93. Sturgeon, R.E., Berman, S.S. and De Sauliniers, I.A.H., *Analytical Chemistry* **52**, 1585 (1980)
94. Guevremont, R., Sturgeon, R.E. and Berman, S.S., *Analytica Chimica Acta* **115**, 1633 (1980)
95. Mykytiak, A.P., Russell, D.S. and Sturgeon, R.E., *Analytical Chemistry* **52**, 1251 (1980)
96. Jaguer, D., *Analytical Chemistry* **50**, 1924 (1978)
97. Sturgeon, R.E., Berman, S.S., De Sauliniers, I.A.H. and Russell, D.S., *Analytical Chemistry* **51**, 2364 (1979)
98. Schmidt, D., *Heligolander Meeresunter-suchungen* **33**, 576 (1980)
99. Lo, J.M., Yun, J.C., Hutchinson, I. and Wal, C.M., *Analytical Chemistry* **54**, 2536 (1982)
100. Brulund, K., Franks, R.P., Knauer, G.A. and Martin, J.H., *Analytica Chimica Acta* **105**, 233 (1979)
101. Sperling, R.K., *Fresenius Zeitschrift für Analytische Chemie* **310**, 254 (1982)
102. Drabach, I., Pheiffer Madsen, P. and Sorensen, J., *Int. J. Environ. Anal. Chem.* **15**, 153 (1983)
103. Kingston, H.M., Barnes, I.L., Brady, T.J. and Rains, R.C., *Analytical Chemistry* **50**, 2064 (1978)
104. Won, C.C., Chiang, S. and Corsini, A., *Analytical Chemistry* **57**, 719 (1985)
105. Peotrowicz, S.R., in Wong, C.S. *et al.* (eds), *Traces Metal in Seawater*, Proceedings of a NATO Advanced Research Institute of Trace Metals in Seawater, 30 March to 3 April 1981, Sicily, Italy, Plenum Press, New York (1981)
106. Scarponi, G., Capodaglio, G., Cescou, P., Cosma, B. and Frache, R., *Analytica Chimica Acta* **135**, 263 (1982)
107. Stein, V. B., Canelli, E. and Richards, H., *Int. J. Environ. Anal. Chem.* **8**, 99 (1980)
108. Slavin, W., *Atomic Spectroscopy* **1**, 66, May–June (1980)
109. Danielsson, L.G., Magnusson, I., Westerlund, I. and Zhang, K., *Analytica Chimica Acta* **144**, 183 (1982)

110. Lee, C., Kim, N.B., Lee, I.C. and Chung, K.S., *Talanta* **24**, 241 (1977)
111. Nakayama, E., Kuwamoto, T., Tokoro, H. and Fujinaka, T., *Analytica Chimica Acta* **131**, 247 (1981)
112. Ishibashi, M. and Shigematsu, T., *Bulletin of the Institute of Chemical Research, Kyoto University*, **23**, 59 (1950)
113. Mullins, T.L., *Analytica Chimica Acta* **165**, 97 (1984)
114. Cheukas, L. and Riley, J.P., *Analytica Chimica Acta* **35**, 240 (1966)
115. Murata, M., Omatsu, M. and Muskimoto, S., *X-Ray Spectroscopy* **13**, 83 (1984)
116. Berger, P., Ewald, M., Liu, D. and Weber, J.H., *Marine Chemistry* **14**, 289 (1984)
117. Yoshimura, K., Nigo, S. and Tarutani, T., *Talanta* **29**, 173 (1982)
118. Murthy, R.S.S. and Ryan, D.E., *Analytica Chimica Acta* **140**, 163 (1982)
119. Sugimae, A., *Analytica Chimica Acta* **121**, 331 (1980)
120. Pellenberg, R.E. and Church, T.M., *Analytica Chimica Acta* **97**, 81 (1978)
121. Berman, S.S., McLaren, J.W. and Willie, S.N., *Analytica Chimica Acta* **52**, 488 (1980)
122. Acebal, S.A., De Luca, O. and Rebello, A., *Analytica Chimica Acta* **148**, 71 (1983)
123. Jagner, D., *Analytical Chemistry* **50**, 1924 (1978)
124. Torsi, G., *Analytica Chemica (Rome)* **67**, 557 (1977)
125. Torsi, G., Oesimoni, E., Palimisano, F. and Sabbatini, L., *Analytica Chimica Acta* **124**, 143 (1981)
126. Tominaga, H., Bansho, K. and Umezaki, Y., *Analytica Chimica Acta* **169**, 171 (1985)
127. Knox, S. and Turner, D.R., *Estuarine and Coastal Marine Science*, **10**, 317 (1980)
128. Carnrick, G.R., Slavin, W. and Manning, D.C., *Analytical Chemistry* **53**, 1866 (1981)
129. Fresnet-Robin, M. and Offmann, F., *Estuarine and Coastal Marine Science* **7**, 425 (1978)
130. Kiriyama, T. and Kuroda, R., *Talanta* **31**, 472 (1984)
131. Nakahara, T. and Chakahorti, C.L., *Analytica Chimica Acta* **104**, 99 (1979)
132. Lee, D.S., *Analytical Chemistry* **54**, 1182 (1982)
133. Tseng, J.H. and Zeitlin, H., *Analytica Chimica Acta* **101**, 71 (1978)
134. Paoulson, A.J., *Analytical Chemistry* **58**, 183 (1986)
135. Motomizu, S., Wakimoto, T. and Toei, K., *Analytical Chemistry* **138**, 329 (1982)
136. Rix, C.J., Bond, A.H. and Smith, J.D., *Analytical Chemistry* **48**, 1236 (1976)
137. Truesdale, V.W., *Marine Chemistry* **6**, 253 (1978)

APPENDIX 2
Radioactive Metallic Elements in Freshwater and Seawater

References to work on the determination of radio nucleides in rivers, groundwaters, rain, potable water, lakes, seawater and river and marine sediments are summarised in Table A2.1.

Table A2.1. Radioactive elements in the environment (references)

	Rivers	Groundwaters	Rain	Potable water	Lakes	Seawater	River sediment	Marine sediment
Naturally occurring								
Actinium-227								154, 170
Actinium-237								154
Carbon-14	28, 35							
Lead-210							173–5	176
Lead-214/Bismuth-214	36		13					
Phosphorus-32	73, 74							
Polonium-210						190		
Polonium-210/lead-210	37, 38					105–116		
Polonium-210/bismuth-210	40							
Protoactinium-231								
Radon-222	41–49	24.25		9, 10		49, 122–30, 133, 136–38	181	
Radium-223	41–49							
Radium-224	41–49							
Radium-226	29, 41–49, 187	22		3–8		49, 122–35, 191	174, 175, 181	176, 180
Radium-228	41–49			3–8		49, 122–35, 192		
Potassium-40	39					104		180
Sodium-22			14, 15					
Sodium-24			14, 15					
Thorium-228					26–27, 188	117–121	183–5	154, 170, 180
Thorium-230					26–27	117–121	183–5	154, 170, 180
Thorium-234					26–27	117–121		170
Thorium-232							183–5	154, 170
Tritium	50–60					93–104		
Uranium-234	85–92					93–104, 190		180, 186
Uranium-235	85–92					93–104, 190		180, 186
Uranium-237			16			93–104, 190		
Uranium-238	85–92					93–104, 190		154, 170
Fall-out products								
Antimony-125	17–21				188			
Barium-140	17–21							

Element						
Beryllium-7			188			
Caesium-137	33, 61–67	17–21	189	159–64	172, 173	171
Cadmium-113	68		188			
Cerium-141	69	17–21	188			
Cerium-144	69	17–21	188	166, 167		171
Cobalt-60	70	12				
Iodine-127	71	12				
Iodine-129						
Iodine-131	71	17–21			1, 2	
Iron-55						
Manganese-54/zinc-85				155		
Niobium-95		17–21	188	168, 169		
Palladium-231						171, 172, 196
Ruthenium-103		17–21	188			171
Ruthenium-106		17–21	188	165		
Strontium-89	30, 31	17–21				
Strontium-90	30, 31 78–83			16–159	23	172
Technetium-99	32, 84				11	
Thallium-204						182
Yttrium-90	80					
Zirconium-95			188			
Zirconium-95/niobium-95	34					
Transuranic elements						
Americium-241						171
Neptunium-237					177	
Plutonium-236				139–56		
Plutonium-238				139–56	177, 178	
Plutonium-239	75–77	17–21		139–56		154, 172
Plutonium-240	75–77	17–21		139–56		154, 172
Plutonium-242				139–56		

REFERENCES

1. Dabosz, E., *Radiochemical and Radioanalytical Letters* **13**, 381 (1973)
2. Morgan, A. and Mitchell, O. UK Atomic Energy Authority Report No. AERE M-1004 (1962)
3. Szy, D. and Lirlam, A., *Mikrochim Acta* **11**, 31 (1971)
4. Higuchi, H., Uesugi, M., Satoh, K., Ohashi, N. and Naguchi, M., *Analytical Chemistry* **56**, 761 (1984)
5. Moron, M.C., Garcia Tenorio, R., Garcia-Montano, E., Garcia-Leon, H. and Madurga, G., *Applied Radiation and Isotopes* **37**, 383 (1986)
6. Friedman, H. and Herneggar, R., *Zeitschrift für Wasser und Abwasser Forshung* **11**, 61 (1978)
7. Baralta, E.J. and Lumsden, E.M., *Journal of Association of Official Analytical Chemists* **65**, 1424 (1982)
8. Mills, W.A., Elliot, W.H. and Sullivan, A.E., *Health Physics* **39**, 1003 (1981)
9. Pritchard, H.M. and Gesell, T.F., *Health Physics* **33**, 577 (1977)
10. Countess, D.J., *Health Physics* **34**, 390 (1978)
11. Garcia-Leon, H., Piazza, C. and Madurga, G., *International Journal of Applied Radiation and Isotopes* **35**, 957 (1984)
12. Murematsu, Y., Ohmomo, Y. and Cristoffers, D., *Journal of Radioactivity and Nuclear Chemistry* **83**, 353 (1984)
13. Michaelis, M.L., *Chemiker Zeitung Chem. Apparat* **93**, 883 (1969)
14. Yasulenis, R.Y., Luyanas, V.Y. and Kekite, V.P., *Soviet Biochemistry* **14**, 673 (1972)
15. Burden, B.A., *Analyst (London)* **93**, 715 (1968)
16. Suzuki, T., Sotobayashi, I., Koyame, S. and Kanda, Y., *Journal of the Chemical Society of Japan, Pure Chemistry Section* **89**, 1084 (1968)
17. Kimura, I. and Hamada, T., *Analytica Chimica Acta* **120**, 419 (1980)
18. Eichholz, G.G. and Galli, A.N., *Radiochemical and Radioanalytical Letters* **4**, 315 (1970)
19. Perkins, R.W., Report of the Atomic Energy Commission, USA, BNWL-1051 Part 2. Procedure for the continuous separation and subsequent direct counting of short-lived cosmic ray produced radio nucleides in rain water (1969)
20. Rambray, O.H., Fisher, E.M. and Salmon, L., Report of the UK Atomic Energy Authority AERE-R5898, Method of collection and radioactivity from distance nuclear test exposures (1970)
21. Ardisson, G., *Trends in Analytical Chemistry* **1**, 281 (1982)
22. Higuchi, H., Uesugi, M., Satoh, K., Ohaski, N. and Noguchi, M., *Analytical Chemistry* **56**, 76 (1984)
23. Alekson, yan, O.M., *Gidrokhim Mater* **53**, 163 (1972)
24. Kobal, I. and Kristan, J., *Radiochemical and Radioanalytical Letters* **10**, 291 (1972)
25. Kobal, I. and Kristan, J., *Mikrochim Acta* **2**, 219 (1973)
26. De Jong, I.G. and Wiles, D.R., *Water, Air and Soil Pollution* **23**, 197 (1984)
27. De Jong, I.G. and Wiles, D.R., *Journal of Radioanalytical Chemistry* **82**, 120 (1984)
28. Fredorak, P.M., Foght, J.M. and Westlake, P.W.S., *Water Research* **16**, 1285 (1982)
29. Conlan, B., Henderson, P. and Walton, A., *Analyst (London)* **94**, 15 (1969)
30. Gregory, L.P., *Analyst (London)* **44**, 2113 (1972)
31. Lapid, J., Munster, M., Farhi, S., Eini, M. and La Louche, L., *Journal of Radioanalytical and Nuclear Letters* **86**, 231 (1984)
32. Robb, P., Warwick, P. and Malcombe-Lawes, D.J., *Journal of Radioanalytical and Nuclear Chemistry* **89**, 323 (1985)
33. Mundschenk, H., *Deutche Gewasserkundliche Mitteilunger* **18**, 72 (1974)

34. Linsalata, P. and Cohen, N., *Health Physics* **43**, 742 (1982)
35. Toeyn, Y.E. and Hanneborg, S., *Marine Biology* **8**, 57 (1971)
36. Hashimato, T., Satoh, K. and Aoyagi, M., *Journal of Radioanalytical and Nuclear Chemistry* **92**, 407 (1985)
37. MacKenzie, A.B. and Scott, R.D., *Analyst (London)* **104**, 1151 (1979)
38. Furnica, G. and Toader, M., *Igiena Bucharest* **18**, 227 (1969)
39. Gertner, A. and Grdinic, V., *Mikrochim. Acta* **1**, 25 (1969)
40. Sill, C.W., *Analytical Chemistry* **50**, 1559 (1978)
41. Kelkan, D.M. and Joshi, P.V., *Health Physics* **17**, 253 (1969)
42. Darrall, K.G., Richardson, P.J. and Tyler, J.F.C., *Analyst (London)* **98**, 710 (1973)
43. Benes, P., Sedlacek, J., Sebesta, F., Sandrik, R. and John, J., *Water Research* **15**, 1299 (1981)
44. Johnson, J.O., US Geological Survey Water Supply Paper 1696G (1971)
45. Michel, J., Moor, W.S. and King, P.T., *Analytical Chemistry* **53**, 1885 (1981)
46. Higuchi, H., Uesugi, H., Satoh, K., Ohashi, N. and Noguchi, N., *Analytical Chemistry* **56**, 761 (1984)
47. Zeigeheim, C.J., Busigin, A. and Phillips, C.R., *Health Physics* **42**, 317 (1982)
48. Pritchard, H.H. and Gesell, T.F., *Health Physics* **33**, 577 (1977)
49. Perkins R.W. Report of the Atomic Energy Commission, USA, BNWL 1051. Radium and radiobarium measurement in seawater and freshwater by sorption and direct multidimensional X-ray spectometry (1969)
50. Noakes, J.E., Neary, N.P. and Spaulding, J.D., *Nuclear Instrumental Methods* **109**, 177 (1973)
51. Files, P., *Radiochemical and Radioanalytical Letters* **15**, 213 (1973)
52. Povinec, P., Chudy, M., Seliga, M., Saro, S. and Szarka, J., *Journal of Radioanalytical Chemistry* **12**, 513 (1972)
53. Tistchenko, S. and Dirian, G., *Bull. Soc. Chim. France* **1**, 16 (1970)
54. Knowles, F.E. and Baratta, E.J., *Radiological Health Data* **12**, 405 (1971)
55. Thordmorsson, P., *Journal of Applied Radiation Isotopes* **25**, 97 (1974)
56. Florkowski, I., Payne, B.R. and Sauzay, G., *International Journal of Radiation and Isotopes* **21**, 453 (1970)
57. Froehlich, K., Herbert, D. and Anreev, A., *Isotopenproxix* **8**, 130 (1972)
58. Wolf, H., *International Journal of Applied Radiation and Isotopes* **24**, 299 (1973)
59. Vinogradov, A.P., Devirts, A.L. and Dobkina, E.I., *Goekhimya* **10**, 1147 (1968)
60. Schello, W.R., Nevissi, A. and Huntamer, D., *Marine Chemistry* **6**, 743 (1978)
61. Rebak, W. and Ubl, G., Report Staatliche Zeitrate für Strahlenschutz S25-1/70 (1970)
62. Stewart, M.L., Pendleton, R.C. and Lords, J.L., *International Journal of Applied Radiation and Isotopes* **23**, 345 (1972)
63. Senegacnik, M. and Paljk, S., *Z. Analyt. Chem.* **244**, 306 (1969)
64. Senegacnik, M. and Paljk, S., *Z. Analyt. Chem.* **244**, 375 (1969)
65. Kleberer, K. and Stuerzer, U., *Gas-u-Wassfach, Wasser-Abwasser* **111**, 29 (1970)
66. Tereda, K., Hayakawa, H., Sawada, K. and Kibu, T., *Talanta* **17**, 955 (1970)
67. Kapustin, V.K., Egorov, A.I. and Eeonov, V.V., *Soviet Journal of Water Chemistry and Technology* **3**, 119 (1981)
68. Palogyi, S., Larsen, R.P. and Tisue, G.I., *Journal of Radioanalytical and Nuclear Chemistry Matters* **96**, 161 (1985)
69. Haberer, K., Stuerzer, U., *Gas-u-Wassfach, Wasser-Abwasser* **113**, 122 (1972)
70. Claassen, H.C., *Analytica Chimica Acta* **52**, 229 (1970)
71. Haberer, K. and Stürzer, U., *Gas-u-Wassfach, Wasser-Abwasser* **109**, 1287 (1968)
72. Palagyi, S., *International Journal of Applied Radiation and Isotopes* **34**, 755 (1983)

73. Milham, R.C., Report of the US Atomic Energy Commission DP.US-73-35 (1973)
74. Furnica, G. and Ionescu, H., *Igena Bucharest* **18**, 105 (1969)
75. Carver, R.D. and Dupzyk, R.J. Report of the Atomic Energy Commission, UK, UCRL-74430 (1973)
76. Golchert, N.W. and Sedlet, J., *Radiochemical and Radioanalytical Letters* **12**, 215 (1972)
77. Kim, U.I., Baumgartner, F. and Schwankren, R., *Fresenius Zeitschrift für Analytische Chemie* **323**, 821 (1986)
78. Gusev, N.G., Ya, U., Margulis, A.N., Marei, A. *et al.*, *Dosimetric and Radiometric Methods* (in Russian), p. 44, Atomizdat, Moscow (1966)
79. Baratta, E.J. and Knowles, F.R., *Journal of the Association of Official Analytical Chemists* **69**, 540 (1986)
80. Mundschenk, H., *Deutsch Gewasserkundliche Mitteilungen* **23**, 64 (1979)
81. Ankudinova, M.M., Mordberg, E.L., Nekhorossheva, M.P. *et al.*, The isolation of strontium-90, collection: Collection or Radiometric and Gamma-spectromic Methods of Analysing Materials in the Environment (in Russian), p. 26 Leningrad (1970)
82. Haberer, K. and Steuzer, U., *Gas-u-Wassfach, Wasser-Abwasser* **121**, 186 (1971)
83. Tesetmale, G. and Leredde, J.L., Centre of Nuclear Studies, Fontenay-aux-Roses, France, Report CEA R-3908 (1970)
84. Golchert, N.W. and Sedlet, J., *Analytical Chemistry* **41**, 669 (1969)
85. Szy, D., Sebessy, L. and Balint, G., *Journal of Radioanalytic Chemistry* **7**, 57 (1971)
86. Szy, D. and Toth, A., *Mikrochimica Acta* **5**, 960 (1970)
87. Gorbushina, L.V., Zhil'tsova, L.Y., Matveeva, E.N., Surganova, N.P., Tenyaev, V.G. and Tyminskii, V.G., *Journal of Radioanalytical Chemistry* **10**, 165 (1972)
88. Takebayashi, T., Matsuda, H. and Umemoto, S., *Talanta* **20**, 892 (1973)
89. Bertine, K.K., Chau, L.H. and Turekian, K.K., *Geochim. Cosmochim. Acta* **34**, 641 (1970)
90. Doerschel, E. and Stolz, W., *Radiochemical and Radioanalytical Letters* **4**, 277 (1970)
91. Fleischer, R.L. and Lovett, D.B., *Geochim. Cosmochim. Acta* **32**, 1126 (1968)
92. Gladney, E.S., Peters, R.J. and Perrin, D.R., *Analytical Chemistry* **55**, 976 (1983)
93. Spencer, R., *Talanta* **15**, 1307 (1968)
94. Bertine, K.K., Chau, L.H. and Turekian, K.K., *Geochimica. Cosmochimica Acta* **34**, 641 (1970)
95. Hashimoto, T., *Analytica Chimica Acta* **56**, 347 (1971)
96. Kim, Y.S. and Zeitlin, H., *Analytical Chemistry* **43**, 1390 (1971)
97. Williams, W.J. and Gillam, A.H., *Analyst (London)* **103**, 1239 (1979)
98. Smith, J. and Grimaldi, O., *Bulletin of the US Geological Society* **1006**, 125 (1957)
99. Leung, G., Kim, Y.S. and Zeitlin, H., *Analytica Chimica Acta* **60**, 229 (1972)
100. Kim, Y.S. and Zeitlin, H., *Analytical Abstracts* **22**, 4571 (1972)
101. Korkish, J. and Koch, W., *Mikrochimica Acta* **1**, 157 (1973)
102. Barbano, P.G. and Rigoli, L., *Analytica Chimica Acta* **96**, 199 (1978)
103. Kim, K.H. and Burnett, W.C., *Analytical Chemistry* **55**, 796 (1983)
104. Bowie, S.H.U. and Clayton, C.G., *Translations of the Institution of Minerals and Metals B* **81**, 215 (1972)
105. Shannon, L.V. and Orren, M.J., *Analytica Chimica Acta* **52**, 166 (1970)
106. Nozaki, Y. and Tsunogai, S., *Analytica Chimica Acta* **64**, 209 (1973)
107. Cowen, J.P., Hodge, V.F. and Folson, T.R., *Analytical Chemistry* **49**, 494 (1977)
108. Tsunogai, S. and Nozaki, Y., *Geochemical Journal* **5**, 165 (1971)
109. Shannon, L.V., Cherry, R.D. and Orren, M.J., *Geochimica and Cosmochimica Acta* **34**, 701 (1970)

110. Hodge, V.F., Hoffman, F.L. and Folsom, T.R., *Health Physics* **27**, 29 (1974)
111. Folsom, T.R. and Hodge, V.R., *Marine Science Communications* **1**, 213 (1975)
112. Folsom, T.R., Hodge, F.V. and Gurney, M., *Marine Science Communications* **1**, 39 (1975)
113. Goldberg, E.D., Koide, M. and Hodge, V.F., Scripps Institution of Oceanography, La Jolla, CA, USA (1976)
114. Flynn, A., *Analytical Abstracts* **18**, 1624 (1970)
115. Reid, D.F., Key, R.M. and Schink, D.R., *Earth Planet Science Letters* **43**, 223 (1979)
116. Nozaki, Y. and Tsunogai, S., *Analytica Chimica Acta* **64**, 209 (1973)
117. Huh, C.A., *Analytical Chemistry* **57**, 2138 (1985)
118. Bacon, M.P. and Anderson, R.F., in Wong, C.S. *et al.* (eds), *Trace Metals in Seawater*, Proceedings of a NATO Advanced Research Institute on Trace Metals in Seawater, 30 March to 3 April 1981, Sicily, Italy, Plenum Press, New York (1981)
119. Spencer, D.W. and Sachs, P.L., *Marine Geology* **9**, 117 (1970)
120. Krishnashwami, S., Lal, D., Somayajulu, B.L.K., Weiss, R.F. and Craig, H., *Earth Planet Science Letters* **32**, 420 (1976)
121. Anderson, R.F., *The Marine Geochemistry of Thorium and Protoactinium*, PhD dissertation, Massachusetts Institute of Technology/Woods Hole Oceanographic Institute, WH01-81-1 (1981)
122. Moore, W.S. and Reid, D.F., *Journal of Geophysical Research* **78**, 8880 (1973)
123. Moore, W.S., *Deep Sea Research* **23**, 647 (1976)
124. Moore, W.S. and Cook, L.M., *Nature (London)* **253**, 262 (1975)
125. US Environmental Protection Agency, Radiochemical Methodology for Drinking Water Regulations EPA 600/4-75-005 (1975)
126. Ku, T.L., Huh, C.A. and Chen, P.S., *Earth Planet Science Letters* **49**, 293 (1980)
127. Chung, Y., *Earth Planet Letters* **49**, 319 (1980)
128. Moore, W.S., *Estuarine and Coastal Shelf Science* **12**, 713 (1982)
129. Reid, D.F., Key, R.M. and Schink, D.R., *Earth Planet Science Letters* **43**, 223 (1979)
130. Michel, J., Moore, W.S. and King, P.T., *Analytical Chemistry* **53**, 1885 (1981)
131. US Environmental Protection Agency National Interim Primary Drinking Water Regulations EPA-57019-79-003 (1976)
132. Key, R.M., Brewer, R.L., Stockwell, J.H., Guinasso, N.L. and Schink, R.D., *Marine Chemistry* **7**, 251 (1979)
133. Broecker, W.S., An application of natural radon to problems in oceanic circulations. In *Proceedings of the Symposium on Diffusion in the Oceans and Fresh Waters*, pp. 116–45, Lamont Geological Observatory, New York (1965)
134. Moore, W.S., Sampling[228] Ra in the deep ocean. *Deep-sea Research* **23**, 647 (1976)
135. Reid, D.F., Key, R.M. and Schink, D.R. Radium extraction from seawater; efficiency of manganese impregnated fibers. *EOS Transactions of the American Geophysical Union*, December (1974)
136. Schink, D., Guinasso, N. Jr, Charnell, R. and Sigalove, J., Radon profiles in the sea—a measure of air–sea exchange. *IEEE Transactions in Nuclear Science* **NS-17**, 184–90 (1970)
137. Chung, Y., *Pacific Deep and Bottom Water Studies Based on Temperature, Radium and Excess-radon measurements*, Dissertation, University of California, San Diego (1971)
138. Lucas, H.F., *Review Scientific Instruments* **28**, 680 (1957)
139. Comar, C.L., *Plutonium, Facts and Interferences*, EPRI EA-43-SR (1976)
140. Livingston, H.D., Mann, D.R. and Bowen, V.T., Analytical methods in oceanography. *Advances in Chemistry*, Series No. 147, ACS (1975)
141. Wong, K.M., *Analytica Chimica Acta* **56**, 355 (1971)

142. Pillai, K.C., Smith, R.C. and Folsom, T.R., *Nature (London)* **203**, 568 (1964)
143. Ballestra, S., Holm, E. and Fukai, R., presented at the Symposium on the Determination of Radionuclides in Environmental and Radiological Materials, Central Electricity Generating Board, London, October 1978
144. Holm, E. and Fukai, R., *Talanta* **24**, 659 (1977)
145. Sakanous, M., Nakamura, M. and Imai, T., Rapid methods for measuring radioactivity in the environment. *Proceedings of the Symposium*, Neuherberg, IAEA, Vienna, p. 171 (1971)
146. Statham, C. and Murray, C.N., *Report of the International Committee of the Mediterranean Ocean* **23**, 163 (1976)
147. Hampson, B.L. and Tennant, D., *Analyst (London)* **98**, 873 (1973)
148. Levine, H. and Lamanna, A., *Health Physics* **11**, 117 (1965)
149. Aakrog, A., *Reference Methods of Marine Radioactivity Studies II*. Technical Report Services No. 169, IAEA, Vienna (1975)
150. Chu, A., *Analytical Abstracts* **22**, 427 (1972)
151. Livingston, H.D., Mann, D.R. and Bowen, U.J., Report of the Atomic Energy Commission, US COO-3563-12, Woods Hole Oceanographic Institute, Massachusetts, USA (1972)
152. Delle Site, A., Marchionni, V. and Testa, C., *Analytica Chimica Acta* **117**, 217 (1980)
153. Testa, C. and Delle Site, A., *Journal of Radioanalytical Chemistry* **34**, 121 (1976)
154. Anderson, R.F. and Fleer, A.I., *Analytical Chemistry* **54**, 1142 (1982)
155. Testa, C. and Staccioli, L., *Analyst (London)* **97**, 527 (1972)
156. Hirose, K. and Sugimura, Y.J., *Radioanalytical and Nuclear Chemistry Articles* **92**, 363 (1985)
157. Silant'ev, A.N., Chumichev, U.B. and Vakulouski, S.M., *Trudy Inst. eksp. Met. glav. uprav gidromet*, Sluzhty Sov. Minist. SSSR **15** (2) (1970) Ref: *Zhur Khim*, 19GD (1) Abstr. No. 1 G209 (1971)
158. Gordon, C.M. and Larson, R.E., *Radiochemical and Radioanalytical Letters* **5**, 369 (1970)
159. Dutton, J.W.R., Report of the Fisheries and Radiobiological Laboratory, FRL 6, Ministry of Agriculture, Fisheries and Food, UK (1970)
160. Lewis, S.R. and Shafrir, H.N., *Nuclear Instrumental Methods* **93**, 317 (1971)
161. Janzer, V.J., *Journal of the US Geological Survey* **1**, 113 (1973)
162. Yamamoto, O., *Analytical Abstracts* **14**, 6669 (1967)
163. Mason, W.J., *Radiochemical and Radioanalytical Letters* **16**, 237 (1974)
164. Morgan, A. and Arkell, O., *Health Physics* **9**, 857 (1963)
165. Kiba, T., Terada, K., Kiba, T. and Suzuki, K., *Talanta* **19**, 451 (1972)
166. Hiraide, M., Sakurai, K. and Mizuike, A., *Analytical Chemistry* **56**, 2851 (1984)
167. Tseng, C.L., Hsieh, Y.S. and Yong, M.H., *Journal of Radioanalytical and Nuclear Chemistry Letters* **95**, 359 (1985)
168. Flynn, W.W., *Analytica Chimica Acta*, **67**, 129 (1973)
169. Stah, S.M. and Rao, S.R., *Current Science (Bombay)* **41**, 659 (1972)
170. Anderson, R.F. and Fleer, A., *Analytical Chemistry* **54**, 1240 (1982)
171. Stanners, D.A. and Aston, S.R., *Estuarine, Coastal and Shelf Science* **14**, 687 (1982)
172. Grummitt, W.E. and Lahaie, G. Report Atomic Energy Commission, Canada, No. AECL 4365. Method for the determination of strontium-90 and caesium-137 in river sediments and soils (1973)
173. Anderson, R.F., Schiff, S.L. and Hesslein, R.H., *Canadian Journal of Fisheries and Aquatic Sciences* **44**, 231 (1987)
174. Binford, M.W. and Brenner, M., *Limnology and Oceanography* **31**, 584 (1986)

175. Appleby, P.G., Nolan, P.J., Oldfield, F., Richardson, F. and Heggit, S.R., *Science of the Total Environment* **69**, 157 (1988)

176. Harvey, B.R. and Young, A.K., *Science of the Total Environment* **69**, 13 (1988)

177. Popplewell, D.S. and Ham, G.J., *Journal of Radioanalytical and Nuclear Chemistry* **115**, 191 (1987)

178. Linsalata, P., Wrenn, M.E., Cohen, N. and Singh, N.P., *Environmental Science Technology* **14**, 1519 (1980)

179. Singh, N.P., Linsalata, P., Gentry, R. and Wrenn, M.E., *Analytica Chimica Acta* **111**, 265 (1979)

180. Bojaniowski, B., Fukai, R. and Holm, E., *Journal of Radioanalytical and Nuclear Chemistry* **110**, 113 (1987)

181. HMSO, Methods for the examination of waters and associated materials. Measurements of alpha and beta activity in water and sludge samples. Determination of radon-222 and radon-226. The determination of uranium, (including general X-ray fluorescence spectrometric analysis) 1985–1986, London (1986)

182. Matthews, A.D. and Riley, J.P., *Analytica Chimica Acta* **48**, 25 (1969)

183. Joshi, S.R., *Analytical Chemistry* **57**, 1023 (1985)

184. Joshi, S.R., *Analytical Chemistry* **57**, 1026 (1985)

185. Joshi, S.R., *Analytical Chemistry* **57**, 1027 (1985)

186. Suttle, A.D., O'Brien, B.C. and Mueller, D.W., *Analytical Chemistry* **41**, 1265 (1969)

187. Elsunger, K.J., King, P.T. and Moore, W.S. *Analytica Chimica Acta* **144**, 277 (1982)

188. Durham, R.W. and Joshi, S.R., *Water Research* **15**, 83 (1981)

189. Hashimoto, T., *Analytica Chimica Acta* **56**, 347 (1971)

190. Cowen, J.P., Hodge, V.F. and Folson, T.R., *Analytical Chemistry* **49**, 494 (1977)

191. Michel, J., Moore, W.S. and King, P.T., *Analytical Chemistry* **53**, 1885 (1981)

192. Key, R.M., Bremer, R.L., Stockwell, J.H., Guinasso, N. L. and Schink, R.D., *Marine Chemistry* **7**, 251 (1979)

APPENDIX 3
Toxicant Concentrations in Sedimentary Matter

In Table A3.1 is reported detailed information on the occurrence of metals in freshwater sediments. Further information on organometallic compounds and organic compounds can be obtained from the references given in Tables A3.2 and A3.3.

Table A3.1. Metals in sediments, mg kg^{-1} dry weight

	River			Lake/pond		
	Location	Concentration	Reference	Location	Concentration	Reference
Aluminium	–	46 200	9	–	26 200–63 800	14
	–	Total: 9890–11 500	12	Lake Ontario, Canada	43 000	13
		acid extractable 522–19 200	12			
Arsenic	River Edisto, USA	0.22–0.63	7		1.9–26	14
	–	1.9–7.1	6			
Antimony					0.01–2.9	14
Barium				–	163–175	14
				Lake Ontario, Canada	2700	13
Bromine				Lake Ontario, Canada	23–96	14
Cadmium	–	0.08–1.22	1	Lake Ontario, Canada	3.5–8.0	–
	River Arno, Italy	1.01–9.6	10	Lake Ontario, Canada	40.0	13
		Total: 0.06–27.5	12			
		Acid extractable 0.1–15.4	12			
Caesium				–	0.5–14.0	14
Calcium				–	12 300–40 000	14
Cerium					53–160	14
Chlorine					20–609	14
Chromium	River Susquehanna, USA	0.48–0.49	8	–	16–50	14
		31.4–1143	2	Lake Ontario, Canada	110	13
	–	108	9			
		450	10			
	River Arno, Italy	Total: 3–368				
		Acid extractable 1.3–128	12			

Element	Location	Concentration	Ref	Location	Concentration	Ref
Cobalt	River Arno, Italy	21.9	10	—	3.9–16.0	14
	—	57	9	Lake Ontario, Canada	200	13
	—	Total: 2.2–5.3 / Acid extractable 1.8–48.9	12			
Copper	—	0.07	12	—		
	River Rideau, Canada	4.2	8	Lake Ontario, Canada	50	13
	River Arno, Italy	59.5–244	9			
	—	1.9–226	10			
		Total: 1–148	11			
		Acid extractable 6.6–74	12			
Dysprosium				—	5.4–74.0	14
Europium					0.77–194	14
Gadolinium				—	6.4–22	14
Gold				—	0.25–19	14
Hafnium				—	1.7–12	14
Indium				—	5.3–19.0	14
Iridium				—	0.5–48	14
Iron	—	16.9–18.4	8		14 700–30 600	14
	—	31 000	9			
		Total: 6960–15 700	12	Lake Ontario, Canada	30 000	13
		Acid extractable 1600–79 800	12			
Lanthanum	—	0.11–0.13	8	—	28–73	14
Lead	River Arno, Italy	60.7–170	10	—	20–180	13
	—	84	9	Lake Ontario, Canada	100	15

continued overleaf

Table A3.1. (*continued*)

	River			Lake/pond		
	Location	Concentration	Reference	Location	Concentration	Reference
Lithium	—	17–59 Total: 51–5060 Acid extractable 5–5160	1 12 12	Lake Ontario, Canada	50	13
Lutecium				—	0.52–1.20	14
Magnesium				— Lake Ontario, Canada	5900–16 800 16 000	14 13
Manganese	River Armo, Italy — — —	0.34 553–704 582 5–3225 Total: 113–9640 Acid extractable 37–9600	8 10 9 11 12	— Lake Ontario, Canada	214–4500 4500	14 13
Mercury	River Arno, Italy River Loire, France River inorganic	0.91–4.4 13.2–46.8 6.5–9.0 Total: 12–21.0	10 4 16	Lake Erie, Canada	1.95–6.79	3
Neodynium	—		10	—	15–137	14
Nickel	River Arno, Italy — —	60.0–79.0 72 Total: 7–238	10 9 12	— Lake Ontario, Canada	1–218 200	14 13

Osmium			—	1–4.5	14
Phosphorus	Acid extractable 1.4–67.6	12	—		
Platinum			—	0.3–8.1	14
Potassium	Total: 675–1870	15	—	5600–22 900	14
Rubidium			—	19–49	14
Ruthenium			—	45–500	14
Samarium			—	7.9–28.0	14
Scandium			—	3.3–9.2	14
Selenium	0.09–0.93	6	—	0.03–1.0	14
Silver	1–5.53	5	—	0.1–1.0	14
			Lake Moira, Canada	1.0–8.05	5
Sodium	9.3	10	River Arno, Italy	3000–9200	14
Strontium			—	10–242	14
Tamerium			—	0.19–0.74	14
Tantalum			—	0.4–1.4	14
Terbium			—	0.95–2.4	14
Thorium			—	4.0–9.4	14
Titanium			—	800–3800	14
Uranium			—	0.78–4.3	14
Vanadium			—	28–68	14
Ytterbium			—	2.34–9.34	14
Zirconium			—	54–488	14

Table A3.2. Organometallic compounds in sediments
(references)

	River	Lake	Marine
Arsenic			17
Lead	18	19	20
Mercury	3, 21–4	42	
Silicon	43		
Tin	44–7		48, 49

Table A3.3. Organic compounds in sediments (references)

Class of organic compound	River	Lake	Marine
Aliphatic hydrocarbons	50–59		58, 60–67
Aromatic hydrocarbons	57–9		68–71
Polyaromatic hydrocarbons	72–8	79	80–83
Phenols	84		
Fatty acids	85		
Phthalate esters	86, 87		
Carbohydrates	43, 88		
Volatile chloroaliphatics	89, 94		
Non-volatile chloroaliphatics	95–102		
Hexachlorobenzene			103
Chlorophenols			104
Chlorinated insecticides			105
Polychlorinated biphenyls	106–13		114, 115
Nitrogen bases			116
Nitrogen containing aromatics			117
Trialkyl and triaryl phosphates		118	
Adenosine triphosphate	119		
Organophosphorus insecticides	120–22		
Dioxins	123		
Humic and fulvic acids	124		125–8
Herbicides	129–31		
Inositol esters		132	
Detergents	133–7		
Priority pollutants (EPA)			137

REFERENCES

1. Sakata, M. and Shimodo, O., *Water research* **16**, 231 (1982)
2. Panko, J.F., Leta, D.P., Lin, J.W., Ohl, S.E., Shim, W.P. and Januer, G.E. *Science of the Total Environment* **7**, 17 (1977)
3. Pillay, K.K.S., Thomas C.C., Sondel, J.A. and Hycke, C.M. *Analytical Chemistry* **43**, 1419 (1971)

4. Fresnet Robin, M. and Ottman, F., *Estuarine, Marine and Coastal Science* **7**, 425 (1978)
5. Lum, K.R. and Edgar, D.G., *Analyst (London)* **108**, 918 (1983)
6. Goulden, P.D., Anthony, D.H.J. and Austen, K.D., *Analytical Chemistry* **53**, 2027 (1981)
7. Sandhu, A.S., *Analyst (London)* **106**, 311 (1981)
8. Zink Neilson, I., *Vatten* **1**, 14 (1977)
9. Agemian, H. and Chau, A.S.Y., *Archives of Environmental Contamination and Toxicology* **6**, 69 (1977)
10. Breder, R., *Fresenius Zeitschrift für Analytische Chemie* **313**, 395 (1982)
11. Lichtfusse, R. and Brummer, G., *Chemical Geology* **21**, 51 (1978)
12. Malo, B.A., *Environmental Science and Technology* **11**, 277 (1977)
13. Agemian, H. and Chau, A.S.Y., *Analyst (London)* **101**, 761 (1976)
14. Nadkarni, R.A. and Morrison, G.H., *Analytica Chimica Acta* **99**, 133 (1978)
15. Aspila, K.I., Agemian, H. and Chau, A.S.Y., *Analyst (London)* **101**, 187 (1976)
16. Minagarva, K., Takizawa, Y. and Fifure, I., *Analytica Chimica Acta* **115**, 103 (1980)
17. Maher, W.A., *Analytica Chimica Acta* **126**, 157 (1981)
18. Chau, Y.K., Wong, P.T.S., Bengert, G.A. and Kramer, O., *Analytical Chemistry* **51**, 186 (1974)
19. Wong, P.T.S., Chau, Y.K. and Luxon, P.L., *Nature (London)* **253**, 263 (1975)
20. Chau, Y.K., Wong, P.T.S., Bengert, S.A. and Kramer, O., *Analytical Chemistry* **51**, 188 (1979)
21. Leong, P.C. and Ong, H.P., *Analytical Chemistry* **43**, 940 (1971)
22. Anderson, D.H., Evans, J.H., Murphy, J.J. and White, W.W., *Analytical Chemistry* **43**, 1511 (1971)
23. Bretthaur, E.W., Moghissi, A.A., Snyder, S.S. and Matthews, N.W., *Analytical Chemistry* **46**, 445 (1974)
24. Feldman, C., *Analytical Chemistry* **46** 1606 (1974)
25. Bishop, J.N., Taylor, L.A. and Neary, B.P., *The Determination of Mercury in Environment Samples*, Ministry of the Environment, Ontario Canada (1973)
26. Jacobs, L.W. and Keeny, D.R., *Environmental Science and Technology* **8**, 267 (1976)
27. *Methods for Chemical Analysis of Water and Wastes*, US Environmental Protection Agency, Cincinnati, Ohio (1974)
28. Iskander, I.K., Syens, J.K., Jacabs, L.W., Keeney, D.R. and Gilmour, J.T., *Analyst (London)* **97**, 388 (1972)
29. Craig, P.J. and Morton, S.F., *Nature (London)* **261**, 125 (1976)
30. Ealy, J.A., Shultz, W.D. and Dean, J.A., *Analytica Chimica Acta* **64**, 235 (1973)
31. Batti, R., Magnaval, R. and Lanzola, E., *Chemosphere* **4**, 13 (1975)
32. Longbottom, J.E., Dressman, R.X. and Lichtenberg, J.J., *Journal of the Association of Official Analytical Chemists* **56**, 1297 (1973)
33. Bartlett, P.D., Craig, P.J. and Morton, S.F., *Nature (London)* **267**, 606 (1977)
34. Uthe, J.F., Solomon, J. and Grift, B., *Journal of the Association of Official Analytical Chemists* **55**, 583 (1972)
35. *Official Methods of Analysis of AOAC* (11th edition) **418** (1970)
36. Nagase, H., Sato, T., Ishikawa, T. and Mitani, K., *Inst. J. Environ, Anal. Chem.* **7**, 261 (1980)
37. Jurka, A.M. and Carter, M.J., *Analytical Chemistry* **50**, 91 (1978)
38. El-Awady, A.A., Miller, R.B. and Carter, M.J., *Analytical Chemistry* **48**, 110 (1976)
39. Cappon, C.J. and Crispin Smith, V.J., *J. Analytical Chemistry* **49**, 365 (1977)
40. Matsunaga, K. and Takahashi, O.A., *Analytica Chimica Acta* **87**, 487 (1976)
41. Langmyhr, F.J. and Aamodt, J., *Analytica Chimica Acta* **87**, 483 (1976)

42. Jensen, S. and Jernelov, M.P., *Nature (London)* **223**, 753 (1969)
43. Pellenberg, R., *Marine Pollution Bulletin* **10**, 267 (1979)
44. Hattori, Y., Kabayashi, A., Takemoto, S., Takami, K., Kuge, Y., Sigimae, A. and Nakemoto, N., *Journal of Chromatography* **15**, 341 (1984)
45. Mueller, M.D., *Fresenius Zeitschrift für Analytiche Chemie* **317**, 32 (1984)
46. Gilmour, C.C., Tuttle, J.H. and Means, J.C., *Analytical Chemistry* **58**, 1548 (1986)
47. Rapsomankis, S., Donard, O.F. and Weber, J.W., *Applied Organometallic Chemistry* **1**, 115, (1987)
48. Randall, L., Han, J.S. and Weber, J.H., *Environmental Technology Letters* **7**, 571 (1986)
49. Unger, M.A., MacIntyre, W.G. and Huggett, R.J., *Environmental Technology and Chemistry* **7**, 907 (1988)
50. Morgan, W.L., *Bulletin of Environmental Contamination and Toxicology* **14**, 309 (1975)
51. Meyers, P.A., *Chemosphere* **7**, 385 (1978)
52. Chesler, S.N., Gump, B.H., Hertz, H.S., May, W.E. and Wise, S.A., *Analytical Chemistry* **50**, 805 (1978)
53. May, W.E., Chesler, S.N., Cram, S.P., Gump, B.H., Hertz, D.S., Enagonio, D.P. and Dyszel, S.M., *Journal of Chromatographic Science* **13**, 535 (1975)
54. Chesler, S.N., Gump, B.H., Hertz, H.S., May, W.E., Dyszel, S.M. and Enagonio, D.P., National Bureau of Standard (US) Technical Note No. 889, Washington, DC (1976)
55. Berthou, F., Yourmelun, Y., Dreano, Y. and Friocourt, M.C., *Journal of Chromatography* **203**, 279 (1981)
56. Mason, P.R., *Marine Pollution Bulletin* **18**, 528 (1987)
57. Vowles, P.D. and Mantoura, R.F, *Chemosphere* **16**, 109 (1987)
58. Blumer, M. and Sass, J., *Marine Pollution Bulletin* **3**, 92 (1972)
59. Farrington, J.W. and Quinn, J.G., *Estuary and Coastal Marine Science* **1**, 71 (1973)
60. Walker, J.D., Calwell, R.R., Hamming, H.C. and Ford, H.T., *Environmental Pollution* **9**, 231 (1975)
61. Bromman, D., Colmsio, A., Ganning, B., Naf, C., Zebuhr, Y. and Ostman, C., *Marine Pollution Bulletin* **18**, 380 (1987)
62. Mark, H.B., *Environmental Science and Technology* **6**, 833 (1972)
63. Zitko, V. and Carson, W.V., Tech. Report Fish Res. Bd, Ottawa, Canada No. 217 (1970)
64. Scarratt, D.J. and Zitko, V.J., Fish Res. Bd, Ottowa, Canada **29**, 1347 (1973)
65. McLeod, W.W., Gennero, D.D. and Brown, D.W., *Analytical Chemistry* **54**, 386 (1982)
66. Hilpert, L.R., May, E.W., Wise, S.A., Chesler, S.N. and Hertz, H.S., *Analytical Chemistry* **54**, 458 (1982)
67. Albaiges, A. and Grimalt, J., *International Journal of Environmental Analytical Chemistry* **31**, 281 (1987)
68. Hennig, H.F.O., *Marine Pollution Bulletin* **10**, 234 (1979)
69. Hargrave, B.T. and Phillips, G.A., *Environmental Pollution* **8**, 193 (1975)
70. Takada, H. and Ishimatari, R., *Journal of Chromatography* **346**, 281 (1985)
71. Krabn, M.M., Moore, L.K., Bogar, R.G., Wigren, C.A., Chau, S.L. and Brown, D.W., *Journal of Chromatography* **437**, 161 (1988)
72. Giger, W. and Schnaffner, C. *Analytical Chemistry* **50**, 243 (1978)
73. Tan, Y.L., *Journal of Chromatography* **176**, 319 (1979)
74. Bjorseth, A., Knutsen, J. and Skei, J. *Science of the Total Environment* **13**, 71 (1979)
75. Garrigues, P. and Ewald, M., *Chemosphere* **16**, 485 (1987)
76. De Laeuw, J.W., De Leer, E.W.B., Damste, J.S.S. and Schuyl, P.J.W., *Analytical Chemistry* **58**, 1852 (1986)

77. Lee, H.K., Wright, G.J. and Swallow, W.H., *Environmental Pollution* **49**, 167 (1988)
78. Marcomini, A., Sfriso, A. and Pavoni, B. *Marine Chemistry* **18**, 120 (1985)
79. Giger, W. and Schnaffner, C., *Analytical Chemistry* **50**, 243 (1978)
80. Readman, J.S.O., Preston, M.R. and Mantoura, R.F.C., *Marine Pollution Bulletin* **17**, 298 (1986)
81. Dunn, B.P. and Stich, H.F., *J. Fish., Res. Bd, Ottawa, Canada* **33**, 2040 (1976)
82. Dunn, B.P., *Environmental Science and Technology* **10**, 1018 (1976)
83. Bates, T.S. and Carpenter, R., *Analytical Chemistry* **51**, 551 (1979)
84. Goldberg, M.C. and Weiner, E.R., *Analytica Chimica Acta* **112**, 373 (1980)
85. Farrington, J.W. and Quinn, J.G., *Geochimica and Cosmochimica Acta* **35**, 735 (1971)
86. Schwartz, H.W., Anzion, G.J.M., Van Vleit, H.P.M., Peerebooms, J.W.C. and Brinkman, U.A.T., *International Journal of Analytical Chemistry* **6**, 133 (1979)
87. Thuron, A., *Bulletin of Environmental Contamination and Toxicology* **36**, 33 (1986)
88. McQuaker, N.R. and Fung, T. *Analytical Chemistry* **47**, 1435 (1975)
89. Charles, M.J. and Simmons, M.S., *Analytical Chemistry* **59**, 1217 (1987)
90. Murray, A.J. and Riley, J.P., *Analytica Chimica Acta* **65**, 261 (1973)
91. Murray, A.J. and Riley, J.P., *Nature (London)* **242**, 37 (1973)
92. Novak, J., Zluticky, J., Kubulka, V. and Mostecky, J., *Journal of Chromatography* **76**, 45 (1973)
93. Amin, T.A. and Narang, R.S., *Analytical Chemistry* **57**, 648 (1985)
94. Teichman, J., Bevenue, A. and Hylin, J.W., *Journal of Chromatography* **151**, 155 (1978)
95. Hallici, J.F., Pinnington, P.F., Handley, A.J., Baldwin, M.K. and Bennett, D., *Analytical Chemistry* **111**, 201 (1979)
96. Bierl, R., *Fresenius Zeitschrift für Analytische Chemie* **230**, No. 4/5 (1988)
97. Wegmann, H., *Deutsche Gewasserkundliche Mitteilungen* **29**, 111 (1985)
98. Lee, H.B. *Journal of Association of Official Analytical Chemists* **71**, 803 (1988)
99. Wegman, R.C.C. and Greve, P., *Science of the Total Environment* **7**, 235 (1977)
100. Schellenberg, K., Leuenberger, C. and Schwarzenbach, R.P., *Environmental Science and Technology* **18**, 652 (1984)
101. Lee, H.B., Stokker, Y.D. and Chau, A.S.Y., *Journal of Association of Official Analytical Chemists* **70**, 1003 (1987)
102. Onuska, F.I. and Terry, K.A. *Analytical Chemistry* **57**, 801 (1985)
103. Beller, H.R. and Simoneit, B.R.T., *Bulletin of Environmental Contamination and Toxicology* **41**, 645 (1988)
104. Xie, T.H., *Chemosphere* **12**, 1183 (1983)
105. Picer, N., Picer, M. and Strohal, P., *Bulletin of Environmental Contamination Toxicology* **14**, 565 (1975)
106. Goerlitz, D.F. and Law, L.M., *Journal of Association of Official Analytical Chemists* **57**, 176 (1974)
107. Kerkhoff, M.R.T., de Vries, A., Wegman, R.C.C. and Hofstee, A.W.M., *Chemosphere* **11**, 165 (1982)
108. Kominar, R.J., Onuska, F.L. and Terry, K.A., *Journal of High Pollution Chromatography and Chromatography Communications* **8**, 585 (1985)
109. Brown, J.F., Bedard, D.L., Brennan, M.J., Carnation, J.C., Feng, H. and Wagner, R.E., *Science* **236**, 709 (1987)
110. Marin, F.A., Noroozian, E., Otter, R.R., Van Dijck, R.C.J.M., De Jong, C.J.J. and Brinkman, T., *Journal of High Resolution Gas Chromatography and Chromatography Communications* **11**, 197 (1988)

111. Alford Stevens, A.L., Budde, W.L. and Bellar, T.A., *Analytical Chemistry* **57**, 2452 (1985)
112. Murtrey, K.D., Wildman, N.J. and Tal, H., *Bulletin of Environmental Contamination and Toxicology* **31**, 734 (1983)
113. Lee, H.B. and Chau, A.S.Y., *Analyst (London)* **112**, 37 (1987)
114. Teichman, J., Bevenue, A. and Hylin, J.W., *Journal of Chromatography* **151**, 155 (1978)
115. Jensen, S., Renberg, L. and Reutergard, L. *Analytical Chemistry* **49**, 316 (1977)
116. Kido, A., Shinohara, R., Eto, S., Koga, M. and Hori, T., *Japan Journal of Water Pollution Research* **2**, 245 (1979)
117. Krone, C.A., Burrows, D.W., Brown, D.W., Robisch, P.A., Friedman, A.J. and Halins, D.C., *Environmental Science and Technology* **20**, 1144 (1986)
118. Ishikawa, S., Taketomi, M. and Shinohara, R., *Water Research* **19**, 119 (1985)
119. Tobin, S.R., Ryan, J.F. and Afghan, B.K., *Water Research* **12**, 783 (1978)
120. Rice, J.R. and Dishberger, H.J., *J. Agric, Food. Chem.* **16**, 867 (1968)
121. Deutsch, M.W., Westlake, M.E. and Gunther, F.A., *J. Agric. Food Chem.* **18**, 178 (1970)
122. Kjolholt, J., *Journal of Chromatography* **325**, 231 (1985)
123. Smith, L.M., Stalling, D.L. and Johnson, J.L., *Analytical Chemistry* **56**, 1830 (1984)
124. Klenke, T., Oskierski, M.W., Poll, K.G. and Reichel, B., *Gas-u-Wassfach Wasser-Abwasser* **127**, 650 (1986)
125. Poutonen, E.L., Morris, R.J., *Marine Chemistry* **18**, 115 (1985)
126. Hayase, K., *Geochimica and Cosmochimica Acta* **49**, 159 (1985)
127. Raspar, B., Nurnberg, H.W., Valentia, P. and Branica, M., *Marine Chemistry* **15**, 217 (1984)
128. Hayase, K., *Journal of Chromatography* **295**, 530 (1984)
129. Spengler, D. and Jumar, A., *Arch. Pflanzenschutz* **7**, 151 (1971)
130. Reeves, R.G. and Woodham, D.W., *J. Agric. Food. Chem.* **22**, 76 (1974)
131. Wauchope, R.D. and Myers, R.S., *Journal of Environmental Quality* **14**, 132 (1985)
132. Weimer, W.C. and Armstrong, D.E., *Analytica Chimica Acta* **94**, 35 (1977)
133. Ambe, Y. and Hanya, T., *Japan Analyst* **21** 252 (1972)
134. Longwell, N. and Maniece, O., *Anal. Abstr.* **2**, 2244 (1955)
135. Sallee, O., *Analytical Chemistry* **28**, 1822 (1956)
136. Ambe, Y., *Environmental Science and Technology* **7**, 542 (1973)
137. Ozretich, R.J. and Schroeder, W.P., *Analytical Chemistry* **58**, 2041 (1986)

APPENDIX 4
Toxicants in Sea Organisms and Phytoplankton

Results obtained in the determination of metals are reviewed in Table A4.1. Further information and references on organometallic compounds and organic compounds in these materials are given, respectively, in Tables A4.2 and A4.3.

Table A4.1. Metals in fish, water creatures other than fish, phytoplankton and weeds

	Fish			Crustacae			Phytoplankton and weeds		
	Location type	Conc. (mg kg⁻¹)	Ref.	Location type	Con. (mg kg⁻¹)	Ref.	Location type	Con. (mg kg⁻¹)	Ref.
Antimony				Oyster tissue	0.4	18			
				Lobster tissue	0.071–0.089	30			
Arsenic	Fish Herring	0.03	1	Oyster tissue	13.4	18	Macro algae	20.0–58.1	27
	Haddock	0.03	1	Lobster	11.9–15.9	195	Algae	20–49	27
	Tuna	0.15	1						
	Smelt	0.41–0.44	3						
	Coho salmon	0.26–0.36	3						
	Plaice	total 24 inorganic 0.02–0.04	19						
	Herring	1.1(tot) 0.02–0.04 (inorg)	19	Canned crab	1.5(tot) 0.06–0.10 (inorg)	19			
	Haddock	2.6(tot) 0.02–0.04 (inorg)	19	Whelk	3.2(tot) 0.06–0.18 (inorg)	19			
	Tuna	2.9(tot) 0.12–0.20 (inorg)	19	Canned lobster	3.6(tot) 0.06–0.08 (inorg)	19			
	Organs dogfish (muscle)	18.7	196	King prawn	14(tot) 0.02–0.04 (inorg)	19			
				Whelk	26(tot) 0.10–0.18 (inorg)	19			
				Lobster	24.6–25.5	20			
				Scallops	7.0–7.8	20			
				Mollusc	2–23.2	27			

Element	Sample	Concentration	Ref
Bismuth	Lobster hepato-pancreas	13.4	194
	Mussel	0.0007–0.0023	71
	Kelp	0.0053	71
Boron	Oyster	0.0042	71
	Carp	up to 1.5	201
Bromine	Lobster	50.6–51.7	195
	Macro cystis	0.0089	71
Cadmium	Oyster	0.0025	21
	Oyster	2.36–2.56	21
	Crab	0.71–0.83	22
	Crab	7.0	23
	Mussel Mediterranean	0.02–0.03	4
	Mussel Port Phillip	0.5	24
	Mussel US West Coast	0.8–20.2	190
	Mussel	0.07–0.40	4
	Shrimp, Crab, Oyster, Mussel	0.07–0.24	4
	Clam	1.3	192
	Lobster	0.5–1.1	23
	Lobster hepato-pancreas	3.5	194
	Fish		
	Catfish	0.039	2
	Rainbow trout	0.10–0.11	3
	White bream	0.04	4
	Sardine	0.02	4
	Gilthead bream	0.03	4
	Grey mullet	0.09	4
	Carp	up to 0.27	201
	Horse mackerel	0.17	4
	Striped mullet	0.02	4
	Crayfish	0.10	5
	Flathead	0.13	5
	Shark	0.08	5
	Organs		
	Muscle	0.075–2.9	6
	Muscle	0.3	7
	Skin	0.14–10.9	6
	Kidney	3.1–5.6	6
	Kidney	7.1	3
	Kidney	4.2	7

continued overleaf

Table A4.1. (*continued*)

	Fish			Crustacae			Phytoplankton and weeds		
	Location type	Conc. (mg kg⁻¹)	Ref.	Location type	Con. (mg kg⁻¹)	Ref.	Location type	Con. (mg kg⁻¹)	Ref.
Cadmium (*continued*)	Gut	0.6–5.6	6						
	Heart	1.5–5.6	6						
	Bone	0.14–3.6	6						
	Gill	0.038	2						
	Gill	0.94–6.67	6, 7						
	Liver	3.1–6.7	3, 6, 7						
	Liver (Perch)	0.17–0.90	8						
	Liver (Pike)	0.17	8						
	Liver	1.5–9.0	8						
	Liver (White fish)	0.19–0.9	8						
	Opercle	9.5	7						
	Blue gill tissue								
	Kidney / Gut / Heart / Gill / Liver	5.6–13.1	6						
	Muscle / Skin / Bone	0.14–1.7	6						
Chromium	*Fish*								
	Cohojack	0.21	3	Lobster	0.75	195	Freshwater plankton	60–70	29
	Rainbow trout	2.2	3				algae	40–630	29
	White bream	0.58	4						
	Sardine	0.28	4						
	Gilthead bream	0.49	4						

Element	Material	Concentration	Reference
	Grey mullet	0.10	4
	Horse mackerel	0.65	4
	Striped mullet	0.14	4
	Carp	up to 2.2	201
	Organs		
	Muscle	0.5–0.8	7
	Gill	4.9–48.2	7
	Opercle	8.3–26.0	7
	Liver	5.6–18.4	7
	Kidney	10.3–27.8	7
Cobalt	Whale heart	0.07	9
	Whale meat	0.07	9
	Whale fat	0.38	9
	Trout	0.14	9
	Lobster	0.34–0.44	195
Copper	*Fish*		
	Rainbow trout	0.53–0.8	3
	Mussel, Shrimp, Crab, Lobster	0.75–2.65	4
	Freshwater plankton	40	29
	Cohojack	1.06	3
	Whale heart	7.9	9
	Lobster Hepato–pancreas	63.0	194
	Algae	56–660	29
	Whale meat	2.29	9
	Whale fat	1.2	9
	Trout	2.6	9
	White bream	1.11	4
	Sardine	2.18	4
	Gilthead Bream	1.20	4
	Grey mullet	1.70	4
	Horse mackerel	0.99	4
	Striped mullet	0.68	4
	Crayfish	3.46	5
	Flathead	0.39	5

continued overleaf

Table A4.1. (*continued*)

	Fish			Crustacea			Phytoplankton and weeds		
	Location type	Conc. (mg kg⁻¹)	Ref.	Location type	Con. (mg kg⁻¹)	Ref.	Location type	Con. (mg kg⁻¹)	Ref.
Copper (*continued*)	Shark	1.01	5						
	Whale	1.2–7.6	9						
	Organs								
	Perch liver	3.7–4.8	8						
	White fish liver	24–62	8						
	Pike liver	11.7	8						
	Liver	6.9	7						
	Liver	1.7	10						
	Gill	5.5	7						
	Gill	0.6	10						
	Opercle	12.4	7						
	Kidney	6.0	7						
	Kidney	0.67	10						
	Blood cell	0.27	10						
	Blood serum	0.57	10						
	Heart	3.0	10						
	Spleen	3.0	10						
	Gut	1.1	10						
	Stomach	0.8	10						
	Skin	0.64	10						
	Muscle	0.22	10						
	Bone	1.6	10						
Iron				Lobster	212–219	195	Freshwater plankton	3700–3800	29
							Algae	340–9720	29

Lead	Fish					
	Rainbow trout	0.92–0.98	3	Crab	2.8	22
				Clam	0.83	2
				Mussel	0.43–0.61	4
	Whale heart	0.62	9	Lobster hepato-pancreas	2.5–12.0	23
	Whale meat	0.45	9	Lobster hepato-pancreas	12.4	194
	Whale fat	1.37	9			
	Carp	up to 2.3	201			
	Trout	0.89	9	Oyster, Crab, Shrimp, Mussel	0.48–0.61	4
	White bream	0.61	4			
	Sardine	0.57	4			
	Gilthead bream	0.68	4	Lobster	0.11–3.2	23
	Grey mullet	1.36	4			
	Horse mackerel	1.05	4			
	Striped mullet	0.12	4			
	Crayfish	0.48	5			
	Flathead	0.92	5			
	Shark	0.57	5			
	Catfish	0.26	2			
	Bluegill	0.32	2			
	Misc. fish	0.12–1.81	11			
	Organs					
	Liver	8.0	3			
	Kidney	36.0	3			
	Muscle	0.12–1.81	11			

continued overleaf

Table A4.1. (*continued*)

	Fish			Crustacae			Phytoplankton and weeds		
	Location type	Conc. (mg kg⁻¹)	Ref.	Location type	Con. (mg kg⁻¹)	Ref.	Location type	Con. (mg kg⁻¹)	Ref.
Manganese	White bream	0.51	4	Lobster hepato-pancreas	17.5	194	Freshwater plankton	230–250	29
	Sardine	1.63	4	Lobster	16, 57	195	Algae	230–4170	29
	Grey mullet	0.33	4						
	Horse mackerel	0.63	4						
	Striped mullet	0.22	4						
Mercury	Pikerel	0.24–1.11	12				Plankton and algae (Lake Erie)	31.2–81.0	13
	Carp	0.23–0.36	13	Oyster	0.14–0.156	2			
	Carp	0.22–2.4	9						
	Carp	0.23–0.36	13	Mussel (Mediterranean)	0.02–0.05	4			
	Shiver	0.28–0.35	9						
	Carp	up to 2.9	201	Shrimp	0.02–0.05	4			
	Chub	0.09–0.16	9	Crab	0.02–0.05	4			
	Buffalo	0.12–0.41	9						
	Blue Cat	0.21–0.27	9						
	Carp	1.5–2.7	197						
	Carp	0.23–0.36	198						
	Walleye	0.33–0.79	198						
	Channel cat	0.26–0.55	90	Lobster	0.31	20			
	Channel cat	0.36–0.42	13	Scallop	0.10	20			
	Crappie	0.09–0.19	9						

349

	Species	Concentration	Ref			
	Crappie	0.09–0.14	197			
	Wallage	0.33–0.79	13			
	Yellow perch	0.29–0.61	13			
	Perch	0.51–0.53	14			
	White bass	0.43–0.72	13			
	Freshwater drin	0.30–0.67	13			
	Coho salmon	0.51–0.69	13			
	White sucker	0.35–0.56	13			
	Gizzard shark	0.20–0.26	13			
	Smallmouth bass	0.55	13			
	Smelt	0.30	13			
	Tuna	0.25–0.58	15			
	Tuna	0.32–0.35	14			
	Canned tuna	0.12–0.13	14			
	Albacore tuna	0.93–0.94	14			
	Barramundi	0.68	14			
	Gemfish	0.32–0.29	14			
	Misc. fish	0.1–0.4	14			
	Misc. fish	0.11–4.01	16			
	Misc. fish	2.6–8.6	17			
	Misc. fish	2.06–7.23	17	Lobster	0.16	195
Molybdenum	Carp	up to 3.6	201			
Nickel	*Fish*					
	Rainbow trout	0.15–0.20	3	Lobster hepato-pancreas	19.4	194
	Whale heart	0.31	9			
	Whale meat	0.17	9			
	Whale fat	0.60	9	Lobster	0.98	195
	Trout	0.34	9			
	Carp	up to 2.2	201			
	Organs					
	Liver	0.92	3			
	Kidney	1.9	3			

continued overleaf

Table A4.1. (*continued*)

	Fish			Crustacea			Phytoplankton and weeds		
	Location type	Conc. (mg kg⁻¹)	Ref.	Location type	Con. (mg kg⁻¹)	Ref.	Location type	Con. (mg kg⁻¹)	Ref.
Plutonium				Mussel	0.3–13.9	—			
Selenium	Misc. fish	0.31–0.55	199	Prawn	4.01	193			
	Misc. fish	0.4–6.6	200	Lobster	2.03–2.70	25			
	Smelt	0.31	3	Oyster	1.7	18			
	Coho salmon	0.38–0.55	3	Lobster	6.2–6.7	20			
	Crayfish	0.17–0.27	5	Scallops	0.71–0.87	20			
	Flathead	0.37	5	Scallops	1.24	193			
	Shark	0.19	5	Oyster	2.26	191			
	Carp	up to 5.5	201	Lobster	2.04–2.21	195			
Scandium				Lobster	0.015	195			
Silver	Whale heart	0.04	9	Lobster	0.86–0.93	195			
	Whale meat	0.02	9						
	Whale fat	0.02	9						
	Trout	0.04	9						
Strontium				Lobster hepato-pancreas	84.9	194			
				Lobster	11.0	195			
Tin							Algae	0.03–1.06	28
Vanadium				White shrimp	0.4–3.05	26			
				Blue crab	1.09–1.84	26			
				Oyster	0.53–1.42	26			

Zinc

	Concentration	[Ref]		Concentration	[Ref]
Fish			Lobster hepato-pancreas	852	[194]
Rainbow trout	10.9–11.8	[3]	Lobster	548–888	[195]
Cohojack	24.6	[3]	algae	20–700	[29]
Whale heart	103	[9]	Plankton		[31]
Whale meat	42	[9]			
Whale fat	26	[9]			
Trout	39	[9]			
White bream	10.6	[4]			
Sardine	6.3	[4]			
Gilthead bream	9.5	[4]			
Grey mullet	12.2	[4]			
Horse mackerel	4.3	[4]			
Striped mullet	6.4	[4]			
Organs					
Liver	12.6	[3]			
Liver	100–150	[8]			
Liver	29.4	[7]			
Muscle	16.4	[7]			
Gill	47.9	[7]			
Opercle	120–2	[4]			
Kidney	57.0	[7]			
Liver (perch)	107–120	[8]			
Liver (white fish)	463–487	[8]			

Plankton

Element	Concentration
Lanthanum	0.15
Cerium	0.14
Neodynium	0.03
Samerium	0.006
Europium	0.003
Gadolinium	0.019
Dysprosium	0.008
Ytterbium	0.001

Table A4.2. Organometallic compounds in fish, crustacae and phytoplankton (references)

Fish	Crustacae	Phytoplankton
27, 32, 33		27, 32–4
35–7		38
33, 39–73	62	
74		
75–8	79	28, 80

Table A4.3. Organic compounds in creatures (references)

	Fish	Crustacae	Lacustrine sediments	Plankton
Aliphatic hydrocarbons	81, 82		83, 84	85
Polyaromatic hydrocarbons		82, 86–90		
Phenols				91
Unsaturated fatty acids			92	93
Phthalate esters	94	94		
Carbohydrates				95
Volatile chloroaliphatics	96–107	108, 109		
Hexachlorobenzene	110			
Chlorophenols	111–115			
α, α α-trifluoro-4-nitro-m cresol	116			
Polychlorostyrenes	117–19			
Polychloronitrobenzene	120			
Chlorinated insecticides	121–38	121, 139–47		148–52
Polychlorinated biphenols	153–7	121, 158–65		
Polychloroterphenyls		166		
Mirex	121, 158, 159			
Toxaphene	167, 168			
Nitrogen bases	169			
Trialkyl and triaryl phosphates	170			
Adenosine triphosphate				171, 172
Organophosphorus insecticides	173, 176	173		171, 172
Organosulphur compounds	177	178, 179		
Dioxins	195, 196	173		
Humic and fulvic acids				180
Sqoxin	181			
Geosmin	182, 183			
Fluridone	184			
Anatoxin A				185
Arsenobetaine		187		
Coprostanol		186		
Sterols			188	
Priority pollutants (E PA)	189			

REFERENCES

1. Brooke, P.J. and Evans W.A., *Analyst (London)* **106**, 514 (1981)
2. Poldoski, J.E., *Analytical Chemistry* **52**, 1147 (1980)
3. Agemian, H., Sturtevant, D.P. and Austen, K.D., *Analyst (London)* **105**, 125 (1980)
4. Ramelow, G., Ozkan, M.A., Tuncel, G., Saydan, C. and Balkas, J.I. *International Journal of Environmental Analytical Chemistry* **5**, 125 (1978)
5. Adeljou, S.B., Bond, A.M. and Hughes, H.C., *Analytica Chimica Acta* **198**, 59 (1983)
6. Blood, C.R. and Grant, G.C. *Analytical Chemistry* **47**, 1439 (1975)
7. Van Hoof, F. and Van Son, M., *Chemosphere* **10**, 1127 (1981)
8. Borg, W., Edin, A., Holm, K. and Skold, E., *Water Research* **15**, 1291 (1981)
9. Armannson, H., *Analytica Chimica Acta* **110**, 21 (1979)
10. Harvey, V.R., *Analytical Chemistry* **50**, 1866 (1978)
11. Pagenkopf, G.K., Neumann, D.R. and Woodruff, R., *Analytical Chemistry* **44**, 2248 (1972)
12. Davidson, J.W., *Analyst (London)* **104**, 683 (1979)
13. Svasankara, N., Pillay, K.K., Thomas, C.C., Sondel, J.A. and Hyche, C.M., *Analytical Chemistry* **43**, 1419 (1971)
14. Louie, H.W., *Analyst (London)* **108**, 1313 (1983)
15. Holak, W., Kruznitz, B. and Williams, J.C., *Journal of Association of Official Analytical Chemists* **55**, 741 (1972)
16. Uthe, J.F., Armstrong, F.A.J. and Tam, K.C. *Journal of Association of Official Analytical Chemists* **54**, 866 (1971)
17. Jones, P. and Nickless, J., *Journal of Chromatography* **89**, 201 (1974)
18. De Oliveire E., McLaren, J.W. and Berman, S.S., *Analytical Chemistry* **55**, 2047 (1983)
19. Branke, P.J. and Evans, W.H., *Analyst (London)* **106**, 514 (1981)
20. Welz, B. and Melcher, M., *Analytical Chemistry* **57**, 427 (1985)
21. Greenberg, R.R., *Analytical Chemistry* **52**, 676 (1980)
22. McLaren, J.W. and Bermann, S.S., *Applied Spectroscopy* **35**, 403 (1981)
23. Topping, G., Report of 6th ICES Trace Metal Intercomparison Exercise for Cadmium and Lead in Biological Tissue. Cooperative Research Report No. 111 (1982)
24. Ashworth, M.J. and Farthing R.H., *International Journal of Environmental Analytical Chemistry* **10**, 35 (1981)
25. Bin Ahmed, R., Hill, J.O. and Magee, R.J., *Analyst (London)* **108**, 835 (1983)
26. Blotcky, A.J., Falcone, C., Medina, V.A. and Rack, E.P., *Analytical Chemistry* **51**, 178 (1979)
27. Maher, W.A. *Analytica Chimica Acta* **126**, 157 (1981)
28. Hodge, V.F., Seidel, S.C. and Goldberg, D., *Analytical Chemistry* **51**, 1256 (1979)
29. Bando, R., Galanti, G. and Varini, P.G., *Analyst (London)* **108**, 722 (1983)
30. Bertine, K.K. and Dong Soo Lee, in Wong, C.S. *et al.* (eds), *Trace Metals in Seawater*, Proceedings of a NATO Advanced Research Institute on Trace Metals in Seawater, 30 March to 3 April 1981, Sicily, Italy. Plenum Press, New York (1981)
31. Edenfield, H. and Greaves, H.J. in Wong, C.S. *et al.* (eds), *Trace Metals in Seawater*, Proceedings of a NATO Advanced Research Institute on Trace Metals in Seawater, 30 March to 3 April 1981, Sicily, Italy, Plenum Press, New York (1981)
32. Fishman, H. and Spencer, R., *Analytical Chemistry* **49**, 1599 (1977)
33. Agemian, H. and Cheam, V., *Analytica Chimica Acta* **101**, 193 (1978)
34. White, J.N.C. and Englar, J.R. *Botanica Marina* **26**, 159 (1983)
35. Chau, Y.K., Wong, P.T.S. and Kramer, O., *Analytical Chemistry* **51**, 186 (1979)

36. Chau, Y.K., Wong, P.T.S. and Goulden, P.D., *Analytica Chimica Acta* **421**, 85 (1976)
37. Binnie, S.E. and Hodges, D.J. *Environmental Technology Letters* **2**, 433 (1981)
38. Wong, P.T.S., Chau, Y.K. and Luxon, P.L., *Nature (London)* **253**, 263 (1975)
39. Jones, P. and Nickless, G., *Analyst (London)* **103**, 1121 (1978)
40. Westoo, G., *Anal. Chem. Scand.* **21**, 1790 (1967)
41. Stainton, M.P., *Analytical Chemistry* **43**, 625 (1971)
42. Magos, L., *Analyst (London)* **96**, 847 (1971)
43. Kopp, J.F., Longbottom, M.C. and Lohring, L.B., *Journal of the American Water Works Association* **64**, 20 (1972)
44. Environmental Protection Agency, *Mercury in Water—Provisional Method and Mercury in Fish—Provisional Method*, Analytical Quality Control Laboratory, Cinncinatti, Ohio (1972)
45. Yamanaka, S. and Ueda, K., *Bulletin of Environmental Contamination and Toxicology* **14**, 409 (1975)
46. Hendzel, M.R. and Jamieson, D.M., *Analytical Chemistry* **48**, 926 (1976)
47. Schultz, C.D., Clear, D., Pearson, J.E., Rivers, J.B. and Hyliu, J.W., *Bulletin of Environmental Contamination and Toxicology* **15**, 230 (1976)
48. Analytical Methods Committee, Chemical Society, London, *Analyst (London)* **92**, 403 (1967)
49. Analytical Methods Committee, Chemical Society, London, *Analyst (London)* **101**, 62 (1976)
50. Friend, M.T., Smith, C.A. and Wishart, D., *Atomic Absorption Newsletter* **16**, 46 (1977)
51. Agemian, H. and Chau, A.S.Y., *Analytica Chimica Acta* **75**, 297 (1975)
52. Matsunaga, K. and Takahashi, S., *Analytica Chimica Acta* **87**, 487 (1976)
53. Analytical Methods Committee, Chemical Society, London, *Analyst (London)* **102**, 769 (1977)
54. Stuart, D.C., *Analytical Chemistry* **96**, 83 (1978)
55. Davies, I.M., *Analytical Chemistry* **102**, 189 (1978)
56. Aspila, K.I. and Carron, J.M., Inter-Laboratory Quality Control Study No. 1. *Total Mercury in Sediments*, Report Series. Inland Waters, Directorate Water Control Board, Special Services Section, Inland Water Directorate, Water Quality Branch, Department of Fisheries and Environment, Burlington, Ontario, Canada (1975)
57. Shum, G.T.C., Freeman, H.C. and Uthe, J.F., *Analytical Chemistry* **51**, 414 (1979)
58. Capelli, R., Fezia, C., Franchi, A. and Zanicchi, P., *Analyst (London)* **104**, 1197 (1979)
59. Collett, D.L., Fleming, D.E. and Tayler, G.E., *Analyst (London)* **105**, 897 (1980)
60. Abo-Rady, M.D.K., *Fresenius Zeitschrift für Analytiche Chemie* **299** 187 (1979)
61. Holden, A.V., *Pesticide Science* **4**, 399 (1973)
62. Uthe, J.F., Solomon, J. and Grift, B., *Journal of Association of Official Analytical Chemists* **55**, 583 (1972)
63. Kamps, L.R. and McMahon, B., *Journal of Association of Official Analytical Chemists* **18**, 351 (1970)
64. Longbottom, J.E., Dressman, R.C. and Lichtenberg, J.J., *Journal of Association of Official Analytical Chemists* **56**, 1297 (1973)
65. Cappon, C.J. and Crispin, Smith V.J., *Analytical Chemistry* **49**, 365 (1977)
66. Bye, R. and Paus, P.E., *Analytica Chimica Acta* **107**, 169 (1979)
67. Callum, G.I., Ferguson, M.M. and Lenihan, J.H.A., *Analyst (London)* **106**, 1009 (1981)
68. Hight, S.C. and Corcoran, M.T., *Journal of Association of Official Analytical Chemists* **70**, 24 (1987)

69. Holak, W., *Analyst (London)* **107**, 1457 (1982)
70. MacCrehan, W.A. and Durst, R.A., *Analytical Chemistry* **50**, 2108 (1975)
71. Pilay, K.K.S., Thomas, C.C., Sondel, J.A. and Hyche, C.M., *Analytical Chemistry* **43**, 1419 (1971)
72. Hight, S., *Journal of Association of Official Analytical Chemists* **70**, 667 (1987)
73. Bache, C.A. and Lisk, D.J., *Analytical Chemistry* **43**, 950 (1971)
74. Wanatabe, N., Yasuda, Y., Kato, K., Nakamura, T., Funasaka, R., Shimokawa, K., Sato, E. and Ose, Y., *Science of the Total Environment* **34**, 169 (1984)
75. Smith, J.D., *Nature (London)* **225**, 103 (1970)
76. McKie, J.C., *Analytica Chimica Acta* **197**, 303 (1987)
77. Short, J.W., *Bulletin of Environmental Contamination and Toxicology* **39**, 412 (1987)
78. Sasuki, Y., Ishizaka, T., Suzuki, T. and Sarito, Y., *Journal of the Association of Official Analytical Chemists* **71**, 360 (1988)
79. Han, J.S. and Weber, J.H., *Analytical Chemistry* **60**, 316 (1988)
80. François, R. and Weber, J.H., *Marine Chemistry* **25**, 279 (1988)
81. Smith, D.H., *Analytical Chemistry* **44**, 536 (1972)
82. Chesler, S.N., Gump, B.H., Hertz, H.S., May, W.E. and Wise, S.A., *Analytical Chemistry* **50**, 805 (1978)
83. Wakeham, S.G., *Environmental Science and Technology* **11**, 272 (1977)
84. Saber, A., Jarosz, J., Martin-Bouer, M., Paturel, I. and Vial, M., *International Journal of Environmental Analytical Chemistry* **28**, 171 (1987)
85. Law, R.J., Fileman, T.W. and Portman, J.E., Ministry of Agriculture and Fisheries and Food, Lowestoft, Suffolk, UK, Report No. 2, Methods of Analysis of Hydrocarbons in Marine and Other Samples (1988)
86. Bjorseth, A., Knutsen, J. and Skei, J., *Science of the Total Environment* **13**, 71 (1979)
87. Dunn, B.P. and Stich, H.F.J., *J. Fish Res. Bd, Ottawa; Canada* **33**, 2040 (1976)
88. Kunte, H., *Arch. Hyg. Bakt.* **151**, 173 (1967)
89. Uthe, J.F. and Musal, C.J., *Journal of Association of Official Analytical Chemists* **71**, 363 (1988)
90. Giam, C.S., Trujillo, D.A., Kira, S. and Hrung, Y., *Bulletin of Environmental Contamination and Toxicology* **25**, 824 (1986)
91. Dallakyan, G.A., Veselovski, V.A., Tarusov, B.N. and Peogosyan, S.I., *Hydrobiology Journal* **14**, 90 (1978)
92. Mendoza, Y.A., Gulcar, F.O., Hu, Z.L. and Buchs, A., *International Journal of Environmental Analytical Chemistry* **31**, 107 (1987)
93. Vairavamurthy, A., Andreae, H.O. and Brooks, J.M., *Analytical Chemistry* **58**, 2684 (1986)
94. Giam, C.S., Chau, H.S. and Reff, G.S., *Analytical Chemistry* **47**, 2225 (1975)
95. Cowie, G.L. and Hedges, J.I., *Analytical Chemistry* **56**, 497 (1984)
96. Itoh, K., Chikuma, M. and Tanaka, H., *Fresenius Zeitschrift für Analytische Chemie* **330**, 600 (1988)
97. Sinex, S.A., Cantillo, A.Y. and Helz, G.R., *Analytical Chemistry* **52**, 2342 (1980)
98. Cantillo, A.Y., Sinex, S.A. and Helz, G.R., *Analytical Chemistry* **56**, 33 (1984)
99. Suhr, N.H. and Ingasmello, C.O., *Analytical Chemistry* **38**, 730 (1968)
100. Sturgeon, R.E., De Sauliniers, J.A.N., Berman, G.S. and Russell, D.S., *Analytica Chimica Acta* **134**, 283 (1982)
101. McQuaker, N.R., Kluckner, P.D. and Chang, G.N., *Analytical Chemistry* **51**, 888 (1979)
102. Walsh, J.N. and Howie, R.A., *Mineral Management* **43**, 967 (1980)
103. Deetman, A.A., Denmeulemeester, P., Garcia, M., Hauck, G., Hollies, J.I., Krock-

enberger, D., Palin, D.E., Prigge, H., Rohrschneider, L. and Schmidthammer, L., *Analytica Chimica Acta* **82**, 1 (1976)

104. Parejko, R.W. and Keller, R., *Bulletin of Environmental Contamination and Toxicology* **14**, 480 (1975)
105. Solomon, J., *Analytical Chemistry* **51**, 186 (1979)
106. De Leon, I.R., Maberry, M.A., Overton, E.B., Roschke, C.K., Remele, P.C., Steele, C.F., Warren, V.L. and Leister, J.L. *J. Chromatog. Sci.* **18**, 85 (1980)
107. Hiatt, H.H., *Analytical Chemistry* **55**, 506 (1983)
108. Murray, A.J. and Riley, J.P., *Analytica Chimica Acta* **65**, 261 (1973)
109. Murray, A.J. and Riley, J.P., *Nature (London)* **242**, 37 (1973)
110. Johnsen, J.L., Stalling, D.L. and Hogen, J., *Bulletin of Environmental Contamination and Toxicology* **11**, 393 (1974)
111. Stark, A., *J. Agric. Food Chem.* **17**, 871 (1969)
112. Rudling, L., *Water Research* **4**, 533 (1970)
113. Renberg, L., *Analytical Chemistry* **46**, 456 (1974)
114. Hoben, H.J., Ching, S.A., Casarett, L.J. and Young, R.A., *Bulletin of Environmental Contamination and Toxicology* **15**, 78 (1976)
115. Sackmauerova, M., Vennigerova, M. and Uhnak, J., *Vodni Hospodarstvi Series B* **31**, 133 (1981)
116. Allen, J.L. and Sills, J.B., *Journal of Association of Official Analytical Chemists* **57**, 387 (1974)
117. Kuchl, D.W., Kopperman, H.L., Veith, G.D. and Glass, G.E. *Bulletin of Environmental Contamination and Toxicology* **16**, 127 (1976)
118. Steinwandter, H. and Zimmer, L., *Fresenius Zeitschrift für Analytische Chemie* **316**, 705 (1983)
119. Ramdahl, T., Carlberg, G.E. and Kolsaber, P., *Science of the Total Environment* **48**, 147 (1986)
120. Steinwandter, H., *Fresenius Zeitschrift für Analytische Chemie* **326**, 139 (1987)
121. Markin, G.P., Hawthorne, J.C., Collins, H.L. and Ford, J.H., *Pesticides Monitoring Journal* **7**, 139 (1974)
122. Frank, R., Armstrong, A.F., Boeleus, R.G., Braun, H.H. and Douglas, C.N., *Pesticides Monitoring Journal* **7**, 165 (1974)
123. Hesselberg, R.J. and Johnsen, J.L., *Bulletin of Environmental Contamination and Toxicology* **7**, 115 (1972)
124. Simal, J., Crous Vidal, J., Maria-Chareo Arias, A., Boado, M.A., Diaz, R. and Vilas D., *An Bromat (Spain)* **23**, 1 (1971)
125. Chau, A.S.Y., *Journal of Association of Official Analytical Chemists* **55**, 519 (1972)
126. Kuem, D.W., *Analytical Chemistry* **49**, 521 (1977)
127. Sackmauerova, M., Pal'Usova, O. and Szokolay, A., *Water Research* **11**, 537 (1977)
128. Langlois, R.E., Stamp, A.R. and Liska, B.J., *Milk Food Technology* **27**, 202 (1954)
129. Luckas, B., Pscheidl, H. and Haberland, P., *Journal of Chromatography* **147**, 41 (1978)
130. Luckas B., Pscheidl, H. and Haberland D., *Nahrung* **20**, K-K2 (1976)
131. Norheim, G. and Okland, M.O., *Analyst (London)* **105**, 990 (1980)
132. Neeley, W.B., *Science of the Total Environment* **7**, 117 (1977)
133. Frederick, L.L., *J. Fish, Res. Bd, Ottawa, Canada* **32**, 1705 (1975)
134. Jan, J. and Malservic, S., *Bulletin Environmental Contamination and Toxicology* **6**, 772 (1978)
135. Olsson, M., Jenson, B. and Reutergard, L., *Ambio* **7**, 66 (1978)
136. Szelewski, M.J., Hill, D.R., Speigel, J. and Tifft, E.C., *Analytical Chemistry* **51**, 2405 (1979)

137. Gaskin, D.E., Smith, G.J.D., Arnold, P.W., Louisy, M.V., Frank, R., Moldrinet, M. and McWade, J.W., *J. Fish, Res. Bd. Ottawa, Canada* **31**, 1235 (1974)
138. Ludke, J.L. and Schmitt, C.J., *Proceedings* of the 3rd USA–USSR Symposium on the Effect of Pollutants upon Aquatic Ecosystems. Theoretical Aspects of Aquatic Toxicology, US Environmental Protection Agency, Duluth, Minnisota, 97–100 (1980)
139. Mills, P.A., Caley, J.F. and Grithen, R.A., *Journal of Association of Official Analytical Chemists* **46**, 106 (1963)
140. Arias, C., Vidal, A., Vidal, C. and Maria, J., *An Bromat (Spain)* **22**, 273 (1970)
141. Kouyoumjian, H.H. and Uglow, R.F., *Environmental Pollution* **7**, 103 (1974)
142. Williams, R. and Holden, A.V. National Institute of Oceanography, Wormley, Godalming, Surrey, UK, private communication
143. Ernst, W., Goerke, H., Eder, G. and Schaefer, R.C. *Bulletin of Environmental Contamination and Toxicology* **15**, 55 (1976)
144. Ernst, W., Schaefer, H., Goerke, H. and Eder, G.Z., *Analytical Chemistry* **227**, 358 (1974)
145. US Environmental Protection Agency Report No. EPA-600-4-74 1974T 108 (1974)
146. Wilson, A.J., Forester, J. and Knight, J., US Fish Wildl, Circ. 355 18–20, Centre for Estuaries and Research, Gulf Breese, Florida, USA (1969)
147. Neudorf, S. and Khan, M.A.Q., *Bulletin of Environmental Contamination and Toxicology* **13**, 443 (1975)
148. Sodergren, A., *Bulletin of Environmental Contamination and Toxicology* **10**, 116 (1973)
149. Sodergren, A., *Journal of Chromatography* **160**, 271 (1978)
150. Picer, N., Picer, M. and Strohal, P., *Bulletin of Environmental Contamination and Toxicology* **14**, 565 (1975)
151. Sackmauerova, M., Pal'Usova, O. and Szokolay, A., *Water Research* **11**, 537 (1977)
152. Sackmauerova, M., Pal'Usova, O. and Hluckan, E. *Vodni Hospodarsevi* **10**, 267 (1972)
153. Sackmauerova, M., Pal'Usova, O. and Szokolay, A., *Water Research* **11**, 551 (1977)
154. Szelwski, M.J., Hill, D.R., Speigel, S.J. and Tifft, E.C., *Analytical Chemistry* **51**, 2405 (1979)
155. Tausch, H., Stehlik, G. and Widlidal, H., *Chromatographia* **41**, 403 (1981)
156. Bush, B. and Barnard, E.L., *Analytical Letters* **15**, 1643 (1982)
157. Tuinstra, L.G.M.T., Driessen, J.J.M., Keukens, H.J., Van Munsteren, T.J., Roos, A.H. and Traag, W.A., *International Journal of Environmental Analytical Chemistry* **14**, 147 (1983)
158. Armour, J.A. and Burke, J.A., *Journal of Association of Official Analytical Chemists* **53**, 761 (1970)
159. Gaul, J. and Cruze-LaGrange, P., *Separation of Mirex and PCBs in Fish*, Laboratory Information Bulletin, Food and Drug Administration, New Orleans, USA (1971)
160. Markin, G.P., Ford, J.H., Hawthorne, J.C., Spence, J.H., Davies, J. and Loftis, C.D., *Environmental Monitoring for the Insecticide Mirex*, USDA APHIS 81–83, November (1972)
161. Butler, P.A., *Biological Science* **19**, 889 (1969)
162. McKenzie, M.D. *Fluctuations in Abundance of the Blue Crab and Factors Affecting Mortalities*, South Carolina Wildlife Research Division. Technical Rep. No. 1 (1970)
163. Mahood, R.K., McKenzie, M.D., Middough, D.P., Bellar, S.J., Davis, J.R. and Spitzbergen, D., *A Report on the Cooperative Blue Crab Studies in South Atlantic States*, US Department of the Interior, Bureau of Commercial Fisheries (Project Nos 2-79-R-1, 2-81-R-1, 2-82-R-1) (1970)

164. Teichman, J., Bevenue, H. and Hylin, J.W. *Journal of Chromatography* **151**, 155 (1978)
165. Tanabe, S., Tatsukawa, R. and Phillips, D.J.H., *Environmental Pollution* **47**, 41 (1987)
166. Freudenthal, J. and Greve, P.A., *Bulletin of Environmental Contamination and Toxicology* **10**, 108 (1973)
167. Hughes, R.A. and Lee, G.F., *Environmental Science and Technology* **7**, 934 (1973)
168. Musial, C.J. and Uthe, J.F., *International Journal of Environmental Analytical Chemistry* **14**, 117 (1983)
169. Chau, Y.K., Wong, D.T.S., Bengert, G.A. and Kramer, O., *Analytical Chemistry* **51**, 156 (1979)
170. Murray, D.A.J., *J. Fish, Res. Bd, Ottawa, Canada* **32**, 457 (1975)
171. Shoaf, W.T. and Lium, B.W., *Journal Research US Geological Survey* **4**, 241 (1976)
172. Martin, G., *Revue Français des Sciences de l'Eau* **2**, 407 (1983)
173. Deutsch M.E., Westlake, W.E. and Gunther, F.A., *J. Agric. Food. Chem.* **18**, 178 (1970)
174. Zakitis, L.H. and McCray, E.M., *Bulletin of Environmental Contamination and Toxicology* **28**, 334 (1982)
175. Lores, E.M., Moore, J.C., Knight, J., Forester, J., Clark, J. and Moody, P., *Journal of Chromatographic Science* **23**, 124 (1985)
176. Szeto, S.Y., Yee, J., Broun, M.J. and Oloffs, P.C., *Journal of Chromatography* **240** 526 (1982)
177. Lindstrom, K. and Schubert, R., *Journal of High Resolution Chromatography and Chromatography Communications* **7**, 68 (1984)
178. Izumi, T. and Ogata, M., *Bulletin of Environmental Contamination and Toxicology* **31**, 518 (1983)
179. Otaga, M., Mujake, Y. and Yamazaki, Y., *Water Research* **13**, 1179 (1979)
180. Gadel, F. and Brucket, A., *Water Research* **21**, 1195 (1987)
181. Kiigemagi, U., Burnard, J. and Terriere, L.C., *J. Agric. Food. Chem.* **23**, 717 (1975)
182. Martin, J.P., McCoy, C.P., Greenleaf, W. and Bennett, L., *Canadian Journal of Fisheries and Aquatic Sciences* **44**, 909 (1987)
183. Perrson, P.E., *Water Research* **14**, 1113 (1980)
184. West, S.D. and Day, W., *Journal of Association of Official Analytical Chemists* **69**, 856 (1986)
185. Wong, S.H. and Hindin, E., *Journal of American Water Works Association* **74**, 528 (1982)
186. Matusik, J.E., Hoskin, G.P. and Sphon, J.A., *Journal of Association of Official Analytical Chemists* **71**, 994 (1988)
187. Francesconi, K.A., Hicks, P., Stockton, R.A. and Irgolic, K.J., *Chemosphere* **14**, 1443 (1985)
188. Dreier, F., Bucks, A. and Gulacar, F.O., *Geochimica et Cosmochimica Acta* **52**, 1663 (1988)
189. Ozretich, R.J. and Schroeder, W.P., *Analytical Chemistry* **58**, 2041 (1986)
190. Goldberg, E.D., Bowen, J.W., Farrington, G., Harvey, J.H., Martin, P.L., Parker, R.W., Risebrough, W. and Robertson, E., *The Mussel Watch Environmental Conservation* **5**, 101 (1978)
191. Ahmed, R.B., Hill, J.O. and Magee, R.J., *Analyst (London)* **108**, 835 (1983)
192. Poldoski, J.E., *Analytical Chemistry* **52**, 1147 (1980)
193. Maher, W.A., *Marine Pollution Bulletin* **16**, 33 (1985)
194. Lansberger, S. and Davidson, W.F., *Analytical Chemistry* **57**, 197 (1985)
195. Chisela, F., Gawlik, D. and Bratter, P., *Analyst (London)* **111** 405 (1986)

196. Beauchemin, D., Bednas, M.E., Berman, S.S., MacLaren, J.W., Siu, K.W.M. and Sturgeon, R.E., *Analytical Chemistry* **60**, 2209 (1988)
197. Thomas, D.J., Hagstrom, R.A. and Kuckar, E.J., *Analytical Chemistry* **44**, 512 (1972)
198. Svasankara-Pillay, K.K., Thomas, C.G. and Sondel, J.A. *Analytical Chemistry* **43**, 1419 (1971)
199. Agemian, H. and Thomason, R., *Analyst (London)* **105**, 902 (1980)
200. Goulden, P.D., Anthony, D.H.J. and Austen, K.D., *Analytical Chemistry* **53**, 2027 (1981)
201. Sakai, M.K. and May, W., *Science of the Total Environment* **74**, 199 (1988)

APPENDIX 5
Composition of Potable Water

Summaries of results obtained in the determination of metals and organics in potable waters from various parts of the world are given in Tables A5.1 and A5.2.

Table A5.1. Metals in potable water

Element	Concentration (μg l^{-1})	Reference
Aluminium	34–38	1
	8–313	2
Antimony	< 13	1
Arsenic	5.4–11.4	3
	< 20	1
Barium	1–2	1
	35–47	4
Beryllium	< 0.01–0.17	4
	< 0.1	1
Bismuth	< 30	1
Boron	< 13	1
Cadmium	< 1	1
Chromium	4.6–18.1	3
	< 1	1
Cobalt	< 1	1
Copper	6–945	3
	< 3	1
Iron	41–45	1
	4.5–1960	3
Lead	< 0.6–99	5
	5–565	6
	19–555	7
	< 7	1
	1.9–9.1	8, 9
Manganese	1.9–9.1	8, 9
	5.9–20.5	3
	< 2	1
Mercury	< 15	1
	0.01	16
	0.001–0.01	10
Molybdenum	< 7	1

continued overleaf

Table A5.1. (*continued*)

Element	Concentration $(\mu g\,l^{-1})$	Reference
Nickel	< 3	1
	0.19	12
	3.7–6.1	3
Selenium	< 10	1
Silver	< 3	1
Tellurium	< 30	1
Thallium	< 60	1
Thorium	< 4	1
Tin	< 13	1
Titanium	< 0.7	1
Uranium	0.4–7	13
Vanadium	7.4–14.8	3
Yttrium	< 0.3	1
Zinc	< 2	1
	394–2043	3
Zirconium	56–56.4	1
Bromine	17.3–743	3
Total iodine	1.1–28.0	14
	0.7–30	15
Free iodine	0.9–28	16

Table A5.2. Organics in potable water

	Haloforms	Concentration $(\mu g\,l^{-1})$	Reference
$CHCl_3$	Chlorinated potable water	21–100	17
		60–80	18
		33.6–40.8	19
		19.4–56.3	20
		23.9–48.7	21
		9.5	22
		29–168	23
		2.1–141.3	24
		20.0–82.9	–
		2.6–17.2	25, 26
		0.57–4.07	27
	ditto ex-Cincinnati	95–128	28
	ditto ex-Durham USA	143–182	28
	ditto ex-Japan	2.6–17.2	29
	Carbon treated potable water	0.5–4.06	19
$Br\,Cl_2CH$	Chlorinated potable water	6–100	17
		6–76	18
		1.7–23.7	20
		22.0–39.9	21
		2.2	22
		2.2–11.1	23

Table A5.2. (*continued*)

Haloforms		Concentration (μg l^{-1})	Reference
		0.2–74.7	24
		11.7–56.3	25, 26, 29
	ditto ex-Japan	1.6–10.5	25, 26, 29
	Carbon-treated potable water	0.08	19
Br$_2$Cl CH	Chlorinated potable water	1.2–100	17
		40	18
		14.5–15.0	17
		0.1–14.8	20
		8.8–24.8	21
		0.1–24.0	23
		0.6	22
		< 0.01–36.8	24
		3.8–20.0	–
		0.8–32	25, 26, 29
		0.16–0.41	27
	ditto ex-Japan	0.6–4.0	25, 26, 29
	Carbon-treated potable water	0.16–15.0	19
Br$_3$CH	Chlorinated potable water	up to 100	17
		< 0.1–8.0	18
		1.72–3.71	17
		1.64–5.55	19
		1.7–2.6	21
		0.16	22
		0.1–5.7	24
		1.8–7.9	–
	ditto ex-Japan	0.3–0.5	25, 26, 29
	Carbon-treated potable water	0.13–0.37	19
CCl$_4$	Chlorinated potable water	0.31–0.68	19
		0.017	22
		1.2	25, 26
		0.02–0.15	27
	ditto ex-Japan	1.2	29
	Carbon-treated potable water	0.02–0.25	19
CH$_2$ClCH$_2$Cl	Chlorinated potable water	up to 100	17
		21.9–24.5	19
CH$_2$ClCHCl	Carbon-treated potable water	0.06–24.5	19
Cl$_2$C=CCl$_2$	Chlorinated potable water	up to 100	17
		0.008	22
	ditto ex-Japan	0.9	25, 26, 29
CH$_3$CCl$_3$	Chlorinated potable water	0.06	22
	ditto ex-Japan	0.5	13, 9, 10
CHCl CCl$_2$	Chlorinated potable water	0.015	22
	ditto ex-Japan	0.2–0.9	25, 26, 29
CHCl$_2$I	Chlorinated potable water	0.007	22
CH$_2$Cl$_2$	Chlorinated potable water	0.3	29
	ditto ex-Japan	0.3	25, 26

continued overleaf

Table A5.2. (*continued*)

Haloforms		Concentration ($\mu g\, l^{-1}$)	Reference
CH$_2$ClCHClCH$_3$	Chlorinated potable water ex-Japan	0.8	25, 26, 29
Total	Chlorinated potable water	28.2	–
Haloforms			
		106–276	18
		72–84.7	19
		22.8–100.3	20
		56.4–116	21
		12.6	22
		31.3–51.9	23
		2.4–258.8	24
		37.3–167.1	24
		6.7–53.5	25, 26, 29
	Carbon-treated potable water	0.97–44.2	19
Polyaromatic hydrocarbons			
Naphthalene		0.005–0.007	30
2-methyl naphthalene		0.002–0.005	
1-methyl naphthalene		0.001–0.002	
Azulene		n.d.	
2-ethyl naphthalene		0.0007–0.002	
2,6-dimethyl naphthalene		0.0007–0.002	
Biphenyl		0.0007–0.001	
1,3-dimethyl naphthalene		0.001–0.002	
2-vinyl naphthalene		n.d.	
2,3-dimethyl naphthalene		0.0007–0.014	
1,4-dimethyl naphthalene		0.0007–0.014	
3-phenyl toluene		0.0002–0.001	
Diphenyl methane		0.001–0.002	
4-phenyl toluene		0.0002–0.004	
Acenaphthylene		0.0005	
Acenaphthylene		0.0002–0.002	
Dibenzyl		0.001–0.002	
1,1-diphenylethylene		> 0.007	
Cis stilbene		> 0.007	
2,2-diphenyl propane		n.d.	
2,3,5-trimethyl naphthalene		0.0006–0.005	
3,3-dimethylbiphenyl		0.0003–0.005	
Fluorene		0.0001–0.002	
4,4-dimethylbiphenyl		0.006–0.007	
4-vinyl biphenyl		n.d.	
diphenylacetylene		0.000 05	
9,10-dihydroanthracene		0.0007	
trans stilbene		0.0005–0.009	
9,10-dihydrophenanthrene		0.0005–0.009	
10,11-dihydro-5-H dibenzo (ad)		0.0004	
cycloheptone phenanthracene		0.0005–0.002	
Anthracene		0.0005–0.002	
1-phenyl naphthalene		n.d.	

Table A5.2. (*continued*)

Haloforms	Concentration (μg l^{-1})	Reference
1-methyl phenanthrene	n.d.–0.011	
2-methyl anthracene	0.0005–0.0007	
9-methyl anthracene	n.d.–0.0007	
9-vinyl anthracene	n.d.	
triphenyl methane	n.d.	
Fluroanthrene	0.0005–0.002	
Pyrene	0.0005–0.002	
9,10-dimethyl anthracene	n.d.–0.0002	
Triphenyl ethylene	n.d.–0.000 08	
p-terphenyl	n.d.	
1,2-benzfluorene	n.d.	
2,3-benzfluorene	n.d.	
Benzyl biphenyl	n.d.	
Triphenylene	0.003	
Benz (a) anthracene	0.003	
Chrysene	0.003	
Xanthene	0.0001–0.0002	
9-fluorenone	0.0009–0.001	
Perinaphthenone	0.0001–0.0003	
Anthrone	n.d.–0.001	
Anthraquinone	0.002–0.004	
Total PAH	0.05–0.13	
Polychlorinated biphenyls	< 0.0001 (as Aroclor 1016)	31
Pentachlorophenol	< 140–340	32
2,4,6-trichlorophenol	< 1	33
Di- and trichloroacetic acid. (chlorinated potable water)	30–160	34
Nitrosamines		
Dimethyl nitrosamine	< 0.001	35
Diethyl-nitrosamines	< 0.001	35
Dipropyl nitrosamine	< 0.001	35
N-Nitrosopiperidine	< 0.001	35
N-Nitrosopynnolidine	< 0.001	35
N-Nitrososarcosinate	< 0.001	35
5-chlorouracil	0.1–14.1	33
5-chlorouridine	0.7–26.7	33
4-chloro resorcinol	1.6–4.7	33
5-chloro salicylic acid	2.3–12.5	33
p-Isopropyldiphenyl-amine	10–15	33
2-hydroxylbenzo-thiazole	5–25	33

REFERENCES

1. Taylor, C.E. and Floyd, T.L., *Applied Spectroscopy* **35**, 408 (1981)
2. Narayaran, A. and Pontany, D.E., *Environmental Technology Letters* **3**, 43 (1982)

3. Saleh, N.S., *Journal of Radioanalytical Chemistry* **74**, 257 (1982)
4. Logos, P., *Analytica Chimica Acta* **98**, 201 (1978)
5. Beetinshaw, H.P., Gelsthorpe, D. and Wheatstone, K.C., *Analyst (London)* **108**, 23 (1981)
6. Vijan, P.N. and Sadana, R.S., *Talanta* **27**, 321 (1980)
7. Sthapit, P.R., Ottaway, J.M. and Fell, G.S., *Analyst (London)* **109**, 1061 (1984)
8. Nikolelis, D.P. and Hadjiiannou, T.P., *Analytica Chimica Acta* **97**, 111 (1978)
9. Nikolelis, D.P. and Hadjiiannou, T.P., *Analyst (London)* **102**, 591 (1977)
10. Temmermon, E., Dumarey, R. and Dams, R., *Analytical Letters* **18**, 203 (1985)
11. Oda, C.E. and Ingle, J.D., *Analytical Chemistry* **53**, 2205 (1981)
12. Pihlar, B., Valenta, P. and Nurnberg, H.W., *Fresenius Zeitschrift für Analytische Chemie* **307**, 337 (1981)
13. Galinier, J.L. and Zikovsky, L., *Eau de Quebec* **14**, 309 (1981)
14. Moxan, R.E., *Analyst (London)* **109**, 425 (1984)
15. Keller, H.E., Doenecke, E., Weidler, K. and Leppla, W., *American New York Academy of Science* **210**, 1 (1973)
16. Minagawa, K., Takizawa, Y. and Kifune, I., *Analytica Chimica Acta* **115**, 103 (1980)
17. Tardiff, R.G. and Dunzer, M., *Toxicity of Organic Compounds in Drinking Water*, Environmental Protection Agency, Cincinnati, USA (1973)
18. Fielding, M., Loughlin, M. and Steele, K., Water Research Centre Enquiry Report ER532, August 1977, Water Research Centre, Stevenage Laboratory, Elder Way, Stevenage, Herts, UK (1977)
19. Von Rensberg J.F.F., Van Huyssteen, J.J. and Hasset, A.J., *Water Research* **12**, 127 (1978)
20. Dietz, E.A. and Singley, K.F., *Analytical Chemistry* **51**, 1809 (1979)
21. Kirschen, N.A., *Varian Instrument Applications* **15**, 2 (1981)
22. Erlund, G., Josefsson, B. and Roos, C., *Journal of High Resolution Chromatography Communications* **1**, 34 (1978)
23. Otson, R., Williams, D.T. and Bothwell, P.D., *Environmental Science and Technology* **13**, 936 (1979)
24. Mieure, J.P., *Journal of the American Water Works Association* **69**, 62 (1977)
25. Fujii, T., *Analytica Chimica Acta* **82**, 117 (1977)
26. Fujii, T., *Bulletin of the Chemical Society of Japan.* **50**, 2911 (1977)
27. Von Rensberg, J.F.T., Van Huyssteen, J.T. and Hasselt, A.J., *Water Research* **12**, 127 (1978)
28. Haeuder, F.E., Jones, R.B., Stevens, A.A., Moore, L. and Haas, J., *Environmental Science and Technology* **12**, 438 (1978)
29. Fujii, T., *Journal of Chromatography* **139**, 297 (1977)
30. Williams, D.T., Benoit, F.H. and Lalid, G.L., *International Journal of Analytical Chemistry* **6**, 277 (1979)
31. Le'Bel, G.L. and Williams, D.T., *Bulletin of Environmental Contamination and Toxicology* **132**, 277 (1977)
32. Morgade, C., Barquet, H. and Pfaffenberger, C.D., *Bulletin of Environmental Contamination and Toxicology* **24**, 257 (1980)
33. Crathorne, B., Watts, C.D. and Fielding, M., *Journal of Chromatography* **185**, 671 (1979)
34. Uden, P.C. and Miller, J.W., *Journal of the American Water Works Association* **75**, 524 (1983)
35. Fine, D.H., Runnbehler, D.P., Hoffman, F. and Epstein, S.S., *Bulletin of Environmental Contamination and Toxicology* **14**, 404 (1975)

Index

Anodic scanning voltammetry,
    determination of,
  Cadmium, 227
  Copper, 227
  Lead, 227
  Mercury, 197, 227
  Organolead, 246, 263
  Selenium, 227
  Tin, 251
  Vanadium, 197
Anthracene, determination in non saline
    water, 157
Antimony, determination in,
  Fish, 226
  Sediments, 196, 197
  Non saline water, 108, 110
Aromatic hydrocarbons, determination in,
  Fish, 238
  Non saline water, 149, 150, 155, 157
  Sediments, 209
Arsenate, determination in non saline
    water, 111
Arsenic, determination in,
  Fish, 223, 226, 249, 250
  Non saline water, 108–110
  Sea water, 124, 125
  Sediments, 195–199
Arsenic, toxicity of, 42–44, 52–58, 107,
    112–115, 119, 120, 188
Arsenobetaine, determination in non saline
    water, 160
Arsenocholine, determination in non saline
    water, 160
Aryl phosphates, determination in fish, 238
Atomic absorption spectrometry,
    determination of,
  Aluminium, 198
  Arsenic, 195, 249
  Bismuth, 195
  Cadmium, 198, 249, 265
  Chromium, 198, 226, 265
  Cobalt, 195, 226
  Copper, 195, 198, 226, 249, 266
  Gold, 108
  Iron, 195, 198, 250, 266
  Lead, 195, 198, 226, 249
  Manganese, 195, 198, 226, 266
  Mercury, 108, 195, 226, 250, 266
  Nickel, 198, 226, 250
  Organoarsenic, 170, 205, 276
  Organolead, 205, 245

  Organomercury, 170, 205, 245
  Organotin, 170, 171, 205, 245
  Platinum, 250
  Selenium, 108, 249
  Siloxanes, 245
  Silver, 226, 249
  Tin, 226, 266
  Titanium, 198
  Uranium, 250
  Zinc, 195, 198, 226, 249, 266
Atrazine, determination in non saline
    waters, 149, 152, 153, 157

Barban, determination in non saline
    water, 149, 152
Barium, determination in,
  Non saline water, 110
  Sea water, 123–125
  Sediment, 195–197
Benomyl, toxic effect, 141
Benzothiazole, determination in non saline
    water, 149, 158
Beryllium, determination in sediment, 199
Best practical environmental option,
    U.K., 31
Bioaccumulation
  Factors affecting, 20–22
  Metals in fish, 22
  Organics, 23
  Organometallics, 23–28
  Tetramethyl lead, 22, 23
  Alkyl mercury, 27, 28
Biomagnification, 19
Bismuth, determination in,
  Non saline water, 108, 110, 263
  Sediments, 188, 195, 198
Bivalvce mollusc, adverse effect of metals,
    93–95, 107, 114, 115
Borate, determination in non saline water,
    110, 111
Borofluoride, determination in non saline
    water, 111
Boron, determination in,
  Fish, 226
  Non saline water, 110
Bromacil, toxicity of, 61, 71, 141
Bromide, determination in,
  Fish, 227, 250
  Non saline water, 111
  Sediment, 186, 197